Foodborne Viral Pathogens

Food Microbiology Series

Series Editor
Dongyou Liu

Foodborne Viral Pathogens, *edited by Peter A. White, Natalie E. Netzler, and Grant S. Hansman* (2017)

Laboratory Models for Foodborne Infections, *edited by Dongyou Liu* (2017)

Food Spoilage Microorganisms: Ecology and Control, *edited by Yanbo Wang* (2017)

Foodborne Viral Pathogens, *edited by Peter A. White, Natalie E. Netzler, and Grant S. Hansman* (2016)

Molecular Biology of Food and Water Borne Mycotoxigenic and Mycotic Fungi, *edited by R. Russell M. Paterson, and Nelson Lima* (2015)

Biology of Foodborne Parasites, *edited by Lihua Xiao, Una Ryan, and Yaoyu Feng* (2015)

FOOD MICROBIOLOGY SERIES

Foodborne Viral Pathogens

Edited by

Peter A. White • Natalie E. Netzler

Grant S. Hansman

CRC Press

Taylor & Francis Group

Boca Raton London New York

CRC Press is an imprint of the

Taylor & Francis Group, an **informa** business

MIX
Paper from responsible sources
FSC® C014174
www.fsc.org

CRC Press
Taylor & Francis Group
6000 Broken Sound Parkway NW, Suite 300
Boca Raton, FL 33487-2742

Printed on acid-free paper
Version Date: 20160711

International Standard Book Number-13: 978-1-4665-7950-7 (Hardback)

Library of Congress Cataloging-in-Publication Data

Names: White, Peter Andrew, editor. | Netzler, Natalie Ellen, editor. |
Hansman, Grant Stuart, editor.
Title: Foodborne viral pathogens / edited by Peter A. White, Natalie E. Netzler,
and Grant S. Hansman.
Other titles: Food microbiology series.
Description: Boca Raton : CRC Press, 2016. | Series: Food microbiology series
| Includes bibliographical references and index.
Identifiers: LCCN 2016025948| ISBN 9781466579507 (hardback : alk. paper) |
ISBN 9781315392301 (ebook)
Subjects: | MESH: Virus Diseases | Foodborne Diseases--virology
Classification: LCC RA601.5 | NLM WC 500 | DDC 615.9/54--dc23
LC record available at https://lccn.loc.gov/2016025948

Visit the Taylor & Francis Web site at
http://www.taylorandfrancis.com

and the CRC Press Web site at
http://www.crcpress.com

Printed and bound in the United States of America by Sheridan

Contents

Preface

Foodborne diseases pose a significant threat to global public health. Viruses are now recognized as the predominant cause of foodborne illnesses worldwide. Enteric and hepatic viruses are frequently transmitted through contaminated food and water, and are responsible for causing widespread disease epidemics. In particular, norovirus and hepatitis A virus are highly infectious and frequently identified as the etiological cause of foodborne gastroenteritis and hepatitis outbreaks, respectively.

Food safety awareness has grown considerably over the last decade; however, foodborne viruses still claim hundreds of thousands of lives each year. These deaths predominantly occur within vulnerable populations, including infants, young children, the elderly, and the immunocompromised, particularly within developing nations. Even within higher-income nations, the seasonal importation of food from around the globe introduces new challenges for food safety. Microbiological monitoring of food is currently focused primarily on bacterial contamination, rather than on viruses. Foodborne viruses are resistant to environmental degradation and have adapted to survive traversal of the digestive tract. Any attempt to destroy microbes within food matrices using various treatments has little or no effect on the viruses present and they often remain infectious. Despite this bleak picture, extensive research is being carried out into the major etiological viral agents of foodborne illnesses and significant breakthroughs are being attained. By raising awareness and promoting hygienic food processing, preparation, and handling practices, global food safety standards can be improved, which will save lives.

Foodborne Viral Pathogens includes details on the molecular biology, pathogenesis, epidemiology, clinical features, diagnosis, potential vaccines, and treatments of all the major foodborne viral pathogens and prions. Each chapter is written by leading virologists and specialists within the field. This book also features techniques used for viral detection and typing, as well as strategies for control and viral risk assessment. *Foodborne Viral Pathogens* is intended as a guide for medical and food microbiology applications and will be a useful resource for anyone interested in foodborne disease.

Peter A. White
University of New South Wales

Natalie E. Netzler
University of New South Wales

Grant S. Hansman
University of Heidelberg

Editors

Peter White is a professor at the University of New South Wales in Australia. He is a molecular virologist with a particular interest in antiviral development, viral evolution, paleovirology, and pandemic noroviruses. He has a breadth of experience in the development of novel molecular assay systems to investigate viral infections. His group also has interests in hepatitis viruses and antibiotic resistance.

Following a BSc (Hons) in biotechnology from King's College London (1992), Peter completed a PhD at University College, London (1996) in molecular microbiology. Prof. White commenced his Postdoctoral research studies at Macquarie University, Sydney, as a recipient of a Royal Society Fellowship and later worked as Hepatitis Group Leader at the Prince of Wales Hospital until joining the University of New South Wales in 2003. He is also an enthusiastic teacher, having convened the third year science course Viruses and Disease for 13 years at UNSW.

Natalie Netzler is a PhD candidate at the University of New South Wales, Australia. Her project focuses on identifying antiviral compounds for norovirus and other caliciviruses. In particular, her interest lies in the discovery of broad-spectrum antivirals with the potential to rapidly treat several clinically significant viruses for which we currently have no antivirals. Natalie is also a tutor on the 2nd year microbiology and 3rd year virology courses at UNSW.

Following a BSc in Biology (1998), Natalie completed an MSc in virology with first class Honors (2000), from the University of Waikato, New Zealand. Following graduation, Natalie worked as a research assistant at the biotechnology company, Genesis Research and Development for several years, investigating ligands involved in plant flowering and siRNA inhibition targeting drug resistance.

Grant Hansman is a researcher and group leader at Heidelberg University and DKFZ, Germany. His group focuses on the norovirus capsid using structural biology (X-ray crystallography and cryo-EM). He is interested in the structural basis for norovirus capsid binding to a known host factor (histo-blood group antigens - HBGAs), deciphering norovirus capsid flexibility and developing norovirus antivirals targeting the capsid.

Following a BSc degree from Macquarie University, he completed an Honors degree at the University of New South Wales, Australia (1999). Grant undertook a PhD at the University of Tokyo, Japan in molecular virology (2005). He had the role of senior scientist at the National Institute of Infectious Diseases, Japan for seven years, in which time he undertook a two year postdoctoral fellowship at the National Institute of Health, USA. Grant Hansman was an editor of the book *Caliciviruses: Molecular and Cellular Virology*, published in 2010.

Contributors

Krisztián Bányai
Institute for Veterinary Medical Research
Center for Agricultural Research
Hungarian Academy of Sciences
Budapest, Hungary

Scott Bowden
Victorian Infectious Diseases Reference Laboratory
Royal Melbourne Hospital at the Peter Doherty Institute for Infection and Immunity
Melbourne, Australia

Rowena Bull
Systems Medicine in Infectious Diseases, Inflammation and Infection Research Center
School of Medical Sciences
University of New South Wales
Sydney, Australia

Michael J. Casteel
Figure Eight, L.L.C
Stanardsville, Virginia

Daniel Cowley
Enteric Virus Group
Murdoch Children's Research Institute
Royal Children's Hospital

and

Department of Paediatrics
University of Melbourne
Victoria, Australia

Simona De Grazia
Department of Health Promotion Sciences and Mother and Child Care
University of Palermo
Palermo, Italy

Erin DiCaprio
Department of Veterinary Biosciences
College of Veterinary Medicine
Ohio State University
Columbus, Ohio

Celeste Donato
Enteric Virus Group
Murdoch Children's Research Institute
Royal Children's Hospital
Parkville, Victoria, Australia

and

Department of Microbiology
La Trobe University
Bundoora, Victoria, Australia

Grant S. Hansman
Department of Infectious Diseases, Virology
University of Heidelberg
Heidelberg, Germany

and

Schaller Research Group at the University of Heidelberg and the DKFZ
Heidelberg, Germany

Andrew G. Kelly
School of Biotechnology and Biomolecular Sciences
Faculty of Science
University of New South Wales
Sydney, Australia

Pattara Khamrin
Department of Microbiology
Faculty of Medicine
Chiang Mai University
Chiang Mai, Thailand

Carl Kirkwood
Enteric Virus Group
Murdoch Children's Research Institute
Royal Children's Hospital

and

Department of Paediatrics
University of Melbourne
Parkville, Victoria, Australia

and

Department of Microbiology
La Trobe University
Bundoora, Victoria, Australia

Eszter Kovács
Institute for Veterinary Medical Research
Center for Agricultural Research
Hungarian Academy of Sciences
Budapest, Hungary

Jianrong Li
Department of Veterinary Biosciences
College of Veterinary Medicine
Ohio State University
Columbus, Ohio

Kun Lee Lim
Department of Pathology
Singapore General Hospital
Singapore

and

School of Biotechnology and Biomolecular Sciences
Faculty of Science
University of New South Wales
Sydney, Australia

Fangfei Lou
Department of Food Science and Technology
College of Food, Agricultural and Environmental Sciences
Ohio State University
Columbus, Ohio

Fabio Luciani
Systems Medicine in Infectious Diseases, Inflammation and Infection Research Center
School of Medical Sciences
University of New South Wales
Sydney, Australia

Jason M. Mackenzie
Department of Microbiology and Immunology
University of Melbourne
Peter Doherty Institute for Infection and Immunity
Melbourne, Australia

Niwat Maneekarn
Department of Microbiology
Faculty of Medicine
Chiang Mai University
Chiang Mai, Thailand

Vito Martella
Department of Veterinary Medicine
University Aldo Moro of Bari
Valenzano, Italy

Kristina D. Mena
University of Texas Health Science Center at Houston
School of Public Health
El Paso, Texas

Natalie E. Netzler
School of Biotechnology and Biomolecular Sciences
Faculty of Science
University of New South Wales
Sydney, Australia

Shoko Okitsu
Division of Microbiology
Department of Pathology and Microbiology
Nihon University School of Medicine
Tokyo, Japan

Takashi Onodera
Research Center for Food Safety
School of Agricultural and Life Sciences
University of Tokyo
Tokyo, Japan

Ashley Predmore
Department of Food Science and Technology
College of Food, Agricultural and Environmental Sciences
Ohio State University
Columbus, Ohio

Peter A. Revill
Victorian Infectious Diseases Reference Laboratory
Royal Melbourne Hospital at the Peter Doherty Institute for Infection and Immunity
Melbourne, Australia

Jason A. Roberts
Victorian Infectious Diseases Reference Laboratory
Royal Melbourne Hospital at the Peter Doherty Institute for Infection and Immunity
Melbourne, Australia

Akikazu Sakudo
Laboratory of Biometabolic Chemistry
School of Health Sciences
Faculty of Medicine
University of the Ryukyus
Nishihara, Okinawa, Japan

Bruce R. Thorley
Victorian Infectious Diseases Reference Laboratory
Royal Melbourne Hospital at the Peter Doherty Institute for Infection and Immunity
Melbourne, Australia

Hiroshi Ushijima
Division of Microbiology
Department of Pathology and Microbiology
Nihon University School of Medicine
Tokyo, Japan

Peter A. White
School of Biotechnology and Biomolecular Sciences
Faculty of Science
University of New South Wales
Sydney, Australia

Section I

New Technologies, Risk Assessment, and Strategies for Control of Viral Foodborne Illness

1 Introduction

Natalie E. Netzler and Peter A. White

CONTENTS

Foodborne disease results in around 600 million cases globally each year, of which approximately one-quarter is attributed to viruses [1]. Mortality resulting from foodborne diseases primarily affects young children in developing nations, while morbidity and economic losses impact both developed and developing countries. In the United States alone, the economic impact of foodborne disease is estimated to cost over US$50 billion per annum [2]. Despite a growing awareness of the importance of food safety within industrialized nations, in developing nations there are a number of factors contributing to foodborne illness. These frequently include living on or below the poverty line, a lack of wastewater infrastructure, the use of contaminated water for drinking and processing food, often accompanied by a lack of education around hygienic practices. This combination of factors often leads to widespread illness and unnecessary deaths.

A predominant cause of foodborne illness is viral contamination, which can be introduced into food during a variety of stages from "paddock to plate." While many foodborne viruses cause self-limiting disease in healthy individuals, populations in developing nations, infants, young children, pregnant women, immunocompromised patients and the elderly are prone to more severe illness and mortality. In developing countries, reasons for increased disease severity include underlying malnourishment, a lack of sufficient health care and close living quarters enabling efficient viral transmission.

Estimates of foodborne disease, including those caused by viruses, are notoriously difficult to quantify accurately, as the majority of affected individuals never report the illness. However, even the most conservative estimates show a significant global disease burden from business closures, time off work and in some cases, death. By raising awareness of the importance of food safety, the morbidity and mortality resulting from viral food contamination can be reduced.

Industrialization leads to increased farming, which can introduce additional challenges for maintaining food safety. These risks include the contamination of field crops and shellfish beds intended for human consumption by animal slurry, the sewage contamination of irrigation water, and farm run-off. In high-income nations, consumers are faced with the additional challenge of an increasingly complex, global food distribution system where seasonal foods are imported. This can introduce foodborne pathogens into the food chain, despite great care taken in food preparation. In terms of the latter, infected individuals can also contaminate food during handling and preparation if basic handwashing practices are not followed. Hepatitis A virus and norovirus outbreaks are well documented for causing illness in this way. Food monitoring for microbes usually only detects bacteria and viral screening is seldom performed, for example, in the oyster industry. Enteric and hepatic viruses are usually resistant to environmental forces, and as a result, attempts at treating food for microbial contamination before distribution have little effect on reducing the level of infectious viruses.

As the methods of its detection have improved and been widely implemented, norovirus has come to be recognized as the most common singular cause of foodborne illness. Norovirus alone causes an estimated 218,000 deaths each year, predominantly in children under 5 years of age from across the developing world [3]. Other significant foodborne viruses include sapovirus, hepatitis

A and E viruses, astrovirus, rotavirus, and adenovirus, along with a raft of novel, emerging viruses such as kobuvirus, picobirnavirus, and torovirus. Additionally, a nonviral, nonbacterial pathogen that can be transmitted in food is the prion, which is also covered in this book.

Foodborne Viral Pathogens is devoted to the role of clinically significant viruses transmitted in food. The book is divided into two main sections. The first section covers key techniques and methodologies used in foodborne viral research and detection; topics include new technologies for viral detection, next-generation sequencing, molecular genotyping, phylogenetics, strategies for the control of foodborne viruses, and viral risk assessment. The second section of the book describes individual foodborne viral pathogens in relation to their classification, morphology, genome structure, transmission, epidemiology, clinical features, pathogenesis, diagnosis, treatment, and prevention. This section specifically covers: norovirus and sapovirus, hepatitis A and hepatitis E viruses, astrovirus, rotavirus, prions, and novel emerging viruses. Adenovirus is covered in Chapters 4 and 5 on viral risks and strategies to control foodborne outbreaks, respectively.

Written by field-specialist scientists, this book presents key molecular techniques and control strategies used in foodborne viral research, providing a state-of-the-art review of the major foodborne viruses. *Foodborne Viral Pathogens* is a reliable roadmap for the future development of improved, innovative methods for analyses of these viral pathogens and a handy textbook for public health professionals, and undergraduate, and graduate students in medical and food microbiology.

REFERENCES

1. World Health Organization (WHO). 2015. WHO estimates of the global burden of foodborne diseases: Foodborne disease burden epidemiology reference group 2007–2015. Switzerland.
2. Scharff RL. 2012. Economic burden from health losses due to foodborne illness in the United States. *J Food Prot* 75:123–131.
3. Patel MM, Widdowson MA, Glass RI, Akazawa K, Vinje J, Parashar UD. 2008. Systematic literature review of role of noroviruses in sporadic gastroenteritis. *Emerg Infect Dis* 14:1224–1231.

2 Next-Generation Sequencing, What Has It Told Us so Far?

Rowena Bull and Fabio Luciani

CONTENTS

INTRODUCTION TO NEXT-GENERATION SEQUENCING

Sequencing techniques have rapidly progressed over the last decade so that they are no longer limited to only DNA and a small number of kilobases generated per day, but now include direct sequencing of DNA, RNA (and even protein), generating up to 600 gigabases (Gb) per day.

Next-generation sequencing (NGS) is the term applied to the techniques that have developed over the last decade and are an advancement of the traditional "Sanger sequencing" method [1]. Sanger sequencing utilized a chain termination sequencing method, whereas NGS utilizes highly parallelized processes resulting in the sequencing of millions of bases at once. A brief description of the NGS technologies that have been applied to study foodborne viruses is outlined in the first two sections.

The ability to rapidly generate sequence data has far-reaching implications and none so more than those in the field of virology. In 2010, there was a 20%–24% increase in the number of viral and phage sequences deposited in GenBank, amounting to just over 1 billion viral and phage bases, and since then there has been an exponential rise. Even the increase in 2010 was well above the average growth for the database as a whole (reviewed in [2]). Interestingly, of the top 20 viruses sequenced that year, three foodborne viruses made the list: human rotavirus A, hepatitis A virus, and hepatitis E virus.

NGS is in part responsible for this increase in sequence generation and since the introduction of NGS, research into virology has changed dramatically and enabled high-resolution analysis investigating drug resistance, immune escape, viral diversity, and epidemiology. In the latter sections of this chapter, a detailed account of how NGS technology has been applied and advanced our understanding of foodborne viruses is provided. In particular, this chapter discusses viral metagenomics (including pathogen discovery, viromics, and quality control assessment), within host-viral population analysis and transmission studies.

CURRENT NGS TECHNOLOGIES

Several NGS technologies have been developed utilizing different sequencing chemistries (Table 2.1). These include Roche 454, Illumina, and SOLiD. In addition, it should also be mentioned that a third wave of sequencing technology is emerging and this has been led by Pacific Biosciences. However, great expectations are flourishing around the development of single molecule sequencing in a chip, such as the Oxford nanopore, which is currently under development (https://www.nanoporetech.com). Several recent reviews have outlined the strengths and weaknesses of these technologies [3].

To date, Roche 454 and Illumina have been the most commonly used methods in all viral studies (reviewed in [4]) and are also the two most popular methods for the sequencing of viral foodborne pathogens (Table 2.2). Roche 454 has been popular due to its longer read lengths, currently up to 1000 base pairs (bp), and is useful for looking at viral variants or the coevolution of single nucleotide polymorphisms (SNPs). However, the higher error rate of Roche 454, particularly spanning homopolymeric stretches, has complicated the detection of SNPs. Illumina, by comparison, has been popular because of its improved error rate and its price efficiency in terms of cost per base. Illumina is useful for SNP detection and the read lengths are rapidly expanding with the use of the paired-end technology.

TABLE 2.1
Representative NGS Sequencing Platforms and Their Characteristics

Platform	Run Time (h)	Read Length (bp)	Throughput per Run (Mb)	Typical Errors	Main Biological Applications
Roche 454 FLX +	23	700, up to 1000	700	Insertions/deletions (indels) at homopolymer regions	Microbial genome sequencing, human genome sequencing, transcriptomics, metagenomics
Illumina					
HiSeq 1000	8	2 × 100	400,000	Indels, especially at the end of reads	Microbial genome sequencing, human genome sequencing, transcriptomics, metagenomics
MiSeq 2000 V3	10	2 × 300	<600,000		
SOLiD 4	12	50 × 35	71,000	End of read substitution errors	Microbial genome sequencing, human genome sequencing, transcriptomics, metagenomics
Ion torrent PGM 318 Chip	3	200	1000	Indels at homopolymer regions	Microbial genome sequencing, human genome sequencing, transcriptomics, metagenomics
Pacific Biosciences	3	8.5 up to 30 Kb	375	Random indel errors	Full-length transcriptomics, discovering large structural variants and haplotypes

Source: Reproduced with permission from Luciani, F., et al., *Trends Biotechnol.*, 30, 443–452, 2012 [5].
Note: Data taken from websites of the NGS companies.

TABLE 2.2
Review of NGS Application to the Study of Foodborne Viruses

Virus	Research Area	NGS Platform	Reference
Norovirus	Within-host evolution	Roche 454	[6]
		Illumina MiSeq	[7]
		Illumina MiSeq	[8]
Picornavirus	Pathogen discovery	Roche 454	[9]
			[10]
			[11]
Orthoreoviruses	Pathogen discovery	Ion torrent	[12]
Astrovirus VA1	Pathogen discovery	Roche 454	[13]
Hepatitis E	Within-host evolution and transmission	Illumina	[14]
Rotavirus and poliovirus	Quality control assessment	Roche 454	[15]

NGS ANALYSIS

Unfortunately, the rapid rate of sequence generation comes with a high error cost. Each technology has different types of errors (Table 2.1) that are dependent on the sequencing chemistry utilized (reviewed in [3]). So while the time and cost of generating millions of sequences has been greatly reduced, the time and skill required to analyze the sequences and differentiate true errors from sequence errors has significantly increased. This is especially true for viral analysis as viral genomes have some of the highest mutation rates, up to 10^{-4} as opposed to 10^{-8} substitutions per site in the human genome [16,17]. Subsequently, large within-host population diversity is naturally observed, particularly in RNA or single-stranded DNA viruses. A comprehensive review outlining the challenges in analyzing viral NGS data has recently been conducted [4].

SAMPLE PREPARATION OPTIONS FOR NGS

NGS can potentially be applied to any nucleic acid of interest. Initially, NGS was limited to samples with high quantities of starting material (nanogram range). Consequently, until recently a polymerase chain reaction (PCR) amplicon-based approach was most commonly utilized, as it was an effortless way to obtain large (microgram) quantities of the target sequence of interest. For the majority of the foodborne viruses that have been discovered using deep sequencing technology (Table 2.2), the most common approach was to use a random PCR approach that exponentially amplified all of the nucleic acid in the sample. The disadvantage of this approach was that it required the virus of interest to represent a significant proportion (>1%) of the total genomic material to ensure detection; the lower the presence of the target sequence, the greater the amount of the total sequence required to detect it. In the past, this approach came at a significant cost to ensure that the novel pathogens that were present in low quantities could be detected. Fortunately, the sensitivity of this approach to detect viral variants present in small quantities has been steadily improving, with a gradual reduction in cost-per-base sequence and an ever-increasing amount of the total data generated from a single run. In addition, the last few years have seen several sequencing preparation kits emerge in the market that are able to prepare sequencing libraries from very small quantities of starting material, even in the picogram range. These developments have been essential, as improvements in sample purity and the concentration of the target enable a simpler and faster bioinformatics workflow.

In regard to the RNA virus field, improvements in the sequencing of RNA directly, without the need for prior reverse transcription-polymerase chain reaction (RT-PCR) amplification, will greatly increase the pace of molecular analysis. To date, there have been two main approaches to preparing viral RNA samples for NGS. The first and simpler approach, which has been successfully applied to full genome sequencing of norovirus (NoV) and hepatitis C virus (HCV), is to use standard RNA extraction protocols to extract RNA from clinical stool specimens. This total RNA is then submitted for NGS [7]. The advantage of this approach is that the sample preparation time upfront is minimal, but the disadvantage is that a large total sequence data yield is required as the majority of the sequence data (98.1%–99.9%) generated will be of human and microbial origin, particularly ribosomal RNA sequences. Generally with this approach, the Illumina NGS pipeline is preferred given the lower cost-per-base ratio. This approach has been able to sensitively detect full-length NoV genomes from a minimum concentration of ~300 copies per microliter when approximately 3 million reads per sample are generated [7]. A more specific approach that has been successfully applied to West Nile virus (WNV), respiratory syncytial virus, and human immunodeficiency virus (HIV), is to have some additional steps upstream of the sequencing that help to remove contaminating cellular material and an additional deoxyribonuclease (DNase) treatment step to remove contaminating DNA [18]. The advantage of this process is that the RNA sample submitted for sequencing is of a higher purity and hence generally represents 2%–70% of the total sequence reads obtained. Although wide ranging, this is a significant improvement on the ~2% when DNase is not used. This approach is particularly useful for samples with low viral loads. The drawbacks are the extra time and cost incurred upstream, and the small starting quantity of RNA requires specialized library preparation kits. The advantage is better sensitivity, a higher return for your sequence expense, and easier bioinformatic analysis due to lower amounts of nonspecific nucleic acid sequences. The preferred option is dependent on the sample and the questions that the researchers wish to ask.

VIRAL METAGENOMICS

Until recently, viral sequencing and genome discovery had been limited to PCR- and cloning-based methods, which required a very specific assay developed on sequence knowledge to look at an individual class of viruses. This often limited viral studies to the examination of known viruses. The advent of NGS has revolutionized viral molecular studies, as it is capable of random amplification and sequencing of all RNA or DNA present in a sample. *Metagenomics is the term often given to the wholesale extraction and sequencing of all nucleic acids in a given sample.* Metagenomics has been an invaluable tool and has been widely applied to the sequencing and identification of a large variety of organisms, in both clinical and environmental settings, as no *a priori* knowledge of the samples sequence(s) is required.

Viral metagenomics is often referred to as *viromics*, and has become an invaluable method in virus discovery and as a sensitive quality control assessment tool [19]. This is particularly important for viruses, as many viruses cannot be cultured. Hence, in the past, detection of new viruses was problematic given that knowledge of the viral sequence was often needed in order to have a sensitive detection method. The application of viral metagenomics as it relates to viral foodborne pathogens is discussed in the following sections.

VIRUS DISCOVERY: RAPID FULL GENOME SEQUENCING OF NEW VARIANTS

NGS technology has massive potential as a diagnostic tool in infections with undetermined etiology. In 2003, a novel coronavirus, termed *severe acute respiratory syndrome* (SARS) virus, was recognized using traditional methods within 1 month of an outbreak of atypical pneumonia. The rapid identification of the novel virus enabled global measures to be promptly introduced in order to contain the spread of the outbreak to around 2000 people. Since the identification of SARS, the advent of NGS technology has become a robust substitute for the previous, more laborious, sequencing

TABLE 2.3
Novel Viruses Identified in Patients Suffering Foodborne Illnesses

Novel Virus	Proposed Viral Family	Specimen	Disease	Region Identified	Reference
Klassevirus/Salivirus	Picornaviridae	Stool	Diarrhea	USA Nigeria Tunisia Nepal Australia	[9]
Orthoreoviruses	Reoviridae		Acute gastroenteritis	Slovenia	[12]
Astrovirus AV1	Astroviridae	Stool	Diarrhea and vomiting	USA	[13]

methods. For example, in 2008, an outbreak of hemorrhagic fever with unexplained etiology was reported in South Africa [20]. Using NGS technology, a novel arenavirus was linked to the outbreak within 2 weeks [20].

In regard to foodborne illness, a large proportion of the causes remain unidentified. In 2011, the Centers for Disease Control and Prevention (CDC) estimated that 80% (~9.4 million illnesses per year) of the cases of reported foodborne illnesses had unspecified causes and 56% (1686 deaths/year) of the deaths associated with foodborne illness had unspecified agents [21]. In developing countries, detection of the etiological agent in cases of gastroenteritis is often omitted, as the disease is seen as "self-limiting" and not life-threatening. Over the last 4 years, an array of NGS techniques have been applied predominantly to nucleic acid material collected from the stools of individuals suffering foodborne-like illnesses with unknown etiology (Table 2.3). To date, a number of novel viruses have been identified. Table 2.3 outlines the specimens these viruses were detected in, their associated diseases, and the regions of detection. It should be noted that for many of these viruses a causative link with disease has been suggested, most have not yet been proved.

It should be noted that the ultrasensitive nature of NGS technology does have drawbacks and care needs to be taken when discovering new pathogens. In 2013, a team of researchers from the National Institute of Health (NIH) thought that they had identified a new pathogen in the blood of 92 Chinese individuals who were suffering from hepatitis, with no known etiology [22]. Another research group at University College San Francisco also independently identified a nearly identical virus (98% nucleotide identity) to the NIH group in a different patient group [23].

They called this virus *parvovirus-like hybrid virus* (PHV). However, Chiu et al. soon realized that this virus had been associated with many other human and viral diseases and they became suspicious. Using NGS technology, they traced the source of PHV virus to the ocean-derived silica that is widely used as part of a nucleic acid extraction method [23], and it became apparent that the samples had been contaminated by the sample preparation procedure.

Quality Control Assessment

To date, many of the effective vaccines against viral diseases utilize live attenuated viruses as the antigen to generate protective immunity. This includes the vaccines for measles, mumps, rubella, yellow fever, and rotavirus. Viral attenuation occurs by genetically adapting the viruses so that they lose their pathogenic potential, but maintain their antigenic profile so as to still generate long-lasting protective immunity. However, there is a risk during the attenuation process that these viral strains could become contaminated by other viruses. Some examples will be described in this chapter.

Metagenomic techniques have been applied to several of the live virus vaccines—including the rotavirus vaccines, Rotarix and RotaTeq—to screen them for viral contaminants [15]. In this study, an NGS analysis using the Roche 454 method identified a single read with 96% identity to simian retrovirus (SRV) in the RotaTeq vaccine. In the Rotarix vaccine, over 6000 reads with 98% identity to porcine circovirus-1 (PCV-1) were also discovered [15]. Further molecular testing indicated that the source of the SRV genome was from the Vero cell line used to generate the live attenuated rotavirus vaccine. However, subsequent analysis indicated that the SRV was not generating viral particles and carried a polymerase mutation that likely made the viral genome defective for replication [15]. It was therefore hypothesized that the SRV DNA had likely integrated into the germ line of the African green monkey host and hence was present in the Vero cell line, from which they were generated. Although low amounts of SRV DNA were present in the analyzed sample, it was considered to be of no risk to people receiving vaccinations, as the genome appeared to be defective [15]. Additionally, two prior infections of humans with replication-competent SRV had no clinical presentations or ongoing viremia [24].

The molecular analysis of the PCV-1 contamination in the Rotarix vaccine was less well defined. PCV-1 is a highly common, nonpathogenic infection in pigs. It was hypothesized that the serum-free media that supplemented the Vero cell line used to grow the vaccine strains had contaminated the cell line [15]. Surprisingly, 41.6% of the total NGS reads generated from the Rotarix vaccine were aligned to the PCV-1 genome. This is a highly significant number of reads. Unfortunately, however, the random nature of the DNA amplification methods used to generate the viral sequences did not enable a direct comparison of the viral load of each virus in the vaccine. The pathogenic potential of PCV-1 has not been well characterized in humans, but the limited studies to date suggest that it is unlikely to be of clinical concern. The highly sensitive and nonspecific nature of NGS techniques has proven them to be a valuable tool for the quality control assessment of biologically derived samples that may be used by and for humans.

WITHIN-HOST POPULATION ANALYSIS

The comparatively small genome size of viruses, their rapid replication rates, and short half-life, mean that viruses can rapidly turn over and generate new viral populations. This is particularly true of single-stranded RNA viruses that also have limited replication fidelity due to the lack of proofreading mechanisms in their RNA-dependent RNA polymerases. In the better-studied human pathogens, such as HIV and HCV, turnover of the viral population is rapid, with an estimated viral half-life of 3–5 h [25]. Rapid turnover is likely essential for viral pathogens associated with chronic disease, as they need to constantly evade the responding immune system and drug therapy to survive.

In the case of acute foodborne viral infections in immunocompetent individuals, limited research has been conducted investigating the viral dynamics within the host (within-host). A study of three NoV-infected immunocompetent individuals [6] suggested that limited viral diversity existed in acutely infected cases. This is also similar to what has been observed in the few acute respiratory viral infections that have been studied [26,27]. Examination of the antigenic domain of the NoV capsid revealed that in these immunocompetent individuals almost 80% of the viral population was dominated by a single viral variant and limited viral evolution was observed over the course of infection (Figure 2.1). This was in contrast to an immunocompromised infant, chronically infected with NoV, who showed a highly heterogeneous viral population with rapid turnover within as little as 3 days. In this case, the dominant variant only represented about 10% of the viral population [6] (Figure 2.1).

Limited within-host diversity was also observed in another NoV study in immunocompetent hosts where an RNA-sequencing (RNA-seq) approach followed by *de novo* assembly was used to rapidly sequence full-length NoV genomes [8]. This same study also used RNA-seq followed by *de novo* assembly to generate viral consensus sequences, which were examined for recombination events within the circulating epidemic variants. However, care should be taken when utilizing longer sequences reconstructed from short NGS reads for analysis, as sample bias or bioinformatic bias may generate *in silico* recombinant sequences that are in fact just biased samplings of a mixed within-host viral population.

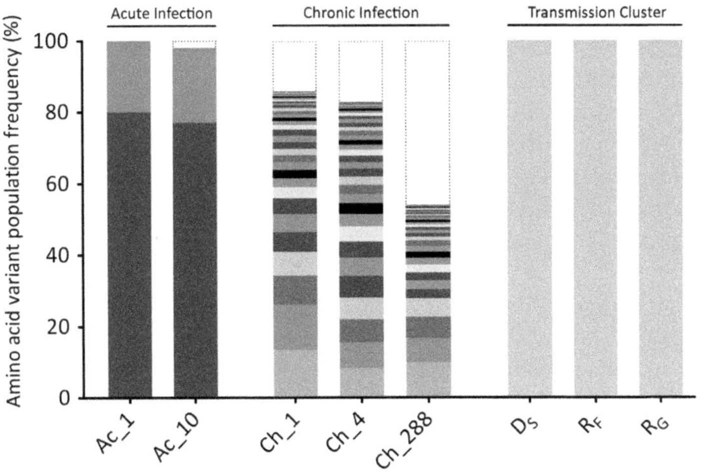

FIGURE 2.1 Comparison of intra-host variation in the NoV capsid sequence. The full-length sequence of the NoV capsid (ORF2) from the whole within-host viral population was reconstructed from NGS data. Sequences were translated to determine the intra-host variant frequency at the amino acid level. The proportion of each unique variant was plotted for each sample point analyzed. In the acutely infected subject (Ac), only two variants were detected and their frequencies of ~79% and ~20%, respectively, remained stable over the 9 days of infection. In contrast, in the chronically infected subject no dominant variant was observed, instead multiple low frequency variants coexisted and their proportion within the population varied over time. The black dotted lines represent the low frequency variants that were under the detection threshold (<2%). (Reproduced with permission from Bull, R.A., et al., *J Virol.*, 86, 3219–3229, 2013.)

UNDERSTANDING VIRAL TRANSMISSION

Viral foodborne pathogens are predominantly transmitted through fecal contamination of food and water; however, infection can also occur through airborne fomites and contact with contaminated surfaces such as door handles. NGS technology has been recently applied to several viral foodborne and bloodborne pathogens to try and understand the process of viral transmission and subsequent infection.

Surprisingly, NGS analysis conducted on samples isolated soon after transmission and subsequent infection by bloodborne pathogens, such as HIV and HCV and more recently the foodborne pathogen NoV, has revealed that despite the likelihood of thousands of viral variants being transmitted only a minor number of these successfully establish an infection [28–32]. This suggests that transmission is an important contributor to interhost viral evolution.

The variants that successfully establish infection are commonly referred to as the *transmitted/founder* (T/F) variant. In HIV and HCV, on average 1–3 T/F variants are identified following blood-blood transmission [28–32]. The strong functional constraints of the transmitted variants are believed to drive this bottleneck event (reviewed in [33]). In HIV, CCR5 usage and glycosylation profiles of the infecting variants are thought to attribute to their success in establishing infection. However, both HIV and HCV are associated with chronic infection and often, following the establishment of infection, will undergo significant immune escape, which includes modifications of their B-cell epitopes. This invariably alters their host cell receptor binding and modifying glycosylation profiles so as to shield the virions from antibody recognition [34]. Alternatively, it has been suggested that, in a HCV transplant recipient, circulating antibodies to the dominant variant prevented infection of the implant. Subsequently, minor variants with resistance to circulating anti-HCV antibodies were able to infect the implant [35].

Interestingly, the number of T/F variants is generally higher via the mucosal transmission of HIV, compared with parenteral transmission. This suggests that variations in the transmission route could alter the number of T/F variants observed in the infection. This is also supported by in-depth viral population analyses of acute respiratory-borne viral infections caused by rhinovirus and equine influenza virus. Here, it was revealed that transmission events were not characterized by strong genetic bottlenecks, but rather by coinfection of a cloud of closely related variants [26,27]. Additionally, in foot-and-mouth disease virus (FMDV), different evolutionary trajectories have been observed in isolated compartments of the same host [36]. Therefore, in the case of NoV as an example, it is possible that the viral population excreted in the vomitus could be genetically different from the virus excreted in the feces. Understanding the transmission event is important for vaccine development and antiviral therapy design.

The only study to date on the human transmission of foodborne pathogens has been conducted on NoV [6]. In this case study, a child suffering gastrointestinal symptoms including vomiting and diarrhea, was suspected of transmitting NoV to two of his primary carers—his grandfather and father—both of whom fell ill within 24 h of the onset of the child's symptoms. NGS sequencing of a stool sample collected from the donor (son) revealed a single variant representing more than 60% of the total population (Figure 2.1). Interestingly, in both recipients (father and grandfather) a novel, minor, distinct variant was transmitted to each recipient (Figure 2.2), rather than the dominant virus found in the host. The T/F variant that became dominant in both of the recipients was present at <0.01% in the donor. Surprisingly, both T/F variants carried amino acid differences in the capsid protein when compared with the donor's dominant variant. Since NoV is known to bind to histo-blood group antigen (HBGA) attachment factors [37–39], it was suggested that NoV may also

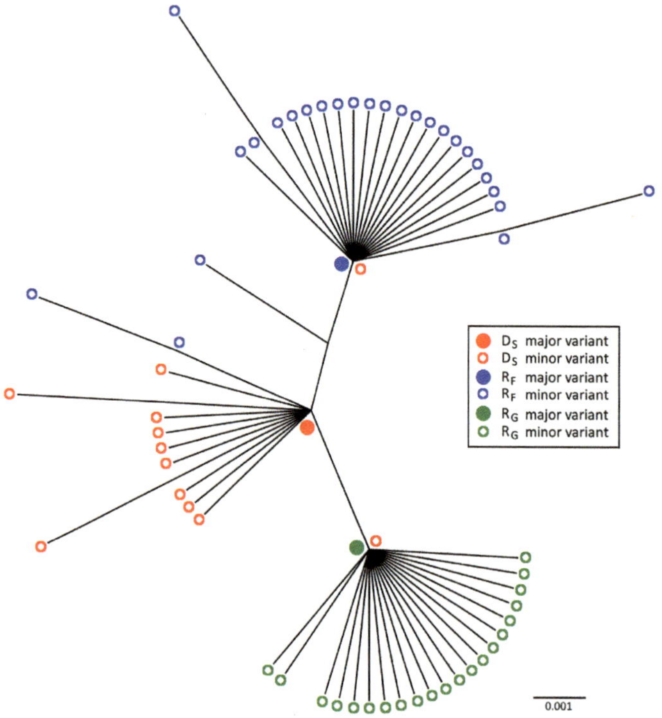

FIGURE 2.2 Phylogenetic analysis of the intra-host NoV population in a three-person transmission cluster. Next-generation sequencing was applied to an amplicon generated from the NoV capsid region (partial ORF2 and partial ORF3) isolated from three individuals. In this high-resolution analysis, each subject's major variant was located at the node of the branch that leads to their minor variants. Furthermore, this analysis revealed that each recipient's major viral variant, R_F and R_G, was found to be 100% identical to a unique minor variant (<0.01%) isolated from the donor (D_S). Surprisingly, the donor's major variant was not identified in the recipient, even at low frequency (<0.01%). The distance scale represents the number of nucleotide substitutions per position. (Reproduced with permission from Bull, R.A., et al., *J Virol.*, 86, 3219–3229, 2013.)

have structural constraints that are imposed during transmission. Alternately, the transmission of a minor variant could just be a random event as NoV has been reported to have a low infectious dose, approximately 18 virus particles [40]. It would be interesting to see if the transmission dynamics of foodborne viral infections are altered when infection occurs via the ingestion of contaminated food or water, compared with fecal–oral transmission.

In another study, NGS was applied to investigate the effect of interspecies transmission on hepatitis E virus (HEV) diversity and host-driven selection [14]. Certain subtypes of HEV (genotype III) are known to be actively circulating between swine and humans [14]. In this study, a human stool containing 2×10^9 copies of subtype 3f HEV RNA was used to orally inoculate two 3-month-old piglets. After inoculation, longitudinal fecal samples and one bile sample were collected 15 days postinoculation. NGS of the HEV inoculum from the human stool sample and the recipient swine samples revealed the transmission of multiple variants (12 of 42). However, the same dominant variant was present in all three subjects, as shown by the conservation of the consensus sequence [14]. From the lack of any observed preferential variant selection, the authors hypothesized that the HEV subtype 3 appears to have adapted to allow for stable transmission between human and swine populations [14].

CONCLUSION

NGS is an invaluable tool that is continuing to improve the time and cost required to investigate viral dynamics in depth. It has been widely applied to answer several questions relevant to foodborne diseases of viral etiology. This includes the identification of novel viral pathogens and investigations of viral genetic traits associated with the transmission and disease state that will help inform vaccine design.

REFERENCES

1. Sanger F, Nicklen S, Coulson AR. 1977. DNA sequencing with chain-terminating inhibitors. *Proc Natl Acad Sci USA* 74:5463–5467.
2. Radford AD, Chapman D, Dixon L, Chantrey J, Darby AC, Hall N. 2012. Application of next-generation sequencing technologies in virology. *J Gen Virol* 93:1853–1868.
3. Liu L, Li Y, Li S, Hu N, He Y, Pong R, Lin D, Lu L, Law M. 2012. Comparison of next-generation sequencing systems. *J Biomed Biotechnol* 2012:251364.
4. Beerenwinkel N, Gunthard HF, Roth V, Metzner KJ. 2012. Challenges and opportunities in estimating viral genetic diversity from next-generation sequencing data. *Front Microbiol* 3:329.
5. Luciani F, Bull RA, Lloyd AR. 2012. Next generation deep sequencing and vaccine design: Today and tomorrow. *Trends Biotechnol* 30:443–452.
6. Bull RA, Eden JS, Luciani F, McElroy K, Rawlinson WD, White PA. 2013. Contribution of intra- and interhost dynamics to norovirus evolution. *J Virol* 86:3219–3229.
7. Batty EM, Wong TH, Trebes A, Argoud K, Attar M, Buck D, Ip CL, et al. 2013. A modified RNA-Seq approach for whole genome sequencing of RNA viruses from faecal and blood samples. *PLoS One* 8:e66129.
8. Wong TH, Dearlove BL, Hedge J, Giess AP, Piazza P, Trebes A, Paul J, et al. 2013. Whole genome sequencing and *de novo* assembly identifies Sydney-like variant noroviruses and recombinants during the winter 2012/2013 outbreak in England. *Virol J* 10:335.
9. Greninger AL, Runckel C, Chiu CY, Haggerty T, Parsonnet J, Ganem D, DeRisi JL. 2009. The complete genome of klassevirus: A novel picornavirus in pediatric stool. *Virol J* 6:82.
10. Li L, Victoria J, Kapoor A, Blinkova O, Wang C, Babrzadeh F, Mason CJ, et al. 2009. A novel picornavirus associated with gastroenteritis. *J Virol* 83:12002–12006.
11. Holtz LR, Finkbeiner SR, Zhao G, Kirkwood CD, Girones R, Pipas JM, Wang D. 2009. Klassevirus 1, a previously undescribed member of the family Picornaviridae, is globally widespread. *Virol J* 6:86.
12. Steyer A, Gutierrez-Aguire I, Kolenc M, Koren S, Kutnjak D, Pokorn M, Poljsak-Prijatelj M, et al. 2013. High similarity of novel orthoreovirus detected in a child hospitalized with acute gastroenteritis to mammalian orthoreoviruses found in bats in Europe. *J Clin Microbiol* 51:3818–3825.

13. Finkbeiner SR, Li Y, Ruone S, Conrardy C, Gregoricus N, Toney D, Virgin HW, et al. 2009. Identification of a novel astrovirus (astrovirus VA1) associated with an outbreak of acute gastroenteritis. *J Virol* 83:10836–10839.
14. Bouquet J, Cheval J, Rogee S, Pavio N, Eloit M. 2012. Identical consensus sequence and conserved genomic polymorphism of hepatitis E virus during controlled interspecies transmission. *J Virol* 86:6238–6245.
15. Victoria JG, Wang C, Jones MS, Jaing C, McLoughlin K, Gardner S, Delwart EL. 2010. Viral nucleic acids in live-attenuated vaccines: Detection of minority variants and an adventitious virus. *J Virol* 84:6033–6040.
16. Sanjuan R, Nebot MR, Chirico N, Mansky LM, Belshaw R. 2010. Viral mutation rates. *J Virol* 84:9733–9748.
17. Nachman MW, Crowell SL. 2000. Estimate of the mutation rate per nucleotide in humans. *Genetics* 156:297–304.
18. Malboeuf CM, Yang X, Charlebois P, Qu J, Berlin AM, Casali M, Pesko KN, et al. 2012. Complete viral RNA genome sequencing of ultra-low copy samples by sequence-independent amplification. *Nucleic Acids Res* 41:e13.
19. Quinones-Mateu ME, Avila S, Reyes-Teran G, Martinez MA. 2014. Deep sequencing: Becoming a critical tool in clinical virology. *J Clin Virol* 61:9–19.
20. Briese T, Paweska JT, McMullan LK, Hutchison SK, Street C, Palacios G, Khristova ML, et al. 2009. Genetic detection and characterization of Lujo virus, a new hemorrhagic fever-associated arenavirus from southern Africa. *PLoS Pathog* 5:e1000455.
21. Centres for Disease Control and Prevention (CDC). 2011. Estimates of Foodborne Illness in the United States. http://www.cdc.gov/foodborneburden/2011-foodborne-estimates.html. Accessed 2014.
22. Xu B, Zhi N, Hu G, Wan Z, Zheng X, Liu X, Wong S, et al. 2013. Hybrid DNA virus in Chinese patients with seronegative hepatitis discovered by deep sequencing. *Proc Natl Acad Sci USA* 110:10264–10269.
23. Naccache SN, Greninger AL, Lee D, Coffey LL, Phan T, Rein-Weston A, Aronsohn A, Hackett J, Jr., Delwart EL, Chiu CY. 2013. The perils of pathogen discovery: Origin of a novel parvovirus-like hybrid genome traced to nucleic acid extraction spin columns. *J Virol* 87:11966–11977.
24. Lerche NW, Switzer WM, Yee JL, Shanmugam V, Rosenthal AN, Chapman LE, Folks TM, Heneine W. 2001. Evidence of infection with simian type D retrovirus in persons occupationally exposed to nonhuman primates. *J Virol* 75:1783–1789.
25. Neumann AU, Lam NP, Dahari H, Gretch DR, Wiley TE, Layden TJ, Perelson AS. 1998. Hepatitis C viral dynamics *in vivo* and the antiviral efficacy of interferon-alpha therapy. *Science* 282:103–107.
26. Cordey S, Junier T, Gerlach D, Gobbini F, Farinelli L, Zdobnov EM, Winther B, Tapparel C, Kaiser L. 2010. Rhinovirus genome evolution during experimental human infection. *PLoS One* 5:e10588.
27. Murcia PR, Baillie GJ, Daly J, Elton D, Jervis C, Mumford JA, Newton R, et al. 2010. Intra- and inter-host evolutionary dynamics of equine influenza virus. *J Virol* 84:6943–6954.
28. Bull RA, Luciani F, McElroy K, Gaudieri S, Pham ST, Chopra A, Cameron B, et al. 2011. Sequential bottlenecks drive viral evolution in early acute hepatitis C virus infection. *PLoS Pathog* 7:e1002243.
29. Fischer W, Ganusov VV, Giorgi EE, Hraber PT, Keele BF, Leitner T, Han CS, et al. 2010. Transmission of single HIV-1 genomes and dynamics of early immune escape revealed by ultra-deep sequencing. *PLoS One* 5:e12303.
30. Keele BF, Giorgi EE, Salazar-Gonzalez JF, Decker JM, Pham KT, Salazar MG, Sun C, et al. 2008. Identification and characterization of transmitted and early founder virus envelopes in primary HIV-1 infection. *Proc Natl Acad Sci USA* 105:7552–7557.
31. Goonetilleke N, Liu MK, Salazar-Gonzalez JF, Ferrari G, Giorgi E, Ganusov VV, Keele BF, et al. 2009. The first T cell response to transmitted/founder virus contributes to the control of acute viremia in HIV-1 infection. *J Exp Med* 206:1253–1272.
32. Salazar-Gonzalez JF, Salazar MG, Keele BF, Learn GH, Giorgi EE, Li H, Decker JM, et al. 2009. Genetic identity, biological phenotype, and evolutionary pathways of transmitted/founder viruses in acute and early HIV-1 infection. *J Exp Med* 206:1273–1289.
33. Keele BF. 2010. Identifying and characterizing recently transmitted viruses. *Curr Opin HIV AIDS* 5:327–334.
34. Go EP, Hewawasam G, Liao HX, Chen H, Ping LH, Anderson JA, Hua DC, Haynes BF, Desaire H. 2011. Characterization of glycosylation profiles of HIV-1 transmitted/founder envelopes by mass spectrometry. *J Virol* 85:8270–8284.

35. Fafi-Kremer S, Fofana I, Soulier E, Carolla P, Meuleman P, Leroux-Roels G, Patel AH, et al. 2010. Viral entry and escape from antibody-mediated neutralization influence hepatitis C virus reinfection in liver transplantation. *J Exp Med* 207:2019–2031.
36. Wright CF, Morelli MJ, Thebaud G, Knowles NJ, Herzyk P, Paton DJ, Haydon DT, King DP. 2011. Beyond the consensus: Dissecting within-host viral population diversity of foot-and-mouth disease virus by using next-generation genome sequencing. *J Virol* 85:2266–2275.
37. Cao S, Lou Z, Tan M, Chen Y, Liu Y, Zhang Z, Zhang XC, Jiang X, Li X, Rao Z. 2007. Structural basis for the recognition of blood group trisaccharides by norovirus. *J Virol* 81:5949–5957.
38. Tan M, Jiang X. 2008. Association of histo-blood group antigens with susceptibility to norovirus infection may be strain-specific rather than genogroup dependent. *J Infect Dis* 198:940–941; author reply 942–943.
39. Shanker S, Choi JM, Sankaran B, Atmar RL, Estes MK, Prasad BV. 2011. Structural analysis of histo-blood group antigen binding specificity in a norovirus GII.4 epidemic variant: Implications for epochal evolution. *J Virol* 85:8635–8645.
40. Teunis PF, Moe CL, Liu P, Miller SE, Lindesmith L, Baric RS, Le Pendu J, Calderon RL. 2008. Norwalk virus: How infectious is it? *J Med Virol* 80:1468–1476.

3 New Technologies for Viral Diagnosis and Detection Using Enteroviruses as an Example

Jason Roberts, Bruce Thorley, Scott Bowden, and Peter A. Revill

CONTENTS

INTRODUCTION

While viruses may be present in food, it is their ability to establish a productive infection upon ingestion that is of public health significance. Viral replication in the gut and subsequent shedding in the feces facilitates laboratory detection, but epidemiological evidence is required to establish a causal association with a specific disease. Norovirus, adenovirus, aichivirus, astrovirus, coronavirus, rotavirus, and sapovirus are all accepted as causative viral agents of foodborne acute gastrointestinal illness [1]. On the other hand, an association between gastrointestinal illness and enteroviruses can be difficult to establish, even though enteroviruses replicate in the gut, since most infections are asymptomatic [2]. Hepatitis A virus and hepatitis E virus are also capable of replicating in the gut but the clinical presentation manifests as hepatitis [1].

In this chapter, we provide a historical perspective of viral diagnostic testing before presenting an overview of recent developments in laboratory testing. Finally, we describe the discovery of a new enterovirus, EV-A120, which did not grow in culture, and how the capsid structure was characterized by protein homology modeling using recent advances in computational biophysics.

AN OVERVIEW OF VIRAL DIAGNOSTIC TESTING

Virus detection has undergone a paradigm shift as a consequence of the development of new technologies. Traditional techniques of virus identification, such as cell culture and negative staining electron microscopy, have not been discarded as they can still be diagnostically useful. However, many laboratories have now joined the molecular revolution and use nucleic acid testing as a routine tool. This shift in technology is exemplified by the procedures used to discover some of the most important foodborne pathogens: norovirus, hepatitis A virus, and hepatitis E virus. Norovirus was initially known as the *Norwalk agent*, named after the city Norwalk, Ohio, where a large outbreak of gastroenteritis was reported among schoolchildren. The causative agent, a 27–35 nm diameter virus-like particle, was visualized using immunoelectron microscopy of the stool filtrates of volunteers [3]. The same technique was used by Feinstone and colleagues to identify hepatitis A virus in extracts of fecal samples taken from people who were experimentally infected with "infectious" hepatitis [4]. In contrast, although epidemics of enterically transmitted non-A, non-B hepatitis had been widely documented in many developing countries, it was not until the molecular cloning of a partial sequence of the causative agent was conducted that it could be definitively defined as hepatitis E virus [5].

Notwithstanding the impetus for molecular assays, serological testing remains a mainstay of viral diagnosis. There is a wide range of serological assays, with the predominant technique being an enzyme immunoassay, where a recombinant viral antigen (or sometimes a polyclonal and/or monoclonal antibody) is immobilized onto a surface and reacts with the test sample. An antigen–antibody complex is detected by a subsequent antibody tagged to a marker, usually an enzyme, which produces a color change when the appropriate substrate is added. These assays offer many advantages: they are suitable for mass screening, compatible for automation, and allow for a rapid turnaround of results. Many serological assays are indirect, detecting the presence of specific antibodies produced by the patient's immune system in response to his or her exposure to an infectious agent, rather than the infectious agent itself. Indeed, serological testing of the blood for a number of pathogens associated with infectious diseases, including hepatitis B, hepatitis C, and human immunodeficiency virus (HIV), represents one of the great improvements in public health. These assays do not necessarily identify whether the infection is past or current and for some infections the window period (i.e., the time for an immune response to become detectable) can be lengthy. Furthermore, with the growing availability of antiviral agents, serological assays are not useful for determining responses to therapy, unlike molecular assays. A different application of serology has been its use in subgrouping and distinguishing related virus types. Using polyclonal sera, monospecific sera, and later monoclonal antibodies, viral subtypes can be distinguished by antibody neutralization. Used originally to separate poliovirus into three serotypes, this technique can be used to type enteroviruses [6], but has been widely replaced by nucleic acid sequence methodology [7,8]. This has led to the identification of more than 50 new types of enteroviruses that can infect humans and is not dependent on the production of specific antisera raised in animals. Viruses with only one serotype, such as hepatitis A or hepatitis E viruses, and those that are difficult to culture are not amenable to characterization by antisera neutralization.

Many of the disadvantages of standard serological assays can be overcome by nucleic acid testing, and the predominant method used is the polymerase chain reaction (PCR), a target amplification technique. As the genetic material of the virus is detected directly, the time to diagnosis can be substantially reduced and by quantification of viral DNA or RNA, the efficacy of antiviral therapy can be assessed. A PCR, as it was first introduced, requires a pair of synthetic oligonucleotides or primers specific to the agent and each hybridizes to one strand of a DNA template (usually double-stranded DNA that has been separated into single-stranded DNA by heating to >90°C). The region spanned by the primers is replicated by a thermostable DNA polymerase, creating a complementary strand of DNA, and the process is repeated for a number of predetermined cycles (usually around 30), exponentially amplifying the target sequence. For RNA targets, complementary DNA (cDNA) is first produced by reverse transcription to allow the PCR to proceed because it requires a DNA substrate (RT-PCR). The detection of the PCR product can be performed by several techniques but most commonly by agarose gel electrophoresis. Rather than relying on the amplified product size as a determinant of specificity, many laboratories employ nested PCR that improves sensitivity as well as specificity. This procedure uses two pairs of primers, one set internal to the other, and two successive rounds of amplification. Other techniques used to demonstrate amplified product specificity include fluorescent dyes and hybridization with specific probes and sequence analysis.

Despite the many advantages afforded by nucleic acid testing in viral diagnosis as outlined thus far, the traditional PCR technique, or conventional PCR, also has some limitations. The detection of the PCR product relies on laborious post-PCR processing, increasing both the time to diagnosis and the chances of amplicon contamination. Moreover, conventional PCR has a poor dynamic range of quantification and for several diseases, including hepatitis B and C, and HIV infections, where viral load monitoring is a necessary component of clinical management. For some chronic infections, viral load has been shown to correlate with a greater risk of disease progression and for diseases where antiviral therapy is available, viral load monitoring can indicate the efficacy of treatment. Therefore, viral load measurements can also be used to assess prognosis. Many of the outlined deficiencies in conventional PCR have been overcome by the development of real-time PCR, in which the target amplification and detection occur concurrently and the accumulating PCR product can be visualized as the reaction progresses (in "real time"). This has been made possible by the labeling of either the amplified sequence or a probe to detect the target sequence with fluorogenic molecules. Appropriate instrumentation can detect the fluorescent signal. When the real-time PCR assay includes standards of a known copy number, the amount of target sequences in a sample can be quantified. The common formats use intercalating fluorescent dyes that bind preferentially to double-stranded DNA, or for even greater specificity, 5′ nuclease or TaqMan® hydrolysis fluorescent probes. These probes work by binding to single-stranded DNA, then as the Taq polymerase synthesizes the nascent strand, the exonuclease activity of the polymerase degrades the probe that has annealed to the template. This probe degradation releases a fluorophore, which had previously been in close proximity to a quencher molecule that had prevented fluorescence. Real-time PCR assays require no postamplification processing, decreasing the time to obtaining a result and substantially minimizing the risk of amplicon carryover contamination. Quantification levels can be generated over a wide dynamic range and accuracy is high, with data generated in the early log-linear phase of amplification when conditions are optimal.

VIRUS DISCOVERY IN THE NEW MILLENNIUM

Most of the human viruses we know today were first identified using filtration, cytopathology, electron microscopy, gel diffusion assays, replication studies in laboratory animals, embryonated eggs or cell culture, or combinations of these techniques [9]. The advent of the molecular biology era has enabled detailed genetic characterization initially through the use of techniques such as restriction fragment length polymorphism (RFLP) and nucleic acid hybridization assays. However, it was the development of PCR in the 1980s, later used in tandem with automated "population-based" Sanger DNA sequence analysis, that facilitated a quantum leap in viral characterization and identification. The ability to rapidly analyze viral sequences without the need for culturing the virus also enabled the development of highly sensitive viral diagnostic and genotyping assays. Whereas previously, with the use of immunoassays, diagnostic assays were predominantly protein-based. The ability to sequence individual viral genomes also enabled advanced phylogenetic analyses to infer evolutionary relationships between individual viruses and virus groups. However, one weakness of traditional population-based Sanger sequencing is its inability to distinguish sequences that are present at a low copy number in the population, with sensitivity limited to sequences representing approximately 10%–20% of the viral pool. That is, if a given virus sequence is present at less than 10% within the population, it will not be distinguished from the dominant virus sequence by

Sanger sequencing. Less prevalent viral strains within a population could only be identified through the cloning of PCR products or infectious assays that depend on the production of a cytopathic effect, such as virus neutralization and plaque purification: a time-consuming and labor-intensive approach. For example, to identify one clone in a population with a sensitivity of 0.5%, it would be necessary to sequence 200 clones. However, new technologies can now overcome these limitations. The advent of next-generation sequencing (NGS)—the most significant advance in the field of viral discovery in the past 20 years—has enabled the detection of viruses with an extremely high level of sensitivity and accuracy, with the identification of organisms representing as little as 0.5% of the viral population routinely reported. This technology is driving viral discovery at previously unimaginable levels. In turn, these technologies have facilitated metagenomic studies, enabling the discovery of new genetic material recovered directly from environmental and tissue samples. In the case of foodborne viruses, these samples may include the food source itself, such as shellfish responsible for outbreaks of hepatitis A virus (HAV) and norovirus [10]; fecal samples; and tissue from the host organisms themselves. This approach has been used to identify novel and previously characterized viruses in fecal and tissue samples from a range of hosts, including bats, foxes, birds, and humans, some examples of which are provided throughout this text. Recently, a metagenomic approach was used to sample aquatic environments and associated sediments, leading to the identification of a completely novel amoeba virus, named *pandoravirus* [11]. The discovery of these and other "megaviruses," with previously unimaginable sequence complexity (pandoravirus genomes range from 1.9 to 2.5 megabases in size [11]), may completely alter our understanding of viral evolution. It remains to be determined what other organisms lie hidden in the murky depths, or even in the food we eat on a daily basis.

METAGENOMICS

The concept of metagenomics was first proposed in 1988 by Handelsmann and colleagues ([12], who characterized the genomes of soil organisms that could not be cultured, by first isolating their DNA and cloning it in culturable organisms (i.e., bacteria). The newly developed PCR technology was well suited to this approach, although it relied on some knowledge of the genetic information of the target organism to enable the design of the primers required for amplification. Circumventing the need for any sequence information, Breitbart and colleagues used a linker amplified shotgun library (LASL) approach to isolate hundreds of viral sequences from 200 L of seawater [13]. This involved first purifying viral DNA by ultracentrifugation, then shearing and ligating adapters to the DNA termini for ligation into a plasmid vector [13]. The DNA could then be sequenced using standard Sanger dideoxy sequencing. The same group subsequently used this approach to perform the first metagenomic analyses of uncultured viruses in human feces, identifying over 1200 viral genotypes [14]. Although ingenious, this approach is highly labor-intensive, requiring the individual screening of thousands of clones. The advent of NGS has enabled sequencing of organisms at previously unheralded depths without any prior knowledge of the genetic information. Although there is still a place for traditional methods such as reverse transcription-polymerase chain reaction (RT-PCR) and PCR, next-generation technology and other technologies, such as mass spectrometry, have enabled us to explore and detect viral populations in ways inconceivable before their introduction. It is not the purpose of this chapter to detail all of the NGS technologies currently available. However, we provide examples of some of these approaches, as well as more traditional methods that have been used to detect known and novel foodborne viruses, and we discuss future trends in this ever-developing field. A summary of the approaches that could be used for the analysis of viral proteins or nucleic acids in a fecal sample is shown in Figure 3.1.

We now describe past and current technologies used to detect viruses, particularly those present in human fecal samples that pose the greatest risk of viral contamination of food and water.

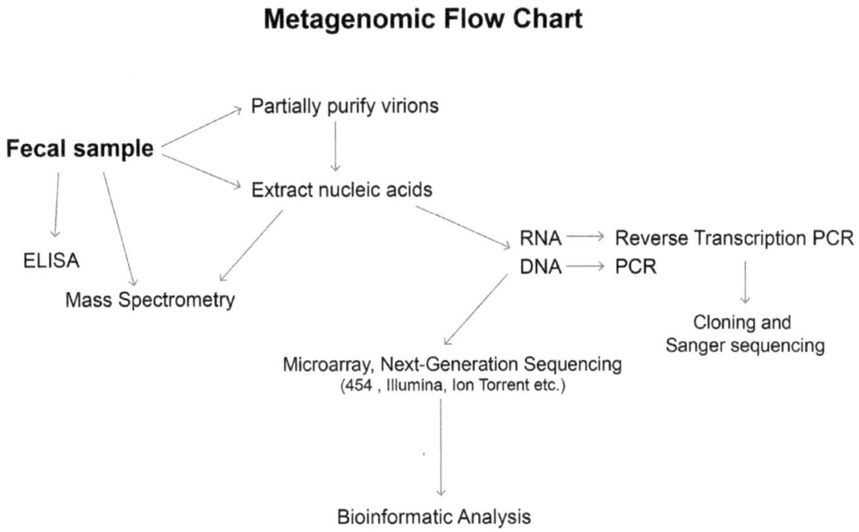

FIGURE 3.1 A flowchart of the methodologies employed for metagenomic analysis of a fecal sample.

PCR and Real-Time PCR

Drexler and colleagues recently demonstrated the power of PCR for virus discovery by using a hemi-nested RT-PCR approach to identify a new strain of hepatitis E virus in bats [15]. Hepatitis E virus is a zoonotic foodborne and waterborne virus that is a common cause of acute hepatitis, particularly in temperate climates. The authors extracted RNA from bat fecal and blood samples, as well as organ tissue from deceased animals, and amplified a hepevirus sequence using degenerate primers designed to amplify all members of the *Hepeviridae*. They also tested human sera for the same viruses and were able to show that the bat viruses were highly divergent from the human strains, suggesting that bats are not a source of hepevirus infection in humans [15].

Mass Spectrometry

Once almost exclusively the domain of protein chemists, mass spectrometry has also become part of the virologist's toolkit. Matrix-assisted laser desorption/ionization time-of-flight (MALDI-TOF) mass spectrometry has been used to identify a range of viruses, including norovirus in human stool samples [16]. This was achieved by an analysis of the norovirus capsid protein, rather than the nucleotide sequence itself, although MALDI-TOF has now been adapted for the analysis of nucleotide sequences (reviewed in [17]).

Microarrays

Oligonucleotide microarrays are another useful method for virus detection. Amplicons derived from viral sequences are tagged, usually with a fluorescent dye, and annealed to a microarray chip, which is then analyzed using the appropriate platform. The advantage of a microarray is that a large number of specific primers can be used without additional effort or high costs. For this reason, a microarray is a particularly suitable method for the detection and subtyping of a panel of viruses based on their diversity. Microarrays have been used for the detection of a range of foodborne viruses, including rotavirus [18], astrovirus, and norovirus [19]. The disadvantage of this technology for viral discovery is that, like PCR, some knowledge of the viral sequence under investigation is required prior to the microarray being performed.

The recent development of panviral microarrays, such as the Virochip [20], which encode thousands of probes specific to over 1500 different viruses on each array, is another important step forward in the detection of new viruses. In terms of virus discovery, this approach is more suited to identifying known viruses in new hosts, rather than completely new organisms. The Virochip has recently been used to detect a novel cardiovirus in respiratory secretions and stool samples from children exhibiting influenza-like symptoms and/or gastrointestinal illness [21]. The study concluded that human Theiler's murine encephalomyelitis virus-like cardioviruses, found primarily in the gastrointestinal tract, were more diverse than previously credited. Further evidence of the suitability of panviral microarrays for the detection of enteric viruses is shown by the recent detection of a novel astrovirus in rabbits, associated with fatal gastroenteritis [22].

Next-Generation Sequencing

For people with an interest in viral discovery, these are the best of times. NGS technologies are identifying new viruses and hosts at previously unimaginable rates. One of the attractions of NGS technology for the identification of new organisms is that, depending on the methodology chosen, there is no requirement (or resultant bias) for *a priori* knowledge of the viral sequences under investigation. In addition, NGS is highly sensitive, producing hundreds of thousands of sequence reads that enable the detection of previously undetectable genomes. Currently, the biggest bottleneck preventing the wider adoption of NGS is analysis of the large amounts of data generated, with intensive bioinformatic analysis required to sort the "wheat from the chaff."

A number of different NGS platforms are available to use, each with inherent advantages and disadvantages. All require the generation of a DNA library, or a PCR amplicon, prior to sequencing. For amplification and sequencing of RNA viruses, the RNA must first be reverse transcribed to cDNA. A detailed description of these technologies is beyond the scope of this chapter and the reader is directed to the following excellent publications to guide their decision on the most suitable platform for their needs [23,24]. A report comparing a metagenomic analysis performed using 454 pyrosequencing and Illumina platforms on the same microbial sample is also informative [25].

Although not a foodborne virus, one of the earliest documented examples of using NGS to detect a novel virus was published by scientists from Columbia University, New York and the Victorian Infectious Diseases Reference Laboratory, Melbourne, who detected a novel arenavirus that had caused febrile-related morbidity and mortality in transplant patients [26]. Using RNA extracted from serum and tissue samples, cDNA was generated and sequenced by pyrosequencing using the 454 FLX platform (Roche). The resultant 100,000 sequences were compared with public sequence databases using basic local alignment search tool (BLAST) algorithms (http://blast.ncbi.nlm.nih.gov/). It is important to note that "hits" with arenavirus sequences were only obtained following the translation of the nucleotide sequences into amino acids using the BLASTX algorithm. The comparison of nucleotide sequences alone did not identify any similarities with known organisms. This conundrum was previously encountered by one of the authors of this chapter (PR), when sequencing an uncharacterized rhabdovirus isolated from taro in Fiji [27]. The amplification of viral sequences using degenerate plant rhabdovirus primers only showed similarity to known rhabdovirus sequences following the translation of the nucleotide sequence [27]. This begs the question, how do we identify completely novel organisms based on sequence information alone if there are no related protein sequences in the public databases?

A 454 pyrosequencing approach was used by Nakamura and colleagues to identify and sequence norovirus genomes in the stored fecal samples of persons who had gastrointestinal illness [28]. In a striking finding, the authors also amplified the genomes of a number of plant viruses from the fecal samples, including pepper mild mottle virus, crucifer tobamovirus, and tobacco mosaic virus. These viruses were likely in the food source of the host. This not only demonstrates the sensitivity of pyrosequencing, but also shows how difficult it would be to differentiate sequences if there was no sequence information in the database with which to compare it. More recently, Phan and colleagues used 454 pyrosequencing to identify a diverse range of enteric viruses and a novel parvovirus in stool samples from West African children suffering acute diarrhea [29]. To enable the isolation of viral nucleic acids from the greater pool of nucleic acids, viral particles were first enriched by filtration and nuclease treatment. Nonspecific PCR amplification was then performed using random primers followed by PCR, using distinctly tagged primers to enable later identification of the samples following the pooling of amplicons. Libraries were generated and sequenced using the 454 TitaniumFLX sequencer and compared with the GenBank protein databases using BLASTX.

A range of other NGS platforms is completely amenable to metagenomic analysis. For example, Yu and colleagues used the Illumina platform to identify a novel polyomavirus (MX-PyV) in fecal samples collected from children with acute diarrheal disease [30] and the Roche Solexa platform has been used to identify a range of viruses in fecal samples obtained from bats [31].

Enteroviruses and Their Association with Gastroenteritis

As mentioned in the introduction, a causal association between gastroenteritis and enterovirus infection can be difficult to determine for sporadic cases. Enteroviruses are known to cause diarrhea and may be shed in the feces for weeks, but are more commonly associated with the clinical manifestation of infection at secondary sites rather than with outbreaks of acute gastroenteritis—these sites include the central nervous system [32]. For example, a category of enteroviruses isolated from human feces during an investigation of polio outbreaks in the 1950s were collectively named *enteric cytopathogenic human orphan* or ECHO viruses, as an association with a specific disease could not be made [33]. In Western Australia, in 1992, symptoms of gastrointestinal illness were reported in 55 of 66 cases of viral meningitis due to echovirus type 6 and in 37 of 59 cases due to echovirus type 9 [34]. Therefore enterovirus can cause gastroenteritis, but this is usually a clinical manifestation that includes other, often more serious, symptoms. Enteroviruses are not usually associated with foodborne illness but part of an investigation into the transmission of poliomyelitis in 1944, determined that food exposed to flies from the homes of human patients suffering the recent onset of poliomyelitis, was capable of causing paralytic polio when fed to primates [35]. A study of poliovirus persistence on soft fruit and salad vegetables when stored in a refrigerator determined that at least 10% of the original inoculum was recoverable after a week, and for some items there was no reduction after 2 weeks, indicating that there is a risk of poliovirus infection with the consumption of contaminated food products [36].

Enterovirus Culture and Typing

Most human enteroviruses can be grown in continuous mammalian cell lines with an observable cytopathic effect within a week of inoculation or serial passage. The few types of enterovirus that could not be propagated by cell culture, such as coxsackievirus A19, were initially identified by growth in suckling mice. The traditional method of identifying human enterovirus isolates was virus neutralization using polyclonal antisera raised in animals. This method was limited by the supply of authenticated antisera and the requirement to raise antiserum against putative new serotypes. The identification of enteroviruses by antisera neutralization also did not provide a means of determining evolutionary similarity between serotypes.

Enteroviruses comprise a single-stranded, positive-sense RNA genome within a nonenveloped capsid. Viral diagnostic laboratories have mostly replaced culture-based methods for enterovirus isolation with pathogen detection, using RT-PCR that targets highly conserved sequences within the 5′ untranslated region (5′-UTR) of all serotypes [37–39]. On the other hand, the lack of variation in the 5′-UTR nucleic acid sequence excludes using this region to reliably differentiate enterovirus types detected in clinical specimens. Instead, nucleic acid-based tests target enterovirus genomic regions with sequence divergence, which encode loop-like protein structures onto the surface of the virus capsid, interspersed between the relatively conserved β-barrel structural motifs of the three major virus capsid proteins: VP1, VP2, and VP3 [40]. The BC loop within VP1 is one of the antigenic determinants on the enterovirus capsid, and enterovirus identification based on the RT-PCR amplification of the BC loop nucleic acid sequence was demonstrated to correlate with antisera neutralization results. Hence, the historical term *serotype* is still often used [8]. The enterovirus type can be ascertained by a comparison of the VP1 nucleic acid sequence with enterovirus prototype sequences. A VP1 nucleic acid sequence with ≥75% identity (85% amino acid identity) to a specific enterovirus prototype would be considered homologous or serotypically identical, so long as the next highest identity with other prototype strains is <70% [2]. Degenerate PCR primers, which anneal to the relatively conserved amino acid motifs in the β-strands of enterovirus VP1 and flank the variable BC loop nucleic acid sequence, were designed using a procedure referred to as *consensus degenerate hybrid oligonucleotide primers* or the CODEHOP method (Figure 3.2) [41,42]. The partial VP1 nucleic acid sequence derived from this assay (approximately 350 nucleotides) can be used to identify enteroviruses directly from original specimens, but the full VP1 sequence (approximately 900 nucleotides) may be required to resolve ambiguous results that do not satisfy the earlier criteria.

From a public health perspective, it is important to detect enteroviruses known to cause a significant disease, such as poliovirus [43], or enterovirus 71, which is commonly associated with hand, foot, and mouth disease, but can also cause neurological illness and fatality [44,45]. Poliovirus is the most well-known enterovirus and the World Health Organization (WHO) global polio eradication program is the largest public health initiative in history. Given that more than 140 WHO polio

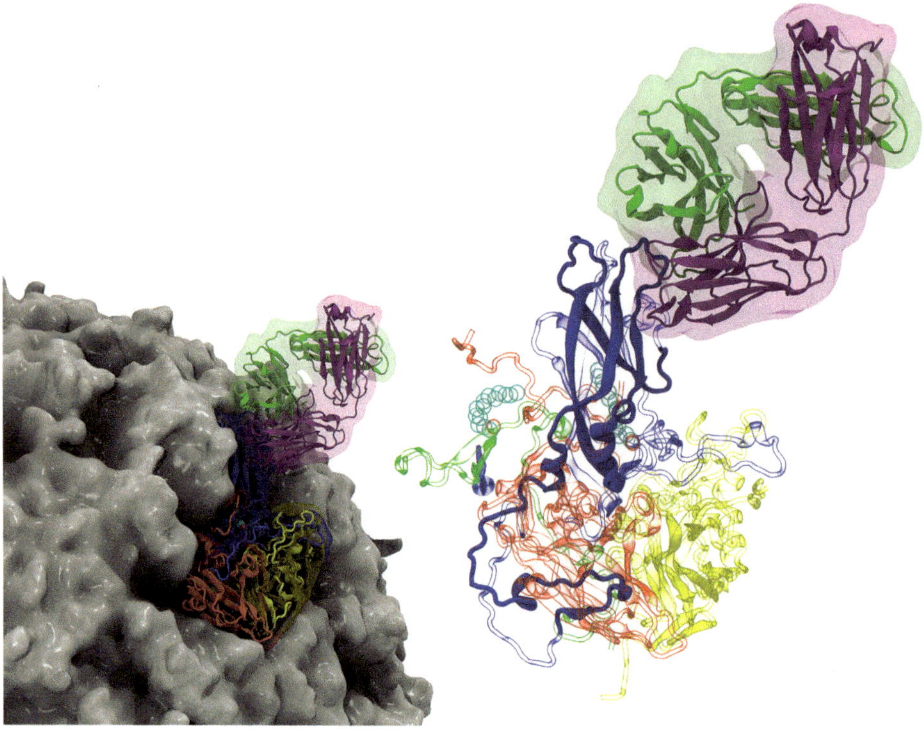

FIGURE 3.2 A representation of a poliovirus capsid (shaded gray) reconstructed from protein data bank x-ray crystallography files 1FPT (From Wien, M.W., et al., *Nat Struct Biol.,* 2, 232–243, 1995) [50] and 1HXS (From Miller, S.T., et al., *J Mol Biol.,* 307, 499–512, 2001) [51] highlighting a single protomer (colored blue—VP1, yellow—VP2 and red—VP3) with an antibody fragment (colored purple and green) bound to the BC-loop of VP1. The breakout figure (right) highlights the areas of interaction between the VP1 and the variable region of the antibody. The solid blue–colored area of the VP1 protein indicates the region amplified by the CODEHOP RT-PCR method.

reference laboratories worldwide routinely test the stool specimens of children with acute flaccid paralysis for enteroviruses, including poliovirus, it is not surprising that the majority of the enterovirus serotypes that have been discovered since 2000 were reported by laboratories involved with the polio eradication program [46–49].

THE DISCOVERY AND PROTEIN HOMOLOGY MODELING OF ENTEROVIRUS A120

As an example of virus discovery and characterization, the steps that led to the classification of a new enterovirus serotype from a case of acute flaccid paralysis, EV-A120, will be presented.

Two fecal specimens from a child presenting with acute flaccid paralysis were processed for virus culture, according to the WHO recommended procedure [52]. No enterovirus was isolated after the inoculation of fecal extracts onto continuous mammalian cell lines (human embryonic lung, Buffalo green monkey kidney, human rhabdomyosarcoma, human larynx carcinoma, and murine epithelial L20B). However, an enterovirus was detected by a seminested real-time panenterovirus RT-PCR performed in parallel with virus culture. Subsequent testing in parallel with a prototype enterovirus control of known cell culture titer using quantitative real-time RT-PCR methods, determined the 50% cell culture infectious dose of the unknown enterovirus to be approximately 0.9 $\log_{10}$. It was hypothesized that the low titer of the virus combined with evidence of toxicity in the cell culture prevented the isolation of EV-A120 from the fecal extract.

Initially, a 297 base pair (bp) fragment of the VP1 genomic region was determined using the CODEHOP method. No significant match was obtained when the fragment was compared with sequences in the public domain. A comparative analysis against known enterovirus prototypes did not fulfill the CODEHOP criteria with the best nucleic acid sequence identity match being $\leq$65.3% nucleic acid identity and $\leq$69.0% amino acid identity to CA-7, indicating the virus to be a putative novel serotype within enterovirus species A [53].

To determine the ancestry of the virus, nucleic acid sequence data of all known species A enteroviruses were aligned with the novel sequence using the Clustal-W algorithm [54], followed by model testing using Molecular Evolutionary Genetics Analysis (MEGA) version 5.2.1 to determine the most appropriate phylogenetic analysis method [55]. The two most suitable models for phylogenetic inference were determined to be the generalized time-reversible (GTR) and the Hasegawa, Kishino, and Yano (HKY) models with a discrete gamma variable incorporating invariant sites (Figure 3.3). No significant differences between the outputs of both the GTR and HKY models were observed when phylogeny was inferred using maximum likelihood methods, performed with 1000 bootstrap pseudoreplicates. The phylogenetic analysis provided strong support that the novel virus was related to a clade incorporating coxsackievirus A7 (CV-A7), coxsackievirus A14 (CV-A14), coxsackievirus A16 (CV-A16), and enterovirus 71 (EV-A71). It is of historic interest to note that CV-A7 was referred to as *poliovirus type 4* during the 1950s, when large outbreaks of poliomyelitis-like illnesses occurred in Russia and the United States [56]. EV-A71 has also been associated with anterior horn cell disease, part of the differential diagnosis for poliomyelitis [57].

Initially, the use of NGS methods was considered to elucidate the putative new enterovirus genome. However, having determined the virus concentration to be only 7.9 infectious virus particles per milliliter of specimen extract, combined with evidence of toxicity, it was predicted that the signal-to-noise ratio would be very low. If successful, a standard RT-PCR-based approach would be less expensive to sequence the approximately 7500 bp enterovirus genome and was chosen as the first option.

To determine whether consensus primers could be designed to elucidate the novel enterovirus genome sequence, the full-length genomes of four enterovirus species A prototypes with the closest sequence matches to the VP1 fragment were aligned using Clustal-W to determine the areas of greatest homogeneity (Figure 3.4). A 44 bp fragment with 100% identity between serotypes was identified within the sequence encoding the 2C proteinase. A reverse orientation PCR primer

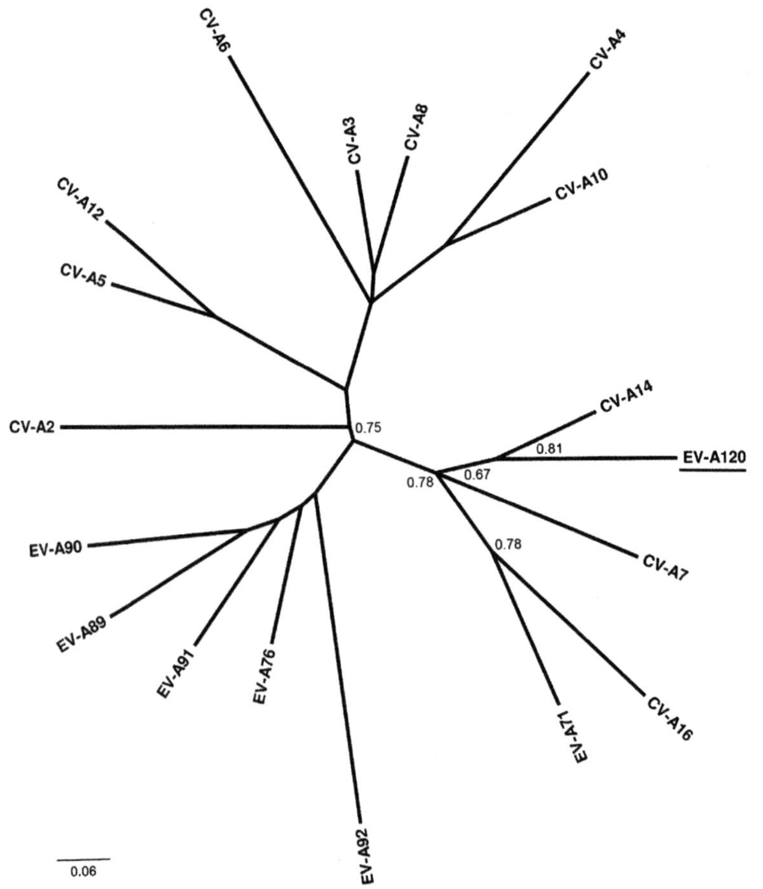

FIGURE 3.3 A comparison of enterovirus species A prototypes and the 297 nucleotide VP1 fragment of EV-A120, represented as a maximum likelihood radial phylogenetic tree (HKY+G+I) with 1000 pseudoreplicates. The scale bar indicates the nucleotide substitutions per site.

was designed for this fragment and amplification of a 3500 bp product of the EV-A120 genome was attempted with EV1 (Figure 3.4) as a forward PCR primer using the SuperScript III one-step RT-PCR kit according to the manufacturer's recommendations, but no product was obtained for sequencing.

To increase sensitivity, a second round of PCR was employed to amplify two fragments using internal forward and reverse primers based on the EV-A120 sequence derived from the CODEHOP fragment (Figure 3.4). The first round product was added to two separate second round PCR reactions. The first PCR used the 5'-UTR forward primer, EV2, and the EV-A120 reverse primer designed from the CODEHOP fragment sequence. The second PCR used the EV-A120 forward primer derived from the CODEHOP fragment with the enterovirus species A 2C proteinase consensus reverse primer used in the first round. This seminested RT-PCR method produced two fragments of 1.8 and 2.0 kb in length, respectively. Sequence data were initially obtained with the primers used to amplify the second round PCR product and the remainder of the two fragments were sequenced by primer walking. The final contiguous sequence was 3700 bp in length and included a fragment of the 5′-UTR, the entire P1 capsid encoding region and part of the P2 nonstructural region extending from 2A to within 2C (GenBank accession No. KM198310).

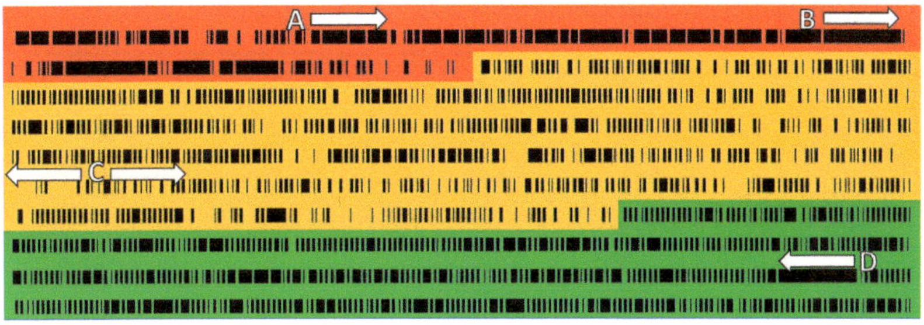

FIGURE 3.4 A schematic diagram of the genomic sequence alignments of four prototype species A enteroviruses closely related to EV-A120; CV-A7, CV-A14, CV-A16, and EV-A71. The black bars indicate areas of 100% identity across all four prototypes: (A) the forward primer EV1, (B) forward primer EV2, (C) forward and reverse primers of the EV-A120 sequence derived from the CODEHOP VP1 fragment, and (D) consensus reverse primer from the 2C proteinase region. Color designation: red—5'NTR, orange—P1 (capsid encoding region), green—P2 (nonstructural protein-encoding regions 2A, 2B and 2C). Arrows indicate the orientation of the primers.

Simplot analysis incorporating bootstrapping methods [58] was performed for the entire capsid region of all prototypic members of the species A enteroviruses, with no significant recombination events detected, indicating that EV-A120 was not a recombinant, but most likely an evolutionary ancestor of CV-A7 and CV-A14.

Having the complete capsid sequence enabled a structural model of the newly discovered enterovirus A120 to be produced using standard comparative protein modeling methods (Figure 3.5). To

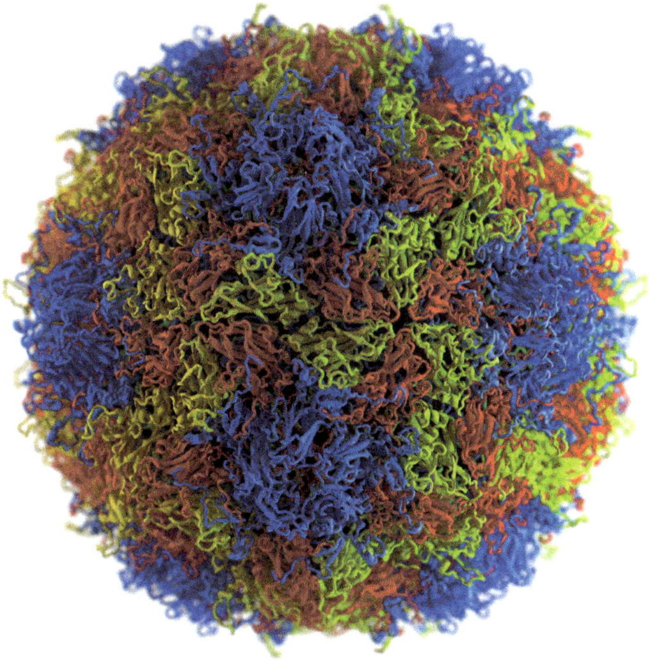

FIGURE 3.5 The EV-A120 virus capsid, reconstructed using comparative protein modeling methods based on the EV-A71 template (PDB:4AED). The secondary structure is represented in "cartoon" form. The structural proteins are designated as blue—VP1, yellow—VP2, red—VP3, green—VP4. VP4 is not readily discerned due to its location within the internal face of the capsid.

do so, an appropriate template must be available in the form of either high-resolution cryo-electron microscopy density maps or x-ray crystallographic data with a reasonably high degree of amino acid sequence identity, preferably greater than 60%. The EV-A120 capsid sequence was queried against SWISS-MODEL to determine the most suitable structural template. The results indicated that x-ray crystallography data were available for EV-A71 with an amino acid identity of 71.9% compared with EV-A120, and cryo-electron microscopy data were available for CV-A7 with an amino acid identity of 79.8%. The x-ray crystallographic data for enterovirus 71 were more complete for the VP1, VP2, VP3, and VP4 chains compared with that available for CV-A7, and were chosen as the basis for comparative modeling.

The EV-A120 capsid was modeled by uploading the amino acid sequence to the SWISS-MODEL website, which produced the four viral proteins in an enterovirus protomer structure based on the EV-A71 x-ray crystallographic data [59]. The EV-A120 protomer was used to construct a complete capsid with the matrix metadata for biological assembly contained within the original EV-A71 x-ray crystallographic data, by a previously described method [60] (Figure 3.5).

The resulting protein homology model is a valuable starting point for comparison with related enteroviruses and subsequent analyses using cryo-electron microscopy methods that require high-resolution three-dimensional density data. When combined with molecular dynamics methods, including free energy perturbation, such models can be used to investigate antiviral drug interactions with the target organism as those described by Vergara-Jaque et al. for influenza virus resistance to the neuraminidase inhibitor, oseltamivir [61].

CONCLUSION

The advent of molecular tools such as cloning, sequencing, PCR, and more recently NGS, has seen an exponential increase in our ability to detect pathogenic organisms. These technologies are directly amenable to metagenomic analysis, including the detection of foodborne viruses. New viruses are being identified and we are also obtaining additional information about the host ranges of previously characterized viruses. The increased sensitivity of NGS is having a positive impact on diagnostic virology, although this can be a "double-edged sword," as the high levels of sensitivity may also result in the detection of organisms that may be incidental to the pathology under investigation. Other technologies such as mass spectrometry, electron microscopy, microarray, and traditional PCR all have their place, but it is likely that further advances in NGS technologies, such as the single molecule amplification techniques under development, will greatly advance the metagenomics field in years to come.

REFERENCES

1. Koopmans M, Duizer E. 2004. Foodborne viruses: An emerging problem. *Int J Food Microbiol* 90:23–41.
2. Pallansch MA, Oberste MS, Whitton JL. 2013. Enteroviruses: Polioviruses, coxsackieviruses, echoviruses and newer enteroviruses. *In* Knipe DM, Howley PM (eds), *Fields Virology*, 6th edn, pp. 490–530. Lippincott Williams & Wilkins, Philidelphia, PA.
3. Kapikian AZ, Wyatt RG, Dolin R, Thornhill TS, Kalica AR, Chanock RM. 1972. Visualization by immune electron microscopy of a 27-nm particle associated with acute infectious nonbacterial gastroenteritis. *J Virol* 10:1075–1081.
4. Feinstone SM, Kapikian AZ, Purceli RH. 1973. Hepatitis A: Detection by immune electron microscopy of a viruslike antigen associated with acute illness. *Science* 182:1026–1028.
5. Reyes GR, Purdy MA, Kim JP, Luk KC, Young LM, Fry KE, Bradley DW. 1990. Isolation of a cDNA from the virus responsible for enterically transmitted non-A, non-B hepatitis. *Science* 247:1335–1339.
6. Hyypia T, Hovi T, Knowles NJ, Stanway G. 1997. Classification of enteroviruses based on molecular and biological properties. *J Gen Virol* 78 (Pt 1):1–11.
7. Oberste MS, Maher K, Flemister MR, Marchetti G, Kilpatrick DR, Pallansch MA. 2000. Comparison of classic and molecular approaches for the identification of untypeable enteroviruses. *J Clin Microbiol* 38:1170–1174.

8. Oberste MS, Maher K, Kilpatrick DR, Pallansch MA. 1999. Molecular evolution of the human enteroviruses: correlation of serotype with VP1 sequence and application to picornavirus classification. *J Virol* 73:1941–1948.
9. Booss J, August MJ. 2013. *To Catch a Virus*. ASM Press, Washington, DC.
10. Bellou M, Kokkinos P, Vantarakis A. 2013. Shellfish-borne viral outbreaks: A systematic review. *Food Environ Virol* 5:13–23.
11. Philippe N, Legendre M, Doutre G, Couté Y, Poirot O, Lescot M, Arslan D, et al. 2013. Pandoraviruses: Amoeba viruses with genomes up to 2.5 Mb reaching that of parasitic eukaryotes. *Science* 341:281–286.
12. Handelsman J, Rondon MR, Brady SF, Clardy J, Goodman RM. 1998. Molecular biological access to the chemistry of unknown soil microbes: A new frontier for natural products. *Chem Biol* 5:R245–R249.
13. Breitbart M, Salamon P, Andresen B, Mahaffy JM, Segall AM, Mead D, Azam F, Rohwer F. 2002. Genomic analysis of uncultured marine viral communities. *Proc Natl Acad Sci USA* 99:14250–14255.
14. Breitbart M, Hewson I, Felts B, Mahaffy JM, Nulton J, Salamon P, Rohwer F. 2003. Metagenomic analyses of an uncultured viral community from human feces. *J Bacteriol* 185:6220–6223.
15. Drexler JF, Seelen A, Corman VM, Fumie Tateno A, Cottontail V, Melim Zerbinati R, Gloza-Rausch F, et al. 2012. Bats worldwide carry hepatitis E virus-related viruses that form a putative novel genus within the family Hepeviridae. *J Virol* 86:9134–9147.
16. Colquhoun DR, Schwab KJ, Cole RN, Halden RU. 2006. Detection of norovirus capsid protein in authentic standards and in stool extracts by matrix-assisted laser desorption ionization and nanospray mass spectrometry. *Appl Environ Microbiol* 72:2749–2755.
17. Revill P, Bowden D, White P. 2010. Viral genotyping and the sequencing revolution. *In* Jerome K (ed), *Lennette's Laboratory Diagnosis of Viral Infections*, 4th edn, pp. 40–58. Informa Healthcare, New York.
18. Chizhikov V, Wagner M, Ivshina A, Hoshino Y, Kapikian AZ, Chumakov K. 2002. Detection and genotyping of human group A rotaviruses by oligonucleotide microarray hybridization. *J Clin Microbiol* 40:2398–2407.
19. Jaaskelainen AJ, Maunula L. 2006. Applicability of microarray technique for the detection of noro- and astroviruses. *J Virol Methods* 136:210–216.
20. Chen EC, Miller SA, DeRisi JL, Chiu CY. 2011. Using a pan-viral microarray assay (Virochip) to screen clinical samples for viral pathogens. *J Vis Exp* 50:945–953.
21. Chiu CY, Greninger AL, Kanada K, Kwok T, Fischer KF, Runckel C, Louie JK, et al. 2008. Identification of cardioviruses related to Theiler's murine encephalomyelitis virus in human infections. *Proc Natl Acad Sci USA* 105:14124–14129.
22. Stenglein MD, Velazquez E, Greenacre C, Wilkes RP, Ruby JG, Lankton JS, Ganem D, Kennedy MA, DeRisi JL. 2012. Complete genome sequence of an astrovirus identified in a domestic rabbit (*Oryctolagus cuniculus*) with gastroenteritis. *Virol J* 9:216.
23. Metzker ML. 2010. Sequencing technologies: The next generation. *Nat Rev Genet* 11:31–46.
24. Radford AD, Chapman D, Dixon L, Chantrey J, Darby AC, Hall N. 2012. Application of next-generation sequencing technologies in virology. *J Gen Virol* 93:1853–1868.
25. Luo C, Tsementzi D, Kyrpides N, Read T, Konstantinidis KT. 2012. Direct comparisons of Illumina vs. Roche 454 sequencing technologies on the same microbial community DNA sample. *PLoS One* 7:e30087.
26. Palacios G, Druce J, Du L, Tran T, Birch C, Briese T, Conlan S, et al. 2008. A new arenavirus in a cluster of fatal transplant-associated diseases. *N Engl J Med* 358:991–998.
27. Revill P, Trinh X, Dale J, Harding R. 2005. Taro vein chlorosis virus: Characterization and variability of a new nucleorhabdovirus. *J Gen Virol* 86:491–499.
28. Nakamura S, Yang CS, Sakon N, Ueda M, Tougan T, Yamashita A, Goto N, et al. 2009. Direct metagenomic detection of viral pathogens in nasal and fecal specimens using an unbiased high-throughput sequencing approach. *PLoS One* 4:e4219.
29. Phan TG, Vo NP, Bonkoungou IJ, Kapoor A, Barro N, O'Ryan M, Kapusinszky B, Wang C, Delwart E. 2012. Acute diarrhea in West African children: Diverse enteric viruses and a novel parvovirus genus. *J Virol* 86:11024–11030.
30. Yu G, Greninger AL, Isa P, Phan TG, Martinez MA, de la Luz Sanchez M, Contreras JF, et al. 2012. Discovery of a novel polyomavirus in acute diarrheal samples from children. *PLoS One* 7:e49449.
31. Ge X, Li Y, Yang X, Zhang H, Zhou P, Zhang Y, Shi Z. 2012. Metagenomic analysis of viruses from bat fecal samples reveals many novel viruses in insectivorous bats in China. *J Virol* 86:4620–4630.
32. Dagan R, Menegus, M. 1995. Nonpolio enteroviruses in the febrile infant. *In* Rotbart H (ed), *Human Enterovirus Infections*, pp. 239–254. ASM Press, Washington, DC.

33. Paul JR. 1971. *A History of Poliomyelitis*. Yale University Press, New Haven, CT.
34. Ashwell MJ, Smith DW, Phillips PA, Rouse IL. 1996. Viral meningitis due to echovirus types 6 and 9: Epidemiological data from Western Australia. *Epidemiol Infect* 117:507–512.
35. Ward R, Melnick JL, Horstmann DM. 1945. Poliomyelitis virus in fly-contaminated food collected at an epidemic. *Science* 101:491–493.
36. Kurdziel AS, Wilkinson N, Langton S, Cook N. 2001. Survival of poliovirus on soft fruit and salad vegetables. *J Food Prot* 64:706–709.
37. Zoll G, Melchers W, Kopecka H, Jambroes G, Van der Poel H, Galama J. 1992. General primer-mediated polymerase chain reaction for detection of enteroviruses: Application for diagnostic routine and persistent infections. *J Cin Micro* 30:160.
38. McIver CJ, Jacques CFH, Chow SSW, Munro SC, Scott GM, Roberts JA, Craig ME, Rawlinson WD. 2005. Development of multiplex PCRs for detection of common viral pathogens and agents of congenital infections. *J Clin Micro* 43:5102–5110.
39. Roberts JA, Thorley BR. 2010. Enterovirus. *In* Schuller M, Sloots TP, James GS, Halliday CL, Carter IWJ (eds), *PCR for Clinical Microbiology*, pp. 229–233. Springer, Dordrecht.
40. Oberste MS, Maher K, Williams AJ, Dybdahl-Sissoko N, Brown BA, Gookin MS, Peñaranda S, Mishrik N, Uddin M, Pallansch MA. 2006. Species-specific RT-PCR amplification of human enteroviruses: A tool for rapid species identification of uncharacterized enteroviruses. *J Gen Virol* 87:119.
41. Nix WA, Oberste MS, Pallansch MA. 2006. Sensitive, seminested PCR amplification of VP1 sequences for direct identification of all enterovirus serotypes from original clinical specimens. *J Clin Microbiol* 44:2698–2704.
42. Rose TM. 2005. CODEHOP-mediated PCR: A powerful technique for the identification and characterization of viral genomes. *Virol J* 2:3763–3766.
43. Stewardson AJ, Roberts JA, Beckett CL, Prime HT, Loh PS, Thorley BR, Daffy JR. 2009. Imported case of poliomyelitis, Melbourne, Australia, 2007. *Emerg Infect Dis* 15:63–65.
44. Khanh TH, Sabanathan S, Thanh TT, Thoa le PK, Thuong TC, Hang V, Farrar J, Hien TT, Chau N, van Doorn HR. 2012. Enterovirus 71-associated hand, foot, and mouth disease, Southern Vietnam, 2011. *Emerg Infect Dis* 18:2002–2005.
45. Zhang Y, Tan X, Cui A, Mao N, Xu S, Zhu Z, Zhou J, et al. 2013. Complete genome analysis of the C4 subgenotype strains of enterovirus 71: Predominant recombination C4 viruses persistently circulating in China for 14 years. *PLoS One* 8:e56341.
46. Oberste MS, Maher K, Michele SM, Belliot G, Uddin M, Pallansch MA. 2005. Enteroviruses 76, 89, 90 and 91 represent a novel group within the species human enterovirus A. *J Gen Virol* 86:445–451.
47. Smura T, Blomqvist S, Paananen A, Vuorinen T, Sobotová Z, Bubovič̌a V, Ivanova O, Hovi T, Roivainen M. 2007. Enterovirus surveillance reveals proposed new serotypes and provides new insight into enterovirus 5′-untranslated region evolution. *J Gen Virol* 88:2520–2526.
48. Brown BA, Maher K, Flemister MR, Naraghi-Arani P, Uddin M, Oberste MS, Pallansch MA. 2009. Resolving ambiguities in genetic typing of human enterovirus species C clinical isolates and identification of enterovirus 96, 99 and 102. *J Gen Virol* 90:1713–1723.
49. Oberste MS, Michele SM, Maher K, Schnurr D, Cisterna D, Junttila N, Uddin M, Chomel J-J, Lau C-S, Ridha W. 2004. Molecular identification and characterization of two proposed new enterovirus serotypes, EV74 and EV75. *J Gen Virol* 85:3205–3212.
50. Wien MW, Filman DJ, Stura EA, Guillot S, Delpeyroux F, Crainic R, Hogle JM. 1995. Structure of the complex between the Fab fragment of a neutralizing antibody for type 1 poliovirus and its viral epitope. *Nat Struct Biol* 2:232–243.
51. Miller ST, Hogle JM, Filman DJ. 2001. *Ab initio* phasing of high-symmetry macromolecular complexes: Successful phasing of authentic poliovirus data to 3.0 A resolution. *J Mol Biol* 307:499–512.
52. WHO. 2004. *Polio Laboratory Manual*. World Health Organization, Geneva, Switzerland.
53. Oberste MS, Maher K, Kilpatrick DR, Flemister MR, Brown BA, Pallansch MA. 1999. Typing of human enteroviruses by partial sequencing of VP1. *J Clin Microbiol* 37:1288–1293.
54. Thompson JD, Higgins DG, Gibson TJ. 1994. CLUSTAL W: Improving the sensitivity of progressive multiple sequence alignment through sequence weighting, position-specific gap penalties and weight matrix choice. *Nucleic Acids Res* 22:4673–4680.
55. Tamura K, Peterson D, Peterson N, Stecher G, Nei M, Kumar S. 2011. MEGA5: Molecular evolutionary genetics analysis using maximum likelihood, evolutionary distance, and maximum parsimony methods. *Mol Biol Evol* 28:2731–2739.
56. Habel K, Loomis LN. Coxsackie A7 virus and the Russian "poliovirus type 4." *Proc Soc Exp Biol Med* 95:597–605.

57. Chen CY, Chang YC, Huang CC, Lui CC, Lee KW, Huang SC. 2001. Acute flaccid paralysis in infants and young children with enterovirus 71 infection: MR imaging findings and clinical correlates. *Am J Neuroradiol* 22:200–205.
58. Lole KS, Bollinger RC, Paranjape RS, Gadkari D, Kulkarni SS, Novak NG, Ingersoll R, Sheppard HW, Ray SC. 1999. Full-length human immunodeficiency virus type 1 genomes from subtype C-infected seroconverters in India, with evidence of intersubtype recombination. *J Virol* 73:152–160.
59. Arnold K, Bordoli L, Kopp J, Schwede T. 2006. The SWISS-MODEL workspace: A web-based environment for protein structure homology modelling. *Bioinformatics* 22:195–201.
60. Roberts JA, Kuiper MJ, Thorley BR, Smooker PM, Hung A. 2012. Investigation of a predicted N-terminal amphipathic α-helix using atomistic molecular dynamics simulation of a complete prototype poliovirus virion. *J Mol Grap Model* 38:165–173.
61. Vergara-Jaque A, Poblete H, Lee EH, Schulten K, González-Nilo F, Chipot C. 2012. Molecular basis of drug resistance in A/H1N1 virus. *J Chem Inf Model* 52:2650–2656.

4 Measuring and Minimizing Health Impacts of Foodborne Viruses

A Targeting Tool for Risk Reduction

Kristina D. Mena

CONTENTS

One afternoon you experience that queasy, uncommonly familiar feeling in your stomach. You wonder if you should lie down or drag yourself to the bathroom. Eventually, vomiting occurs—along with fever and chills—and some diarrhea. Do you go to the doctor? *Probably not.* You wonder why it happened. Did you "catch" something at the office? *Maybe.* Was it something you ate for lunch? *Probably not.* Could it be that delicious dinner you had at your friend's house the night before? *Possibly.*

The aforementioned scenario is something we all have experienced. A range of nonspecific symptoms—symptoms that make us uncomfortable at best, or debilitate us resulting in missed work and unmet responsibilities. Often, these symptoms are related to consuming food compromised with pathogens, such as enteric viruses, bacteria, or protozoa. However, we typically never learn the causative agent of our illness, primarily because we do not typically see a physician when such symptoms occur, or if we do, tests are not conducted that identify the responsible agent. For example, in the United Kingdom, only 1 in 250 people suffering gastroenteritis are actually tested for pathogens. It has been estimated that approximately 48 million foodborne illnesses occur each year in the United States, with only 20% linked to a specific pathogen [1,2]. In the United States, the economic costs associated with foodborne disease have reached more than $50 billion annually [3].

Quantitative microbial risk assessment (QMRA) is a mechanism to comprehensively address the environmental, economic, and human host factors influencing foodborne infections and

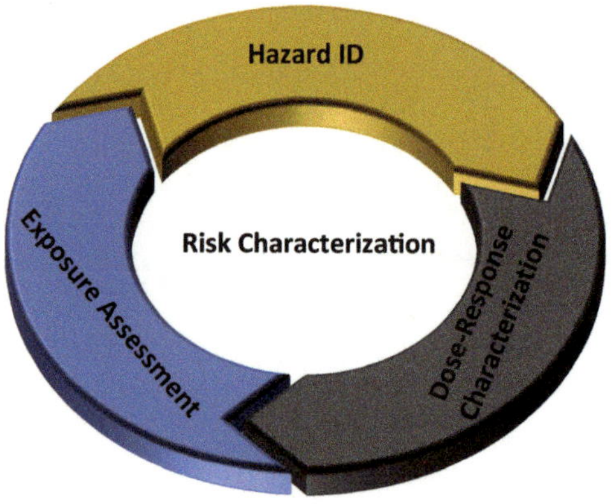

FIGURE 4.1 The four components of quantitative microbial risk assessment (QMRA).

illnesses. The four-component process includes hazard identification, dose–response characterization, exposure assessment, and risk characterization (Figure 4.1). The components are iterative and dynamic, with each step potentially providing input for another. QMRA tells the story of how a microorganism—such as an enteric virus—can impact human lives through an environmental source, such as food. The process can be both qualitative (e.g., *What can happen?*) and quantitative (e.g., *How likely is it to happen?*), and usually contains an array of risk management applications. Specifically, QMRA can be used to *predict* human health impacts associated with foodborne pathogens and subsequently *prevent* infections and illnesses.

Risk is defined as the possibility, likelihood, or probability of an adverse effect due to consequences from a specific hazard or set of hazards. For 30 years, the risk assessment method has been applied to evaluate chemical hazards in the environment [4]. During the 1980s, the method was explored for its application to microbial hazards. In particular, dose–response studies of various pathogens were evaluated to identify best-fit models that could be used in microbial risk assessments [5]. This vital paper paved the way for risk assessment as a tool to address the health consequences associated with exposure to pathogens in the environment, such as water. Since then, numerous studies have conducted risk assessments on groups of pathogens—or specific microorganisms such as viruses—to explore their human health impact, not only through water, but also food [6–19]. In addition, risk management approaches to address food safety along the food production chain have also been explored using these same principles [20–22].

QUANTITATIVE MICROBIAL RISK ASSESSMENT

Identifying Foodborne Viral Hazards

Characterizing microbial agents as environmental health hazards is the first step of the QMRA process. As mentioned at the start of the chapter, the incidence of gastroenteritis within populations is underestimated, since most people do not seek medical treatment from a physician when experiencing diarrhea [23,24]. Therefore, associating diarrheal cases with specific sources or causative agents is difficult. Foodborne disease is typically attributed to bacterial pathogens; however, enteric viruses can also impact health through the foodborne route. Foodborne viruses can cause health consequences ranging from diarrhea, vomiting, and hepatitis to encephalitis [25]. Noroviruses and hepatitis A virus, in particular, are leading causes of foodborne disease throughout Europe and the United States [25–29]. Other viruses that can be transmitted through food include adenovirus, astrovirus, hepatitis E virus, sapovirus, some enteroviruses, and rotavirus.

Several approaches for attributing specific pathogens and different types of food to foodborne disease have been employed [30–32], although most focus primarily on bacteria. The methods involve epidemiological study designs, disease outbreak investigation data, expert elicitation, molecular subtyping, and exposure assessments. In 2012, Batz et al. [31] addressed 168 food–pathogen combinations and ranked them based on their associated costs of illness and quality-adjusted life years (QALYs). Most of the combinations studied involved bacterial pathogens; however, norovirus was part of the analysis. When evaluating the economic and health burden associated with exposure to norovirus in foods containing multiple ingredients (e.g., sandwiches, salads, and dressings), the food–pathogen combination ranked fifth among the other combinations addressed.

In the United States, noroviruses cause approximately 23 million gastroenteritis cases annually [28] and are suspected to be the cause of most adult diarrheal illness [33]. Noroviruses are transmitted through both water and food, and can survive on environmental surfaces [28]. In addition, noroviruses have a relatively low infectious dose (10–100 viral particles) [34,35]. While some infections are asymptomatic, clinical illness results in diarrhea and vomiting. Viral particles in feces and vomitus, along with virion survivability in the environment, aid in the transmission of noroviruses. Person-to-person transmission is common in closed environments; noroviruses have been the responsible agents in numerous gastroenteritis outbreaks throughout populations on cruise ships and in elder care facilities. A variety of foods have been implicated in outbreaks including shellfish, deli sandwiches, and salads [2].

Hepatitis A virus is also transmitted by both water and food, as well as person-to-person transmission. Food—or an infected food handler—is often found to be the source of outbreaks [36,37]. Fruits and vegetables, as well as shellfish, may be contaminated with hepatitis A, and all have been implicated in documented outbreaks [29,38–41]. One of the largest hepatitis A virus outbreaks occurred in 1988 in China, and impacted >300,000 people who consumed hepatitis A virus-laden clams [42]. Like noroviruses, hepatitis A virus is an unenveloped, naked virus and is capable of surviving in food and on environmental surfaces for lengthy time periods, which contributes to person-to-person transmission [43]. In addition, hepatitis A virus is shed in the feces of infected individuals before the onset of clinical symptoms, making it difficult to prevent transmission.

Other viruses, such as rotavirus and hepatitis E virus, are more commonly associated with waterborne transmission; however, these viruses can also cause foodborne disease [28]. Rotaviruses cause gastroenteritis and are a major cause of hospitalization and death among children less than 5 years of age worldwide [44], although the introduction of the rotavirus vaccine has likely reduced this number. These viruses commonly occur in wastewater as they are excreted in large numbers by infected individuals [26]. A microbial risk assessment concluded rotavirus to be one of the most infectious waterborne agents known [44]. Documented outbreaks have involved educational settings (schools and universities) in particular [28,45,46]. When identified, implicated foods have included chicken and tuna salad sandwiches.

Adenoviruses have been implicated in recreational waterborne outbreaks resulting in gastroenteritis and respiratory infections [17,47]. Therefore, it is likely that these viruses are not only transmitted through the fecal–oral route, but also through aerosolized droplets [28]. Like rotavirus, adenovirus can be found in wastewater. Surveillance studies have also detected adenovirus in shellfish [48–50]. One study detected adenovirus in almost half (47%) of the shellfish samples tested [49].

CHARACTERIZING DOSE RESPONSE

Dose–response characterization explores the relationship between the amount of the hazard (virus) in an exposure and the human health consequence(s). Dose–response information based on human volunteer studies is available for many microorganisms, including enteric viruses [35,51–54]. Dose–response models have been evaluated for their ability to represent dose–response relationships between human hosts and a variety of foodborne pathogens [5,10,35,55]. Goodness-of-fit criteria

FIGURE 4.2 Sequence of probability of events following pathogen exposure.

are used to assess model applicability. Simpler models with fewer parameters are favored over more complex models.

In a QMRA, the dose–response characterization predicts what happens after there is an exposure(s) to a pathogen; it typically does not address the *likelihood* of exposure occurring in the first place (exposure is assumed). The models predict the chance of infection, and can also address the probabilities of subsequent illness and death by incorporating morbidity and case-fatality ratios. Figure 4.2 shows the chain of events (based on probability) following exposure to a pathogen. Each arrow represents a probability distribution.

Two dose–response models in particular—the exponential model and the beta-Poisson model—have been shown to best fit microbial dose–response data in most cases [5]. The exponential model can be used to predict the probability of infection from exposure to several food/waterborne pathogens [10]. It is a simple model in that it includes only one dose–response parameter that reflects the specific pathogen of interest:

$$P_{\text{infection}} = 1 - \exp(-rd) \tag{4.1}$$

where:

$P_{\text{infection}}$ equals the probability of infection
r is the dose–response parameter representing the fraction of surviving microorganisms in the dose that are capable of initiating infection
d is the number of microorganisms in the exposure

The model assumes the microorganisms in the exposure are pathogenic, and are randomly distributed.

The beta-Poisson model is a modification of the exponential model in that it has two dose–response parameters:

$$P_{\text{Infection}} = 1 - \left[1 + \frac{d}{N_{50}}\left(2^{1/\alpha}\right)\right]^{-\alpha} \tag{4.2}$$

where:

$P_{\text{infection}}$ equals the probability of infection
d is the number of microorganisms in the exposure
N_{50} is the dose that causes infection in half of the exposed population
α is a dose–response parameter representing the slope of the dose–response curve

As α gets bigger, the beta-Poisson model approaches the exponential model. The beta-Poisson model assumes more heterogeneity between the pathogen and host interaction. Both the exponential and the beta-Poisson models assume that as little as one pathogen in an exposure is capable of initiating an infection.

Annual (and lifetime) infection risks can also be predicted by using the following equation:

$$P_{annual} = 1 - (1 - P_{infection})^{n} \tag{4.3}$$

where:

P_{annual} is the yearly risk of infection
$P_{infection}$ is the (daily) probability of infection
n is the number of days of exposure in a year

Risks of illness and death can be estimated by multiplying $P_{infection}$ by the appropriate morbidity and mortality ratios for that pathogen.

ASSESSING EXPOSURE

Exposure assessment incorporates numerous factors that make this component the greatest source of variability and uncertainty in the QMRA process. This step considers the amount of infectious microorganisms (e.g., viral particles) in the exposure, the amount of the exposure source (e.g., food) consumed, and the frequency of exposure over time. The quality and the accuracy of the input data are critical in that these factors directly impact risk assessment output. When considering Equations 4.1 and 4.2, the factors described in the exposure assessment help define model parameter *d* (the number of microorganisms in the exposure). The following equation shows the relationship between *d* and exposure-related factors associated with exposure to viruses in food:

$$d = C \times R^{-1} \times 10^{-DR} \times I \times A \tag{4.4}$$

where:

C is the concentration of microorganisms per serving of food (viral particles/g)
R^{-1} is the recovery efficiency of the detection method used to determine C
10^{-DR} considers the decimal reduction of viruses that may result from food processing
I is the fraction of microorganisms in the exposure capable of initiating infection (i.e., percentage infectious)
A equals the amount of food (g) per serving consumed

The concentration (C) can be estimated from food monitoring data or information gleaned from disease outbreak investigations. A range of infection risks can be estimated by assuming lower and upper levels of viral particles in the food, or by estimating the average exposure concentrations. When concentration data are laboratory generated, the recovery efficiency (R^{-1}) of the detection method used needs to be considered. In addition, as with any environmental microbiology data generated, careful consideration should be given to those samples with negative results—perhaps the detection method employed was not sensitive enough to detect the low numbers of viruses present in the sample. In addition, if food quality is assessed prior to any processing (such as heating through cooking), the potential reduction in virus concentration due to food processing needs to be addressed (10^{-DR}).

Viral inactivation through food preparation is variable, and depends on the specific virus, the type of food matrix, and the extent of food processing [16]. Parameter I can be defined based on the detection method used or reasonably estimated from infectivity studies. For protective infection risk estimate outputs, I is typically assumed to be 1 (or 100%); this assumes that all viral particles in the exposure are able to infect the host. The last parameter, A, incorporates the amount of food (mass in g) consumed. This information can be found in the published literature on what constitutes a serving (g) for a specific type of food, or can be obtained from disease outbreak investigations where the amount of food consumed is known.

When considering the four components of the QMRA process (hazard identification, dose–response characterization, exposure assessment, and risk characterization), exposure-related factors drive risk; if there is no exposure, there is no risk. The importance of appropriately defining the parameters in Equation 4.4 to address exposure cannot be overestimated. These exposure-related variables that determine *d* directly impact risk output that may then be used to inform risk mitigation strategies. Improving laboratory detection methods to increase the sensitivity and specificity for viruses found in the environment is critical in QMRA. In addition, the amount of (contaminated) food consumed also has an integral role in the QMRA process. Therefore, the ability to appropriately estimate the amount of a food source consumed—based on weight and frequency—is also important. This factor is dependent on the food type and host.

Besides virus characteristics, factors related to the host (consumers) should also be considered. Consumers who are immunocompromised have a greater probability of severe health consequences associated with foodborne viruses, so defining the host (pregnant women, the elderly, children, etc.) should be considered. Some host-related factors may be directly reflected in Equation 4.4 when determining *d*. For example, adults may be more likely to consume larger portions of certain foods than children; therefore, *A* would be different when considering these different subpopulations. Further, some foods may be more likely to be consumed by adults (such as shellfish perhaps) than children.

A recent review of the literature suggests that some subpopulations within Europe and the United States may be more susceptible to foodborne disease [56]. Specifically, studies indicate that the minority of low-income populations have greater risks of foodborne disease than other groups [56,57]. This may be due to limited access to high-quality foods and/or repeated exposure to foodborne pathogens. In addition, culture may play a role in some subpopulations by influencing food consumption patterns or food handling practices [56]. For example, subpopulations who routinely consume foods at high risk of microbial contamination may experience greater rates of foodborne illness.

TRANSMISSION

Food may become compromised with viral pathogens through different mechanisms. The number of pathways that enteric viruses can follow to compromise food and be transmitted to humans can complicate a QMRA exposure assessment of foodborne viruses. Since some foodborne viruses are also commonly associated with water, defining the true exposure route in such a risk assessment is challenging. Some "foodborne" illnesses may actually be due to the consumption of food—such as fresh produce—that was irrigated, washed, processed, or prepared with virus-laden water [14]. The environmental stressors imposed on viruses as they move along those transmission pathways impact the virus concentration in the environmental source, and thereby dose (*d*), reaching the consumer.

Consuming "ready-to-eat" foods or foods that are minimally processed increases the likelihood of exposure to enteric viruses. Many food commodities—such as vegetables and fruits—are susceptible to contamination in the preharvest environment, especially from viral-laden irrigation water. When consumed raw, produce potentially pose greater health risks than other types of food. In addition, the trend in choosing healthier food options continues to rise with more people consuming fresh fruits and vegetables [58]. An analysis of reported foodborne illnesses during outbreaks occurring from 1998 to 2008 attributed almost half of the foodborne diseases to produce [59]. Specifically, leafy vegetables were linked to foodborne illnesses more than any other type of food. Interestingly, half of all foodborne norovirus outbreaks occurring between 2001 and 2008 was associated with contaminated produce [60].

IRRIGATION WATER

Irrigation water can harbor pathogens, including enteric viruses, which can be transferred to food crops [38,61,62]. Whether it is reclaimed water or surface water, water used for irrigation is subject to contamination with a variety of pathogens. QMRA has been used to address the role of irrigation water quality on food and subsequent foodborne transmission of pathogens [22,23,63,64]. Studies have evaluated the microbial transfer rate and retention of pathogens, including viruses, on food crops [65,66]. These rates can then be used in the QMRA process when considering microbial contamination of food crops, which informs the exposure assessment. Studies have shown a transfer rate ranging from 0.055% to greater than 4% involving coliphage PRD-1 and lettuce, cantaloupe, and bell peppers [62,66].

Survival studies evaluating irrigation water quality have shown that enteric viruses, such as hepatitis A virus, survive longer than bacteria on produce, and may survive postharvest [62,66]. Other studies have also addressed the impact of irrigation water quality on food crops based on the type of irrigation method employed and the type of crop [62,65,66]. Recently, pesticides have been identified as a possible source of norovirus contamination of produce crops [64,67,68]. Pesticides may become contaminated with the virus when mixed with virus-laden water. The data generated from these studies regarding irrigation water quality and type of application, virus transfer rates and survivability, and type of food crop, can be useful for the exposure assessment of viruses on food associated with poor irrigation water quality.

FOOD HANDLERS

Food handlers are another source of enteric viruses that may compromise food. The fecal–oral route of virus transmission, the virus shedding patterns from infected individuals (including those who are asymptomatic), and the survival mechanisms of these viruses, enhance the possibility of transmission from infected food handlers to consumers [69–71]. Transmission of hepatitis A virus, norovirus, and adenovirus in particular, has been associated with infected food handlers [71]. The fact that these viruses are particularly stable in the environment and lack an envelope increases their chance of survival and, therefore, transmission by food handlers. In addition, their low infectious dose also contributes to their success in initiating infection in exposed host(s). For norovirus-infected individuals, for example, high numbers of viruses may be excreted, yet the median infectious dose is only 18 virions [35]. In the absence of proper handwashing, the possibility of an infected food handler contaminating his or her environment is likely, and the food service arena provides an optimal opportunity for transmission to others. Norovirus can survive in food and on various surfaces, such as tabletops and utensils, for several days, making routine disinfection of surfaces at food service venues an important part of a cleaning regimen to prevent disease transmission.

The food service culture also contributes to the likelihood of food handler transmission of viruses. Food service workers may not be allowed, encouraged, or financially able to take time off from work if ill with gastroenteritis. Foodborne outbreaks associated with food preparers and servers who were ill onsite have been well documented [72]. In addition, if a food handler does take time away from work for the duration of a gastrointestinal illness, he or she may return to work while still shedding viruses, as viral shedding may continue for a few weeks once symptoms have dissipated. Further, up to 30% of infected individuals can be asymptomatic when infected with norovirus, thereby excreting viruses without knowing their infection status [34]. A person infected with norovirus could possibly shed the virus for as long as 4 weeks, be asymptomatic, and unknowingly transmit the virus to others.

HOME ENVIRONMENT

Although it is not uncommon for food handlers at food service establishments to be implicated in foodborne outbreaks, studies have shown that much of foodborne disease transmission occurs within the home environment [73,74]. Unlike restaurants, there is no regulatory oversight in the home and there is often a lack of awareness of food safety practices. As with disease outbreaks associated with public food services, the transmission of foodborne illnesses within the home is typically associated with the same factors: an infected person preparing the food, improper food holding temperatures (both hot and cold), and cross-contamination. However, unlike with restaurant personnel, those preparing and serving food in the home are often not cognizant of their potential role as transmitters of foodborne pathogens. This is likely due to the lack of oversight within household kitchens, and the media focus on food service establishments and food manufacturers during disease outbreak investigations and public announcements.

Although compromised food in the home does not lead to widespread disease outbreaks and media attention, it does cause sporadic foodborne illness that contributes to the overall health and economic burden of foodborne disease [73]. It is speculated that as much as 40% of foodborne illnesses can be linked to the home [75]. Studies have shown that consumers come in contact with a number of foodborne agents in the home environment, including viruses such as hepatitis A virus and rotavirus [76,77]. The combination of a lack of knowledge or awareness of food safety practices, and the habits and behaviors of consumers as they prepare food in their kitchens, emphasizes the need to target this environment in food safety efforts [78–80].

Studies have shown that the transportation and storage of food by consumers is a major contributor to foodborne disease transmission within the home environment [75,78,81,82]. While survey studies suggest that consumers are aware of the important role that temperature plays in maintaining food quality, shoppers' decisions do not consistently reflect that belief [75,82]. While many will selectively pull cold foods from the shelves as close to the time of purchase as possible, many shoppers will still complete other errands before returning home from the grocery store. In addition, few will use coolers to transport cold food. Moreover, surveys show that many consumers are not knowledgeable of proper refrigeration temperatures or consistently use a refrigerator thermometer [82].

Cross-contamination during food preparation is also a factor in the transmission of foodborne disease in the home. Studies indicate that consumers do not use separate cutting boards for raw meat and other foods, or disinfect cutting boards between use [75,78]. In addition, the lack of handwashing during food preparation in home kitchens is not uncommon. While studies underscore the need for food safety education that reaches a broad audience, it has also been recognized that food safety awareness and acceptance may not easily change consumer behavior, that is, food preparation habits [79].

A microbial risk assessment approach has been undertaken to incorporate consumer choices and food preparation behaviors in the home [83]. Unlike other microbial risk assessments that assume exposure is imminent, this risk assessment addresses the issue of *likelihood* as it uses Rasch modeling to predict the likelihood of particular food-related consumer behaviors. Since different, specific consumer behaviors will impact microorganism occurrence and survival in foods, information gleaned will inform—and thus improve—exposure assessments of foodborne pathogens. In addition, risk assessments using this technique can also lead to effective risk management and risk communication. The human health impact(s) concluded from such an assessment can be linked to specific consumer behavior, which can then identify appropriate risk mitigation targets. This information can also directly inform risk communication that will convey to the public the food safety practices that can be implemented in the home.

OTHER TRANSMISSION SOURCES

Since enteric viruses are transmitted via the fecal–oral route, they are commonly found in sewage and may or may not be inactivated through wastewater treatment. Bivalve mollusks may come in contact with (un)treated sewage and become contaminated with viruses such as hepatitis A virus and norovirus. Zoonotic foodborne transmission is also possible when considering hepatitis E virus in wildlife, such as pigs and deer, and the fact that hepatitis E virus has been shown to infect humans through this transmission route [84]. Hepatitis A virus and norovirus may also be transmitted by animals on rare occasions [71]. Figure 4.3 summarizes the factors that impact exposure to foodborne viruses.

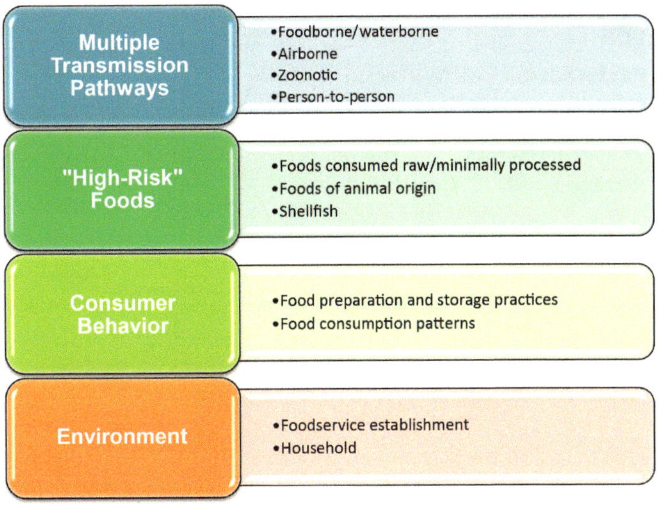

FIGURE 4.3 Factors impacting human exposure to foodborne viruses.

CHARACTERIZING RISK: WHAT ARE THE CHANCES?

Risk characterization integrates the information obtained from the other QMRA steps and packages it in a way that is both qualitative and quantitative. All assumptions made when data were lacking are explicitly stated here in order to ensure appropriate interpretations of the QMRA output. When estimating human health risks, point estimates of risk can be determined, or ranges of health risks can be calculated.

Simple spreadsheets using software (such as Microsoft Excel) can be used for point estimates of risk; programs, such as @RISK (Palisade Corporation, Newfield, New York) and Crystal Ball (Decisioneering Inc., Denver, Colorado) can be incorporated—along with a Monte Carlo approach—to compute distributions of risk, as described in Haas et al., 1999 [10]. Since a certain degree of uncertainty exists in all risk assessments, a Monte Carlo analysis can also be conducted to address this issue.

Case Study I

Norovirus is a common etiological agent of both foodborne and waterborne diseases worldwide. As discussed previously, norovirus is excreted in high numbers by infected individuals, has a relatively low infectious dose, and can survive environmental stressors making it easily transmitted to others [35]. Assuming exposure, the onset of infection and subsequent illness depend on a variety of factors related to the host–virus interaction. Within sensitive (immunocompromised) populations, severe health consequences may occur—such as dehydration—leading to death. These patients are also susceptible to an ongoing chronic norovirus infection. Distinct epidemiological patterns have been observed within vulnerable populations in health-care settings, reflecting the impact of host factors on health risks [85].

Previous dose–response characterizations and QMRAs of norovirus have shown the beta-Poisson model (Equation 4.2) to be appropriate in predicting the probability of infection following exposure [35,86]. In a simple case study of norovirus exposure and health effects, d (dose) can be assumed to be a certain number of virus particles in the exposure whether it is via water, food, or air. It can be assumed that all of the noroviruses in the exposure are capable of initiating infection in a susceptible host. Defined parameters for this case study are listed in Table 4.1.

TABLE 4.1
Parameters for a Norovirus Case Study

Parameter	Value	Reference
Concentration of norovirus	Five viral particles per exposure	Assumption
Dose–response model	beta-Poisson $\alpha = 0.0545$ $N_{50} = 5.24 \times 10^6$	[35,86]
Infectivity	100%	Assumption
Consumer	General population	Assumption
Morbidity ratio	0.50	[87]
Number of exposures per year	Three exposures	Assumption

Assuming five viral particles in each exposure, and assuming $\alpha = 0.0545$ and $N_{50} = 5.24 \times 10^6$ (35, 86), Equation 4.2 can be rewritten as

$$P_{\text{infection}} = 1 - \left[1 + \frac{5 \text{ viruses}}{5.24 \times 10^6}\left(2^{1/0.0545} - 1\right)\right]^{-0.0545} \quad (4.5)$$

with a resulting probability of infection ($P_{\text{infection}}$) of 1.5×10^{-2}, which equals 1.5/100 or 1.5:100. This means that for every 100 exposures (at $d = 5$ viruses), it can be expected that about two will result in infection. This does not necessarily translate to individuals experiencing clinical illness, but rather reflects individuals who are infected and may be excreting viruses.

A systematic review of over 900 norovirus outbreaks described in the published literature from 1993 to 2011 found that the highest attack rates (median = 50%, interquartile range = 31%–71%) were observed during foodborne outbreaks of norovirus when compared with outbreaks involving transmission by water, person-to-person, and environmental sources [87]. If we assume that the median attack rate of 50% reported from this study represents an appropriate morbidity ratio for this case study, a risk of illness can be estimated using the following equation:

$$P_{\text{illness}} = P_{\text{infection}} \times \text{morbidity ratio} \quad (4.6)$$

where $P_{\text{infection}}$ equals 1.5×10^{-2} and the morbidity ratio equals 0.50. This results in an estimated probability of norovirus illness of 7.5×10^{-3} (7.5/1000). (This illness risk estimate is based on the exposure parameters assumed when calculating the risk of infection.)

Annual risks of infection can be predicted using Equation 4.3. If it is assumed that an individual is exposed to norovirus three times per year at an exposure concentration of five viral particles, Equation 4.3 can be rewritten as

$$P_{\text{annual}} = 1 - \left(1 - P_{\text{infection}}\right)^3 \quad (4.7)$$

where $P_{\text{infection}}$ equals 1.5×10^{-2}. This scenario results in an estimated yearly risk of infection of 4.4×10^{-2} (4.4/100). Yearly, microbial risks are used by the United States Environmental Protective Agency (USEPA) to inform the development of health-based standards for drinking water quality. The USEPA recommends that annual microbial risks of infection for drinking water should not exceed 1.0×10^{-4} (1/10,000) [6,88]. If this case study involved drinking water as the transmission route for norovirus, the estimated yearly infection risk of 4.4/100 does not meet this guideline. In addition, in typical QMRAs of drinking water microbial hazards, a frequency of 365 days of exposure is assumed. If that assumption is applied here, the estimated annual risk of infection is 9.9/10, which greatly exceeds the USEPA 1/10,000 yearly infection risk recommendation. This scenario, however, assumes that an individual encounters an exposure of norovirus containing five viral particles every day for an entire year. Although it is likely that this assumption results in an overestimate of risk, it could be argued that overestimating leads to a conservative risk estimate that is protective of the most vulnerable subpopulations.

Case Study II

Hepatitis A virus has been implicated in numerous foodborne outbreaks, as described in the section 'Identifying Foodborne Viral Hazards'. In this case study, assume you are investigating the potential for hepatitis A virus to cause a foodborne outbreak associated with lettuce within a community of 100,000 residents. You are tasked with predicting the risks of infection associated with lettuce that has become contaminated with hepatitis A virus. Equation 4.4 summarizes the data you obtain from the published literature and the information you assume (due to lack of data) to help you estimate daily and yearly risks of infection. You assume one viral particle per 1000 g of lettuce (a recovery efficiency of 100%), and assume an actual serving weight of lettuce to be 5.9 g (adjusted for loss at retail and consumer levels) [89]. To be conservative, you assume no loss of viruses due to food processing, and that all hepatitis A viruses in the exposure are infectious.

After considering the exposure-related factors, *d*—the number of hepatitis A viruses in the exposure—is estimated to be 0.0059 viruses. By applying the exponential model and defined parameter as listed in Table 4.2, the probability of infection is estimated to be 3.23×10^{-3}, which equals about 3/1000. This translates to a prediction of approximately 300 infections throughout this community of 100,000 residents, assuming a one-time exposure to a lettuce serving (at 5.9 g) containing one infectious hepatitis A virus per 1000 g lettuce. The risk output for this scenario may be an overestimate due to the conservative assumptions made in the QMRA process; however, the calculated risk may be underestimated since secondary transmission was not considered.

As with the norovirus case study, annual risks of infection can be predicted using Equation 4.3. Assuming n equals 104 exposures over a 1 year time frame (Table 4.2), annual risks of infection are estimated at 2.86×10^{-1} (almost 3/10). If annual risks are estimated assuming four lettuce servings per month at the same virus exposure level, yearly risks equal 1.4/10, or about 1/10. When considering the USEPA recommendation that microbial risks of infection for drinking water should not exceed 1/10,000 per year [88], the virus contamination levels presented in this case study regarding a food commodity translate to annual microbial infection risks that do not meet this goal.

TABLE 4.2
Parameters for a Hepatitis A Virus Case Study

Parameter	Value	Reference
Concentration of hepatitis A virus	1 virus/1000 g lettuce	Assumption
Dose–response model	Exponential $r = 0.5486$	[10]
Recovery efficiency (R^{-1})	100%	Assumption
Decimal reduction	0	Assumption
Infectivity	100%	Assumption
Consumer	General population	Assumption
Amount consumed per serving	5.9 g	[89]
Number of servings per year	104 servings in a year (average of twice per week)	Assumption

QMRA APPLICATIONS AND IMPACTS

QMRA is the first step in the risk analysis process of risk assessment, risk management, and risk communication [90]. QMRA output may be both qualitative and quantitative, and provides the foundation for mitigating risks and discussing prevention strategies with the public. Specifically, human health risk estimates can be used to evaluate risk reduction programs by associating probabilities of health outcomes from exposure to environmental hazards before and after an intervention, for example. The qualitative nature of the hazard identification step can inform the public of critical information regarding clinical symptoms associated with foodborne viral illnesses and foods that may pose greater risks than others. In addition, health risk estimates can be linked to specific consumer behaviors that can then inform a risk communication dialogue. Further, regulators can utilize health risk estimates from QMRAs to inform policy for the food industry [22].

From a risk management perspective, information generated from QMRAs can be used in Hazard Analysis and Critical Control Point (HACCP) programs to identify those places in the food production and processing chain that may be particularly susceptible to hazard contamination or persistence. HACCP is a seven-step process that includes: (1) hazard analysis; (2) critical control point identification; (3) guideline development for critical control points; (4) the creation of a monitoring regimen for critical control points; (5) a defined remediation plan; (6) a record-keeping procedure; and (7) the development of a verification plan [91]. QMRA offers a proactive approach to identifying "critical control points" during a food manufacturing process that can then be targeted in the implementation and application of hazard control measures. QMRA quantitative output can reflect how specific critical control points impact health risk(s), and how remediation mitigates these risk(s). QMRA complements HACCP as they both provide tools to prevent foodborne disease. Figure 4.4 shows the relationship between QMRA's identification of critical control points and HACCP, a risk management application.

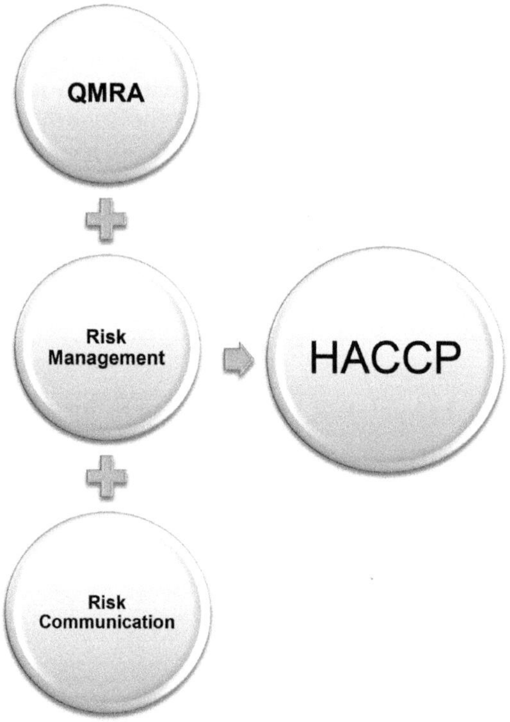

FIGURE 4.4 The interface of quantitative microbial risk assessment (QMRA) that identifies critical control points (CCPs), and hazard analysis critical control point (HACCP) program within the risk analysis framework.

The last component of the risk analysis framework is risk communication. Risk communication integrates the activities of risk assessment and risk management, and involves a dialogue among risk managers and various stakeholders. Risk perception directly impacts risk communication efforts, and has the potential to create misunderstandings. Knowing the goal(s) of the communication, as well as the participants (audience), is critical for effective risk communication [92]. For example, conveying health risk information related to foodborne pathogens can provide useful information regarding food safety and illness prevention. However, foodborne illness rates have been linked to specific consumer food preparation behaviors that may be related to one's culture, so connecting health risks to potentially culture-based food handling practices may be seen as criticism, rather than helpful information [92].

CONCLUSION

Quantitative microbial risk assessment gives science a voice in the policy arena. It ensures that critical components of complex problems are considered during regulatory decision-making and policy development. Minimizing foodborne disease is an international challenge. With the globalization of food distribution, the "farm to fork" labyrinthine continuum continues to expand, becoming more convoluted. The flexibility of the QMRA framework makes it possible to interactively investigate foodborne viral disease so that associated impacts can be predicted and, ultimately, infections can be prevented.

REFERENCES

1. Scallan E, Griffin PM, Angulo FJ, Tauxe RV, Hoekstra RM. 2011. Foodborne illness acquired in the United States—Unspecified agents. *Emerg Infect Dis* 17:16.
2. Gould LH, Walsh KA, Vieira AR, Herman K, Williams IT, Hall AJ, Cole D. 2013. Surveillance for foodborne disease outbreaks—United States, 1998–2008. *MMWR Surveill Summ* 62:1–34.
3. Scharff RL. 2012. Economic burden from health losses due to foodborne illness in the United States. *J Food Prot* 75:123–131.
4. National Research Council (NRC). 1983. *Risk Assessment in the Federal Government: Managing the Process*. National Academy Press, Washington, DC.
5. Haas C. 1983. Estimation of risk due to low doses of microorganisms: A comparison of alternative methodologies. *Am J Epidemiol* 118:573–582.
6. Regli S, Rose JB, Haas CN, Gerba CP. 1991. Modeling the risk from Giardia and viruses in drinking water. *J Am Water Works Assoc* 83:76–84.
7. Gerba CP, Rose JB. 1993. Estimating viral disease risk from drinking water. *In* Cothern C (ed), *Comparative Environmental Risks*, pp. 117–134. Lewis Publishers, Boca Raton, FL.
8. Haas CN, Rose JB, Gerba C, Regli S. 1993. Risk assessment of virus in drinking water. *Risk Anal* 13:545–552.
9. Rose JB, Sobsey MD. 1993. Quantitative risk assessment for viral contamination of shellfish and coastal waters. *J Food Prot* 56:1043–1050.
10. Haas CN, Rose JB, Gerba CP. 1999. *Quantitative Microbial Risk Assessment*. John Wiley & Sons, New York.
11. Mena KD. 2002. Environmental exposure to viruses: A human health risk assessment methodology. *In* Bitton G (ed), *Risk Assessment of Environmental Exposure to Viruses*, Vol 5, pp. 2741–2753, *Encyclopedia of Environmental Microbiology*. John Wiley & Sons, New York.
12. Mena KD, Gerba CP, Haas CN, Rose JB. 2003. Risk assessment of waterborne coxsackievirus. *J Am Water Works Assoc* 95:122–131.
13. Mena KD, Rose JB, Gerba CP. 2004. Addressing food safety issues quantitatively: A risk assessment approach. *In* Beier RC, Pillai SD, Phillips TD (eds), *Preharvest and Postharvest Food Safety: Contemporary Issues and Future Directions*, pp. 415–426. Blackwell Publishing, Ames, IA.
14. Mena KD. 2006. Produce quality and foodborne disease: Assessing water's role. *In* James J (ed), *Microbial Hazard Identification in Fresh Fruit and Vegetables*, pp. 95–114. John Wiley & Sons, Hoboken, NJ.
15. Mena KD. 2007. Waterborne viruses: Assessing the risks. *In* Bosch A (ed), *Human Viruses in Water*, vol 17, pp. 163–175. Elsevier, Amsterdam, the Netherlands.

16. Havelaar A, Rutjes S. 2008. Risk assessment of viruses in food: Opportunities and challenges, *In* Koopmans M, Cliver D, Bosch A, Doyle M (eds), *Food-Borne Viruses*, pp. 221–236. ASM Press, Washington, DC. doi: 10.1128/9781555815738.ch10.
17. Mena KD, Gerba CP. 2009. Waterborne adenovirus. *Rev Environ Contam Toxicol* 198:133–167.
18. Espinosa AC, Jesudhasan P, Arredondo R, Cepeda M, Mazari-Hiriart M, Mena KD, Pillai SD. 2011. Sensitivity of poliovirus type 1 chat strain and rotavirus SA-11 to e-beam irradiation on iceberg lettuce and spinach: Quantifying the reduction in potential health risks. *Appl Environ Microbiol* 78:988–993.
19. Soto-Beltran M, Mena KD, Gerba CP, Tarwater PM, Reynolds K, Chaidez C. 2013. Risk assessment of Listeria monocytogenes in queso fresco in Culiacan, Mexico. *J Micro Research* 3:111–116.
20. Brown M, Stringer M. 2002. *Microbiological Risk Assessment in Food Processing*. CRC Press LLC, Boca Raton, Florida.
21. Dennis SB, Kause J, Losikoff M, Engeljohn DL, Buchanan RL, Schaffner D. 2008. Using risk analysis for microbial food safety regulatory decision making. *In* Schaffner DW (ed), *Microbial Risk Analysis of Foods*, pp. 137–175. ASM Press, Washington, DC.
22. Mena K, Pillai S. 2008. An approach for developing quantitative risk-based microbial standards for fresh produce. *J Water Health* 6:359–364.
23. Petterson SR, Teunis PF, Ashbolt NJ. 2001. Modeling virus inactivation of salad crops using microbial count data. *Risk Anal* 21:1097–1108.
24. Scallan E, Jones TF, Cronquist A, Thomas S, Frenzen P, Hoefer D, Medus C, Angulo FJ. 2006. Factors associated with seeking medical care and submitting a stool sample in estimating the burden of foodborne illness. *Foodborne Pathog Dis* 3:432–438.
25. Koopmans M. 2012. Food-borne viruses from a global perspective. *In* Choffnes ER, Relman DA, Olsen L, Hutton R, Mack A (eds), *Improving Food Safety through a One Health Approach: Workshop Summary*, pp. 225. Institute of Medicine, National Academies Press, Washington, DC.
26. Koopmans M, Duizer E. 2004. Foodborne viruses: An emerging problem. *Int J Food Microbiol* 90:23–41.
27. McCabe-Sellers BJ, Beattie SE. 2004. Food safety: Emerging trends in foodborne illness surveillance and prevention. *J Am Diet Assoc* 104:1708–1717.
28. Vasickova P, Dvorska L, Lorencova A, Pavlik I. 2005. Viruses as a cause of foodborne diseases: A review of the literature. *Vet Med* 50:89–104.
29. Bouwknegt M, Verhaelen K, de Roda Husman A, Rutjes S. 2013. Quantitative risk profile for viruses in foods. RIVM report 330371008.
30. Flint JA, Van Duynhoven YT, Angulo FJ, DeLong SM, Braun P, Kirk M, Scallan E, Fitzgerald M, Adak GK, Sockett P. 2005. Estimating the burden of acute gastroenteritis, foodborne disease, and pathogens commonly transmitted by food: An international review. *Clin Infect Dis* 41:698–704.
31. Batz MB, Hoffmann S, Morris Jr JG. 2012. Ranking the disease burden of 14 pathogens in food sources in the United States using attribution data from outbreak investigations and expert elicitation. *J Food Prot* 75:1278–1291.
32. Pires SM, Vieira AR, Perez E, Lo Fo Wong D, Hald T. 2012. Attributing human foodborne illness to food sources and water in Latin America and the Caribbean using data from outbreak investigations. *Int J Food Microbiol* 152:129–138.
33. Mead PS, Slutsker L, Dietz V, McCaig LF, Bresee JS, Shapiro C, Griffin PM, Tauxe RV. 1999. Food-related illness and death in the United States. *Emerg Infect Dis* 5:607.
34. Centers for Disease Control and Prevention (CDC). 2003. Norovirus activity—United States, 2002.
35. Teunis PF, Moe CL, Liu P, Miller SE, Lindesmith L, Baric RS, Le Pendu J, Calderon RL. 2008. Norwalk virus: How infectious is it? *J Med Virol* 80:1468–1476.
36. Chironna M, Lopalco P, Prato R, Germinario C, Barbuti S, Quarto M. 2004. Outbreak of infection with hepatitis A virus (HAV) associated with a foodhandler and confirmed by sequence analysis reveals a new HAV genotype IB variant. *J Clin Microbiol* 42:2825–2828.
37. Acheson D, Fiore AE. 2004. Hepatitis A transmitted by food. *Clin Infect Dis* 38:705–715.
38. Dentinger CM, Bower WA, Nainan OV, Cotter SM, Myers G, Dubusky LM, Fowler S, Salehi ED, Bell BP. 2001. An outbreak of hepatitis A associated with green onions. *J Infect Dis* 183:1273–1276.
39. Coelho C, Heinert A, Simões C, Barardi C. 2003. Hepatitis A virus detection in oysters (Crassostrea gigas) in Santa Catarina state, Brazil, by reverse transcription–polymerase chain reaction. *J Food Prot* 66:507–511.
40. Shieh Y, Khudyakov Y, Xia G, Ganova-Raeva L, Khambaty F, Woods J, Veazey J, Motes M, Glatzer M, Bialek S. 2007. Molecular confirmation of oysters as the vector for hepatitis A in a 2005 multistate outbreak. *J Food Prot* 70:145–150.
41. Pinto RM, Costafreda MI, Bosch A. 2009. Risk assessment in shellfish-borne outbreaks of hepatitis A. *Appl Environ Microbiol* 75:7350–7355.

42. Halliday ML, Kang L-Y, Zhou T-K, Hu M-D, Pan Q-C, Fu T-Y, Huang Y-S, Hu S-L. 1991. An epidemic of hepatitis A attributable to the ingestion of raw clams in Shanghai, China. *J Infect Dis* 164:852–859.
43. Goyal SM. 2004. Viruses in food. *In* Beier RC, Pillai SD, Phillips TD (eds), *Preharvest and Postharvest Food Safety: Contemporary Issues and Future Directions*, pp. 101–117. Blackwell Publishing, Ames, IA.
44. Gerba CP, Rose JB, Haas CN, Crabtree KD. 1996. Waterborne rotavirus: A risk assessment. *Water Res* 30:2929–2940.
45. Matsumoto K, Hatano M, Kobayashi K, Hasegawa A, Yamazaki S, Nakata S, Chiba S, Kimura Y. 1989. An outbreak of gastroenteritis associated with acute rotaviral infection in schoolchildren. *J Infect Dis* 160:611–615.
46. Mikami T, Nakagomi T, Tsutsui R, Ishikawa K, Onodera Y, Arisawa K, Nakagomi O. 2004. An outbreak of gastroenteritis during school trip caused by serotype G2 group A rotavirus. *J Med Virol* 73:460–464.
47. Crabtree K, Gerba C, Rose J, Haas C. 1997. Waterborne adenovirus: A risk assessment. *Wat Sci Tech* 35:1–6.
48. Pina S, Puig M, Lucena F, Jofre J, Girones R. 1998. Viral pollution in the environment and in shellfish: Human adenovirus detection by PCR as an index of human viruses. *Appl Environ Microbiol* 64:3376–3382.
49. Muniain-Mujika I, Calvo M, Lucena F, Girones R. 2003. Comparative analysis of viral pathogens and potential indicators in shellfish. *Int J Food Microbiol* 83:75–85.
50. Myrmel M, Berg E, Rimstad E, Grinde B. 2004. Detection of enteric viruses in shellfish from the Norwegian coast. *Appl Environ Microbiol* 70:2678–2684.
51. Ward R, Krugman S, Giles JP, Jacobs AM, Bodansky O. 1958. Infectious hepatitis: Studies of its natural history and prevention. *N Engl J Med* 258:407–416.
52. Couch R, Cate T, Gerone P, Fleet W, Lang D, Griffith W, Knight V. 1966. Production of illness with a small-particle aerosol of Adenovirus type 4. *Bacteriol Reviews* 30:517–528.
53. Katz M, Plotkin SA. 1967. Minimal infective dose of attenuated poliovirus for man. *Am J Public Health* 57:1837–1840.
54. Ward RL, Bernstein DI, Young EC, Sherwood JR, Knowlton DR, Schiff GM. 1986. Human rotavirus studies in volunteers: Determination of infectious dose and serological response to infection. *J Infect Dis* 154:871–880.
55. Holcomb DL, Smith MA, Ware GO, Hung YC, Brackett RE, Doyle MP. 1999. Comparison of six dose-response models for use with food-borne pathogens. *Risk Anal* 19:1091–1100.
56. Quinlan JJ. 2013. Foodborne illness incidence rates and food safety risks for populations of low socioeconomic status and minority race/ethnicity: A review of the literature. *Int J Environ Res Public Health* 10:3634–3652.
57. Henley SC, Stein SE, Quinlan JJ. 2012. Identification of unique food handling practices that could represent food safety risks for minority consumers. *J Food Prot* 75:2050–2054.
58. Sivapalasingam S, Friedman CR, Cohen L, Tauxe RV. 2004. Fresh produce: A growing cause of outbreaks of foodborne illness in the United States, 1973 through 1997. *J Food Prot* 67:2342–2353.
59. Painter JA, Hoekstra RM, Ayers T, Tauxe RV, Braden CR, Angulo FJ, Griffin PM. 2013. Attribution of foodborne illnesses, hospitalizations, and deaths to food commodities by using outbreak data, United States, 1998–2008. *Emerg Infect Dis* 19:407.
60. Hall AJ, Eisenbart VG, Etingüe AL, Gould LH, Lopman BA, Parashar UD. 2012. Epidemiology of foodborne norovirus outbreaks, United States, 2001–2008. *Emerg Infect Dis* 18:1566.
61. Steele M, Odumeru J. 2004. Irrigation water as source of foodborne pathogens on fruit and vegetables. *J Food Prot* 67:2839–2849.
62. Gerba CP, Choi CY. 2006. Role of irrigation water in crop contamination by viruses. *In* Goyal S (ed), *Viruses in Foods*, pp. 257–263. Springer, New York.
63. Stine SW, Song I, Choi CY, Gerba CP. 2005. Application of microbial risk assessment to the development of standards for enteric pathogens in water used to irrigate fresh produce. *J Food Prot* 68:913–918.
64. Stine SW, Song I, Choi CY, Gerba CP. 2011. Application of pesticide sprays to fresh produce: A risk assessment for hepatitis A and salmonella. *Food Environ Virol* 3:86–91.
65. Choi C, Song I, Stine S, Pimentel J, Gerba C. 2004. Role of irrigation and wastewater reuse: Comparison of subsurface irrigation and furrow irrigation. *Wat Sci Tech* 50:61–68.
66. Stine SW, Song I, Choi CY, Gerba CP. 2005. Effect of relative humidity on preharvest survival of bacterial and viral pathogens on the surface of cantaloupe, lettuce, and bell peppers. *J Food Prot* 68:1352–1358.
67. Potera C. 2013. Pesticide spraying may spread norovirus. *Environ Health Perspect* 121:a148.

68. Verhaelen K, Bouwknegt M, Rutjes SA, de Roda Husman AM. 2013. Persistence of human norovirus in reconstituted pesticides: Pesticide application as a possible source of viruses in fresh produce chains. *Int J Food Microbiol* 160:323–328.
69. De Wit M, Koopmans M, Kortbeek L, Wannet W, Vinje J, Van Leusden F, Bartelds A, Van Duynhoven Y. 2001. Sensor, a population-based cohort study on gastroenteritis in the Netherlands: Incidence and etiology. *Am J Epidemiol* 154:666–674.
70. Amar C, East C, Gray J, Iturriza-Gomara M, Maclure E, McLauchlin J. 2007. Detection by PCR of eight groups of enteric pathogens in 4,627 faecal samples: Re-examination of the English case-control Infectious Intestinal Disease Study (1993–1996). *Eur J Clin Microbiol Infect Dis* 26:311–323.
71. Duizer E, Koopmans M, Koopmans M, Cliver D, Bosch A. 2008. Emerging food-borne viral diseases. *In* Koopmans MPG, Cliver DO, Bosch A (eds), *Food-Borne Viruses: Progress and Challenges*, pp. 117–145. ASM Press, Washington, DC.
72. Moe CL. 2009. Preventing norovirus transmission: How should we handle food handlers? *Clin Infect Dis* 48:38–40.
73. Scott E. 1996. Foodborne disease and other hygiene issues in the home. *J Appl Bacteriol* 80:5–9.
74. Scott E. 2003. Food safety and foodborne disease in 21st century homes. *Can J Infect Dis* 14:277.
75. Langiano E, Ferrara M, Lanni L, Viscardi V, Abbatecola AM, De Vito E. 2012. Food safety at home: Knowledge and practices of consumers. *J Public Health* 20:47–57.
76. Jay LS, Comar D, Govenlock LD. 1999. A video study of Australian domestic food-handling practices. *J Food Prot* 62:1285–1296.
77. Scott E. 2001. The potential benefits of infection control measures in the home. *Am J Infect Control* 29:247–249.
78. Redmond EC, Griffith CJ. 2003. Consumer food handling in the home: A review of food safety studies. *J Food Prot* 66:130–161.
79. McIntosh WA. 2004. Food safety risk communication and consumer food-handling behavior. *In* Beier RC, Pillai SD, Phillips TD (eds), *Preharvest and Postharvest Food Safety: Contemporary Issues and Future Directions*, pp. 405–414. Blackwell Publishing, Ames, IA.
80. Unusan N. 2007. Consumer food safety knowledge and practices in the home in Turkey. *Food Control* 18:45–51.
81. Kagan LJ, Aiello AE, Larson E. 2002. The role of the home environment in the transmission of infectious diseases. *J Commun Health* 27:247–267.
82. Godwin S, Coppings R. 2005. Analysis of consumer food-handling practices from grocer to home including transport and storage of selected foods. *J Food Dist Research* 36:55–62.
83. Fischer AR, Frewer LJ, Nauta MJ. 2006. Toward improving food safety in the domestic environment: A multi-item Rasch scale for the measurement of the safety efficacy of domestic food-handling practices. *Risk Anal* 26:1323–1338.
84. Tei S, Kitajima N, Takahashi K, Mishiro S. 2003. Zoonotic transmission of hepatitis E virus from deer to human beings. *Lancet* 362:371–373.
85. Lopman BA, Adak GK, Reacher M, Brown DW. 2003. Two epidemiologic patterns of norovirus outbreaks: Surveillance in England and Wales, 1992–2000. *Emerg Infect Dis* 9:71.
86. CAMRA. Center for Advancing Microbial Risk Assessment. 2013. Case Studies: Quantitative Microbial Risk Assessment (QMRA) Wiki. http://qmrawiki.msu.edu/. Accessed November 2013.
87. Matthews J, Dickey B, Miller R, Felzer J, Dawson B, Lee A, Rocks J, Kiel J, Montes J, Moe C. 2012. The epidemiology of published norovirus outbreaks: A review of risk factors associated with attack rate and genogroup. *Epidemiol Infect* 140:1161–1172.
88. United States Environmental Protection Agency (USEPA). 1989. National Primary Drinking Water Regulations: Filtration, Disinfection; Turbidity, Giardia lamblia, Viruses, Legionella and Heterotrophic Bacteria. Final Rule.
89. United States Department of Agriculture (USDA). 2013. USDA/Economic Research Service Loss Adjusted Food Availability Data. http://www.ers.usda.gov/data-products/food-availability-(per-capita)-data-system/. http://www.ers.usda.gov/data-products/food-availability-(per-capita)-data-system.aspx. Accessed October, 2013.
90. National Research Council (NRC). 1994. *Science and Judgment in Risk Assessment.* National Academy Press, Washington, DC.
91. Buchanan RL, Whiting RC. 1998. Risk assessment: A means for linking HACCP plans and public health. *J Food Prot* 61:1531–1534.
92. Hallman WK, Schaffner D. 2008. Communicating about microbial risks in foods. *In* Schaffner DW (ed), *Microbial Risk Analysis of Foods*, pp. 205–262. ASM Press, Washington, DC.

5 Strategies for Control of Food- and Waterborne Viruses

Erin DiCaprio, Fangfei Lou, Ashley Predmore, and Jianrong Li

CONTENTS

INTRODUCTION TO FOOD- AND WATERBORNE VIRUSES

A foodborne virus is defined as any virus that may be transmitted by food and is able to cause illness via the fecal–oral route. Similarly, a waterborne virus is usually transmitted by contaminated water. These food- and waterborne viruses usually infect the human digestive system, and are subsequently shed and dispersed by stool and/or vomit. The most common food- and waterborne human viruses are norovirus, sapovirus, astrovirus, human rotavirus (HuRV), adenovirus, poliovirus, enteroviruses, hepatitis A virus (HAV), and hepatitis E virus (HEV). Among these foodborne viruses, human norovirus has long been considered the most prominent cause of viral-associated foodborne illness, causing a significant proportion of all nonbacterial acute gastroenteritis worldwide. Approximately 23 million people suffer from norovirus-associated gastroenteritis each year in the United States [1]. Unlike bacteria, viruses are strict intracellular parasites and are only capable of replicating inside a living host. Therefore, viruses will never amplify in food or water during transportation, processing, and storage. Importantly, all known food- and waterborne viruses are nonenveloped viruses and are therefore extremely stable in food, water, and the environment. These viruses can survive on hands, in dried human and animal feces, on kitchen surfaces, floors, carpets, and even hospital lockers for long time periods. Additionally, only a few virus particles are capable of causing illness and, following infection, a large number of virus particles can be shed in the stools of afflicted individuals (up to 10^{11}). Most food- and waterborne viruses are much more resistant to heat, acidic pH, dehydration, and disinfection compared with bacteria (or even enveloped viruses), thus the current procedures used to control bacteria in food processing, preservation, and storage may not be fully effective against viral pathogens. Another challenge is that many important food- and waterborne viruses (such as human norovirus and sapovirus) cannot be grown robustly in cell culture. In fact, the replication of these noncultivable viruses is still poorly understood. Thus, the study of human norovirus relies on cultivable surrogates from the *Caliciviridae* family, including feline calicivirus (FCV), murine norovirus (MNV), and Tulane virus (TV). In recent years, many strategies have been developed to control food- and waterborne viruses. This chapter focuses on the most recent progress in nonthermal, thermal, and chemical processing technologies, preventive measures, regulatory efforts, and vaccination programs to control viruses in humans, food, water, and the environment.

NONTHERMAL PROCESSES AGAINST FOODBORNE VIRUSES

INTRODUCTION

A nonthermal process is a "cold" process, which can be used for decontamination, pasteurization, and sterilization. One of the key attributes of the nonthermal processed product is its excellent quality, wherein the products maintain a "fresh" characteristic. The advantages include better nutritional value (e.g., vitamins, enzymes, and protein), better sensory (e.g., texture and color) and microbiological quality, and minimal or no use of preservatives [2,3]. Commonly known nonthermal technologies include high-pressure processing (HPP), irradiation, ultraviolet (UV) light, ultrasound, pulsed electric field, and cold plasma. The application of a particular nonthermal processing technology depends on the type of matrix targeted, the location of the pathogens, and the effectiveness of the technology. For viruses located within the food matrix, technologies targeting food surface areas, such as UV, ultrasound, and cold plasma, are not applicable. Some technologies, such as pulsed electric field, are only suitable for liquid matrices. Other technologies, such as gamma irradiation or the electron beam (e-beam), have limited penetration, thus, the efficacy may be compromised. To date, HPP is considered to be the most effective, promising, and practical nonthermal technology to inactivate food- and waterborne viruses.

High Pressure Processing

For HPP, food samples (or other matrices) need to be prepackaged in flexible pouches/containers, due to the possible volume decrease under pressure, and then loaded in a pressure vessel filled with a pressure-transmitting fluid (water or oil) prior to operation. Once the vessel is filled and sealed, pressure can be generated by pumping additional fluid into the vessel. The pressure is subsequently applied uniformly throughout the product. HPP is an energy-efficient process, because it requires no additional energy once the desired pressure is reached. HPP units are available in a variety of dimensions, capacities, and throughputs, ranging from small research-scale sizes (2 L) to commercial sizes (up to 687 L). Extensive research efforts have shown that HPP is capable of efficiently inactivating many food- and waterborne viruses and their research surrogates (Table 5.1). Significant viral inactivation (≥5-log reduction) was reported for a variety of nonenveloped food- and waterborne viruses following the application of ≤600 MPa of pressure, including human norovirus surrogates such as FCV, MNV, HAV, and HuRV [2,4–7]. However, viruses can differ widely in their sensitivities to HPP. For example, poliovirus is very resistant to HPP, with no significant reductions in infectivity reported after relatively harsh treatments, such as 600 MPa at 20°C for 60 min [8].

Mechanism of Virus Inactivation by HPP

Lou et al. [7] performed a systematic study in 2011 to determine the mechanism of inactivation of norovirus by HPP using MNV as a model. After pressure treatment, the virion structure changed from a discrete, small round shape to a large amount of undefined protein debris as visualized by electron microscopy (EM) (Figure 5.1). The primary mechanism of norovirus inactivation is the disruption of the virion capsid structure. However, the capsid proteins of MNV were not degraded and were still capable of reacting with antibodies, indicating that the primary and secondary structures

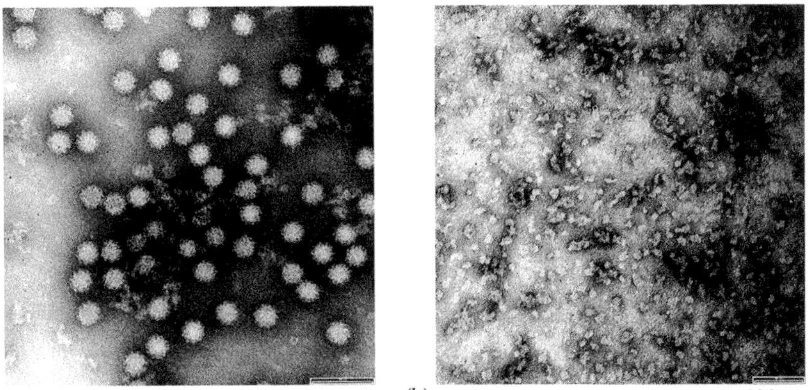

FIGURE 5.1 HPP disrupts MNV particles after treatment at 600 MPa at 4°C for 2 min. Treated and untreated virus particles were negatively stained with 1% ammonium molybdate and visualized by EM. (a) Untreated MNV virions. (b) MNV virions treated at 600 MPa. (Adapted from Lou, F. et al., *Appl Environ Microbiol*, 77, 1862–1871, 2011.)

of viral proteins remained intact, although the quaternary and tertiary structures of viral capsid proteins had become completely distorted [7]. In a subsequent study in 2012, Lou et al. [9] showed that HPP also impaired the capacity of human norovirus to bind to the histo-blood group antigens (HBGAs), which are considered attachment factors for human norovirus infections. This was also supported by another study showing a significant decrease in the ability of human norovirus to bind to HBGAs after treatment at 400 MPa and 5°C for 5 min [10]. However, HPP does not degrade viral genomic RNA, because high pressure usually does not break covalent bonds at the level applied for food processing [7,11].

Factors That Affect the Effectiveness of Virus Inactivation by HPP

To effectively inactivate pathogens, it is critical to optimize the conditions for pressure treatment. As shown in Table 5.1, the effectiveness of HPP is influenced by many factors such as processing parameters (e.g., applied pressure, holding time, and initial temperature) and nonprocessing parameters (e.g., virus structure, food matrix, and pH) [2,6,7,12,13].

Processing Parameters

Generally, the extent of virus inactivation increases commensurate with the pressure level and holding time. However, it was discovered that virus inactivation is more pressure dependent and a first-order relationship was exhibited between pressure levels and inactivation outcome [10]. As for treatment time, the HPP inactivation curves of FCV, MNV, and HAV showed variation in virus titer to be a function of treatment time, exhibiting pronounced tailing; indicating that longer holding times did not significantly enhance inactivation [5,7,12,21].

The initial temperature at which pressure is applied can also have a significant impact on the pressure inactivation of viruses, whereby temperature can either work synergistically or antagonistically with pressure. It is reported that pressure inactivation of MNV was favored at refrigerated temperatures [7]; a 350-MPa treatment held for 5 min at 30°C inactivated only 1.2 logs of virus, while the same treatment achieved a reduction of 5.6 logs at 5°C [21]. This is consistent with the temperature sensitivity of another recently developed a human norovirus surrogate, TV, during HPP treatment [22]. In contrast, other studies have shown that certain viruses were more sensitive to HPP at room temperature than at lower temperatures. For HAV, pressure inactivation was enhanced as temperatures increased above 30°C compared with temperatures ranging from 5°C to 30°C [5]. FCV was found to be minimally affected by pressure at room temperature and inactivation was enhanced at either below or above 20°C [20]. These observations suggest that the optimal temperature for pressure inactivation is specific to the virus of interest.

Intrinsic Parameters

The pH of the suspending medium or food is an important consideration in the pressure inactivation of viruses. HAV is known to be pH stable at atmospheric pressure; however, its inactivation is significantly enhanced in an acidic environment (pH 3) compared with a neutral environment (pH 7) at 400 MPa [18]. This synergistic effect of pH and HPP could be beneficial for processing selected acidic food products such as salsa and strawberry puree [16]. On the other hand, it was shown that the MNV, TV, FCV, and HuRV were more easily inactivated at a neutral pH compared with an acidic pH [6,7,12,20,22]. As surrogates of the enteric human norovirus, MNV and TV both exhibit high stability at low pH [7,22]. For MNV, an 8.1-log reduction of virus titer was achieved at 350 MPa for 2 min at pH 7.0, whereas only a 6.0-log virus reduction was achieved at pH 4.0 at the same pressure level and holding time [7]. A treatment of 350 MPa for 2 min at 21°C resulted in a 3.8-log reduction of TV at a neutral pH, but the same treatment only reduced TV by 2.4 logs at pH 4 [22]. The titer of HuRV was reduced by 3.4 logs, following exposure to 250 MPa of pressure at

TABLE 5.1
Effectiveness of High Pressure Processing on the Inactivation of Food- and Waterborne Viruses

Virus					
Family	Virus Name[c]	Pressure Treatment	pH	$\log_{10}$ Reduction (Matrix)	Reference
Reoviridae	Rotavirus	300 MPa, 4°C, 2 min	7	4.1 (medium)	[7]
		300 MPa, 4°C, 2 min	4	1.9 (medium)	[7]
Picornaviridae	HAV	450 MPa, ambient temp,	NA[a]	7 (medium)	[14]
		5 min	NA	3 (oyster)	[15]
		400 MPa, 9°C, 1 min	3.67	4.3 (mashed strawberries)	[16]
		375 MPa, 21°C, 5 min	5.12	4.7 (sliced green onions)	[16]
		375 MPa, 21°C, 5 min	NA	3.23 TCID50/mL	[17]
		500 MPa, 4°C, 5 min	NA	(sausages)	[18]
		400 MPa, 50°C, 1 min	NA	4.0 (medium)	[18]
		400 MPa, 50°C, 1 min,		0.4 (medium)	
		6% NaCl			
	Poliovirus	600 MPa, ambient temp,	NA	no reduction (medium)	[14]
		5 min	NA	no reduction (medium)	[8]
		600 MPa, 20°C, 60 min			
	Aichivirus	600 MPa, ambient temp,	NA	no reduction (medium)	[19]
		5 min			
Caliciviridae	FCV	275 MPa, ambient temp,	NA	>6 TCID50/mL (medium)	[14]
		5 min	NA	5 (medium)	[12]
		200 MPa, −10°C, 4 min	NA	0.3 (medium)	[12]
		200 MPa, 20°C, 4 min	6.0	4.1 (medium)	[20]
		250 MPa, 20°C, 1 min	NA	0.7 (medium)	[20]
		250 MPa, 20°C, 5 min,	NA	5 TCID50/mL (medium)	[13]
		12% NaCl	NA	2.89 TCID50/mL (sausage)	[17]
		300 MPa, ambient temp,			
		3 min			
		500 MPa, 4°C, 5 min			
	MNV-1	400 MPa, 5°C, 5 min	NA	4.05 (oyster)	[21]
	MNV-1	450 MPa, 20°C, 5 min	NA	6.85 (medium)	[21]
	TV	250 MPa, 4°C, 2 min	NA	2.9 (oyster)	[22]
	Norwalk virus	250 MPa, 4°C, 2 min	NA	0.3 (oyster)	[22]
	Human NoV	600 MPa, 6°C, 5 min	NA	4GEC[b] (human volunteer)	[23]
	GII-4	450 MPa, 45°C, 15 min	NA	<0.5GEC	[24]
Leviviridae	MS2	600 MPa, 21°C, 5 min	NA	3.3 (medium)	[25]
	MS2	500 MPa, 4°C, 5 min	NA	1.46 (sausage)	[17]

[a] Not applicable.
[b] Genomic equivalent copies.
[c] Abbreviation of virus name. HAV, hepatitis A virus; FCV, feline calicivirus; MNV-1, murine norovirus 1; TV, Tulane virus; Human NoV, human norovirus; MS2, bacteriophage MS2.

pH 7.0, but only 1.2 logs at pH 4.0 [6]. Currently, the mechanism by which pH influences the pressure inactivation of other nonenveloped viruses remains unknown. It is likely that the susceptibility of a virus to pH is dependent on the nature of the viral capsid proteins. Furthermore, it seems that the influence of pH on the efficiency of pressure inactivation is also dependent on the magnitude of the pressure applied. For instance, a study carried out by Lou et al. [7] demonstrated that vesicular stomatitis virus (VSV; an enveloped virus) appeared to be more sensitive to HPP at an acidic pH (5.7-log reduction) than at a neutral pH (4.0-log reduction) at 550 MPa, but not at 450 MPa.

A baroprotective effect from salt was first suggested due to the observation of an increased resistance of HAV to pressure in diluted seawater [14]. A similar phenomenon was later reported for FCV, where the addition of sodium chloride during HPP at 200 MPa at 4°C or 250 MPa at 20°C diminished the pressure inactivation of FCV in medium from over 6 logs to approximately 2–3 logs [20]. The mechanism underlying this protective effect from salt is still under investigation. One hypothesis is that NaCl, added exogenously, helps to stabilize the solvation cage surrounding the hydrophilic virus proteins and maintains the void volumes by displacing water molecules, thus altering the solution density and compressibility under HPP.

Finally, the composition of a medium or food matrix is another important factor that affects viral inactivation during pressure treatment. It has been reported that carbohydrates, fats, salts, proteins, ions, and other food constituents can protect viruses from inactivation [18,20,26–28]. For instance, sucrose (40%) provided a baroprotective effect against HPP inactivation of FCV and resulted in <1-log inactivation at 250 MPa and 20°C for 5 min [20]. Murchie et al. [29] pressure-treated several viruses including FCV in shellfish, seawater, and culture medium, and the viruses were found to be most resistant when treated in oysters and mussels. Comparisons of the effect of HAV inactivation in oyster homogenates and 0.3% NaCl solutions at a similar pH revealed an increased pressure resistance in oyster homogenates compared with salt solutions, suggesting the baroprotective characteristic of other oyster components [18]. Consistent results were observed in a study on MNV, demonstrating that viral reduction was higher in medium (8 logs) than in strawberries (5.8 logs) and in strawberry puree (4.7 logs) under 450 MPa for 2 min at 4°C [7]. These observations demonstrate that the food matrix confers protection from the inactivation of viruses by HPP. Thus, it is necessary to optimize the processing parameters for each product since the efficiency of viral inactivation varies with the food matrix.

Evidence That Human Norovirus Is More Resistant to HPP Than Its Surrogates

As the leading causative agent for foodborne illnesses, the stability of human norovirus to HPP is poorly understood, since it is still widely noncultivable. However, its surrogate viruses (FCV, MNV, and TV) can be effectively inactivated by HPP. A human volunteer study conducted by Leon et al. in 2011 [23], provided convincing evidence on the stability of human norovirus to HPP. It is reported that HPP at 600 MPa at 6°C for 5 min completely inactivated 4-log genomic copies of the Norwalk virus (human norovirus genogroup I genotype 1; GI.1) in seeded oysters, based on the lack of infection and virus shedding in the challenged volunteers. However, treatment at 400 MPa (at 6°C or 25°C) for 5 min, which completely inactivated human norovirus surrogates (FCV and MNV), was insufficient to prevent human subjects from norovirus infection and shedding, indicating that human norovirus may be more stable than surrogates. In another study, human norovirus GII.4 virus-like particles (VLPs) were used as a norovirus model to evaluate the stability of human norovirus because VLPs possess the authentic antigenicity, structure, and attachment factors binding the activities of human norovirus [9]. Similarly, it was shown that pressures of 500–600 MPa were not sufficient to disrupt human norovirus VLPs or impair their receptor binding functions. In fact, treatments at 700 MPa for 30 min, 800 MPa for 15 min, and 900 MPa for 2 min were required for the complete disruption of human norovirus VLPs [9]. Despite the fact that the Norwalk virus tested in the volunteer study and human norovirus GII.4 VLPs differ in pressure susceptibilities, these two studies consistently demonstrate that the authentic human norovirus could be more pressure resistant than the widely used surrogate viruses. The difference between the Norwalk virus and VLPs may result from the differences in the virus strain, particle conformation, and the susceptibility of the human volunteers to human norovirus. Overall, although these studies provide useful insights about the survival of human norovirus, ultimately, the effectiveness of HPP in inactivating human norovirus depends on the successful cultivation and further study of this virus.

Application of HPP in Processing High-Risk Foods

Fresh produce is a frequent high-risk food for norovirus infections, as human norovirus can attach tightly and be subsequently internalized and disseminated in fresh produce [30–33]. HPP can effectively inactivate surface and internal viral contaminants while preserving the texture and organoleptic characteristics of the food after treatment, making HPP a promising technology to process fresh fruits and vegetables. Under 350 MPa at 4°C, 2.4-, 2.2-, and 2.4-log virus reductions of MNV were achieved in fresh lettuce, fresh-cut strawberries, and strawberry puree, respectively [7]. Kingsley et al. [16] also demonstrated that HPP was an effective means of reducing HAV in produce, including green onions and strawberry puree that had pH values of 5.12 and 3.67, respectively. Cascarino [34] found that HAV and FCV inoculated in salsa could be completely inactivated after 1 min of treatment at a minimum pressure level of 250 MPa [34]. Adverse sensorial changes may still occur in fresh produce after HPP treatment; therefore, HPP may be practical to process fresh produce–related products, such as purees, sauces, and juices, as well as fruits intended for frozen storage. In recent years, HPP has had commercial success with various fruit and vegetable product categories. Pressure-treated fruit and vegetable-based products that have been introduced on the market include salsa (Fresherized Foods), chopped Spanish onions (Winsoms of Walla Walla), green peppers, apple sauce (Leahy Orchards), and various fruit blends and purees [35].

Shellfish, such as oysters, mussels, and clams, are another high-risk food for enteric virus contamination. Hence, the application of HPP to eliminate foodborne viruses in shellfish has been an active area of research. Kingsley et al. [21] showed that a 400 MPa treatment for 5 min at 5°C was sufficient to inactivate 4.1 logs of MNV in oyster tissue. HPP was also used to treat oysters with a pressure of 400 MPa for 1 min (9°C) and induced a 3-log reduction of HAV [15] and a 4-log reduction of MNV at a treatment temperature of 5°C [21]. In addition, Gogal et al. [36] tested the inactivation efficacy of HPP on viruses *in vivo*. The author artificially contaminated oysters with MNV, pressurized the oysters at 400 MPa for 5 min, and subsequently orally fed them to immunocompromised mice. HPP was found to successfully prevent MNV infection of mice, suggesting that this processing method may be an effective food processing intervention for norovirus-contaminated shellfish. Currently, HPP is employed for the commercial processing of shellfish, particularly oysters. Not only does HPP significantly reduce pathogen levels in oysters, but it also causes the oysters to be more voluminous, more flavorful, and more pleasantly textured compared with untreated oysters [37–40]. Moreover, HPP treatment causes the adductor muscle of oysters to detach from the shell, thus opening (shucking) the oyster. This "self-shucking" aspect of HPP can drastically cut down on labor requirements associated with manual shucking, thus constituting an important economic advantage for oyster processors [29,38].

Other Nonthermal Processing Technologies

Irradiation

Irradiation (or "ionizing radiation") can be generated by either cobalt-60 or cesium-137 radioisotopes, or high-energy e-beams. Food irradiation technologies (gamma irradiation and e-beam) have been shown to be effective nonthermal processing technologies to inactivate bacteria, fungi, insects, and other pests in foods without posing health risks [41]. However, viruses are generally considered to be more resistant to irradiation than bacteria and fungi due to their small sizes and stable structures. In fact, studies have shown that both e-beam and gamma irradiation methods are not effective against foodborne viruses at the dosage approved by the Food and Drug Administration (FDA) for inactivating bacteria in fresh iceberg lettuce and spinach (up to 4.0 kGy). For example, Sanglay et al. [42] found that less than a 1-log MNV reduction was observed in either cabbage or strawberries after 4 kGy of e-beam irradiation. An approximate 3-log reduction of simian rotavirus SA-11 (a surrogate for HuRV) on lettuce or spinach was observed when exposed to 10 MeV, 18 kW e-beam irradiation at the dosage of 4 kGy. Whereas poliovirus exhibited greater resistance to e-beam irradiation compared with rotavirus in the same study [42,43]. As for gamma irradiation, only a 1.7- to 2.4-log MNV reduction was observed on fresh produce (strawberries, romaine lettuce, and spinach) by a gamma irradiation dose of 5.6 kGy [44]. In addition to showing that viral nucleic acid was damaged by gamma irradiation, which is the well-described mechanism by which the ionizing radiation inactivates microorganisms, Feng et al. [44] also discovered that gamma irradiation disrupted virion structures and viral capsid proteins. As a comparison, it has been shown that a 1.25 kGy radiation dose is effective at inactivating 5 logs of *Escherichia coli* O157:H7 in refrigerated meat [45]. Viruses could also potentially survive the 12D irradiation process used for *Clostridium botulinum* in meat products unless previously damaged by other methods, such as heating. Viruses are considered a minor target for ionizing radiation and may potentially survive the conditions of radappertization (i.e., "radiation's commercial sterility" or approximately 30 kGy). Overall, irradiation may not be feasible as a technology to eliminate viruses from high-risk foods at the current FDA approved dosages. It is believed that implementing the hurdle concept using an additional technology (such as heating) in combination with food irradiation would ensure food safety.

Ultraviolet Light

UV light treatment is a technology currently used to process foods to inactivate microorganisms. The radiation can be emitted in either continuous mode or pulsed mode. Continuous shortwave UV light, with a wavelength of 254 nm, has a broad germicidal spectrum that includes bacteria, viruses, protozoa, fungi, yeasts, and algae in many types of materials [46,47]. Human norovirus surrogates, MNV and FCV, were shown to be susceptible to UV. The titers of MNV and FCV were both reduced by approximately 3 logs at a dosage of 26 mJ/cm^2 UV light [48,49], while a 5-log inactivation of FCV was achieved at a dose of 50 mJ/cm^2 [46]. It is known that UV light is absorbed by the DNA in the cells and causes cross-linking of thymine. If the UV damage is severe then repair mechanisms are impaired and this leads to genetic mutations that can prevent the cellular function and reproduction of the microorganisms [50].

Pulsed UV light is a non-thermal technology that has been developed in recent years. An intense pulse of light is generated by high-power electrical energy and therefore is more effective than continuous UV light in terms of the pathogen inactivation [51]. The infectivity of poliovirus was completely eliminated (>5-log reduction) in phosphate buffered saline (PBS) after 25 UV pulses, corresponding to a dose of 28 mJ/cm^2 [52]. Jean et al. [53] found that 2 seconds of pulsed UV light treatment (59 mJ/cm^2) was sufficient to completely inactivate (>5-log reduction) MNV and HAV, both in liquid media (PBS) and when dried on surfaces (polyvinyl chloride disks).

UV light appears to be an effective alternative or complement to conventional sterilization techniques (e.g., thermal treatment) to eliminate viruses. Unfortunately, the major limitation of UV technology is its low penetration capacity. For example, UV is unable to inactivate viruses that have been internalized in fresh produce. Additionally, UV cannot be used on contaminated shellfish due to the protection that the shells and tissues impart to the pathogens. As a result, the applications for UV technology in the food industry are limited to the disinfection of water or clear liquid food, and the decontamination of surfaces in restaurants or food processing pilot plants [46,54].

Ultrasound

High intensity ultrasound, with a frequency ranging from 20 kHz to 2 MHz and power between 100 and 500 W/cm^2, can be used for various processing applications including microorganisms, enzyme inactivation, and biocomponent separation. At present, ultrasound is only used to process liquid foods. As a recently developed food processing technology, very limited research has been conducted to address its efficacy at reducing the infectivity of foodborne viruses. Su et al. [55] investigated the effectiveness of high intensity ultrasound in eliminating human norovirus surrogates (bacteriophage MS2, FCV, and MNV). In this study, an ultrasound probe was directly placed into virus suspensions to avoid dissipating any ultrasound energy. When MS2, FCV, and MNV were suspended in PBS, a 4-log reduction was found after 5, 10, and 30 min exposure to ultrasound, respectively. However, the extent of inactivation was significantly reduced when the viruses were suspended in orange juice, indicating that the food matrix protected the viruses from inactivation. This result also suggests that ultrasound may not be sufficient to effectively inactivate human norovirus surrogates in foods by ≥5-log pathogen reductions to meet the FDA standard [56]. Therefore, the combination of ultrasound with other techniques such as heat, pressure, or antimicrobials should be used to ensure food safety [55].

THERMAL PROCESSES AGAINST FOODBORNE VIRUSES

INTRODUCTION

Thermal processing is defined as the combination of the temperature and time required to eliminate a desired number of microorganisms. The thermal inactivation of foodborne viruses is complex and the efficacy of thermal treatment is influenced by several factors. A summary of the inactivation of foodborne viruses by heat treatment can be found in Table 5.2. The first, most evident factor is the target virus itself, as different families and even strains of viruses have the potential to behave differently under heat treatment. Treatment parameters such as temperature and holding time must be established for each foodborne virus independently. Another component that needs to be addressed is the sample matrix in which the target virus is suspended. Even small amounts of protein, salts, and carbohydrates in a liquid medium can confer protection to foodborne viruses during thermal treatment. The complex nature of most foods will undoubtedly affect the thermal inactivation of viral contaminants.

The inactivation of foodborne viruses by thermal processing rarely occurs at a constant rate and therefore does not follow a log-linear (monophasic reduction) thermal inactivation kinetics model [58,74]. This complicates the establishment of appropriate D-values for food processing operations. Recently, alternative models are being applied to predict viral thermal inactivation. The biphasic reduction model can be used when there are two rates of inactivation observed [58,75]. The presence of two rates of inactivation may be due to different fractions of the virus population having increased/decreased heat resistance, an aggregation of viral particles, or differing experimental conditions (i.e., suspension media). The Weibull model may be used to describe viral inactivation, which varies due to increased treatment and is not constant over time [58,76]. The application of these statistical models can be beneficial for estimating the survival of viruses during food production; however, the application relies on the establishment of comparable data on the thermal stability of different viruses [58].

THERMAL INACTIVATION OF VIRUSES IN LIQUID MEDIUMS

A systematic comparison of the thermal stability of several foodborne viruses, including adenovirus type 5, poliovirus Sabin 1, parechovirus 1, MNV, and human norovirus genogroup II, genotype 4 (GII.4), was conducted at "hot bar" holding temperatures (56°C) and pasteurization temperatures (73°C) [58]. All viruses were found to be stable at a treatment temperature of 30°C for 20 min in both Dulbeco's Modified Eagle Medium (DMEM) and a 1% stool suspension. The time to first log (TFL) reduction was then determined at 56°C and 73°C for each virus in both media. At 56°C, parcehovirus 1 was found to be the most stable with a mean TFL of 27 min in DMEM and no reduction in 1% stool suspension. MNV had a mean TFL of 4.21 min and 3.2 min in DMEM and 1% stool suspension, respectively. Both adenovirus type 5 and poliovirus Sabin 1 had an average TFL of <0.30 min in DMEM and the TFL was reduced to 0.1 min in 1% stool suspension. At 73°C, all viruses in 1% stool suspension were inactivated in <0.73 min and in DMEM in <0.53 min, with the exception of MNV, which was inactivated in 1.06 min in DMEM. The amount of viral nucleic acid after heat treatment was determined using a real-time reverse transcriptase polymerase chain

TABLE 5.2
The Efficacy of Thermal Treatment for the Inactivation of Water- and Foodborne Viruses

Virus	Treatment	Matrix	Log_{10} Reduction	Reference
MNV-1, TV	55°C, 2 min	Medium	<2	[57]
Human NoV GII.4	56°C, 30 min	Medium	0.2 log RNA	[58]
MNV-1	56°C, 4.21 min	Medium	1	[58]
MNV-1	56°C, 3.2 min	1% stool suspension	1	[58]
Parechovirus 1	56°C, 27.0 min	Medium	1	[58]
Parechovirus 1	56°C, 30 min	1% stool suspension	No reduction	[58]
Human NoV	60°C, 30 min	Stool filtrate	Infectious to human	[59]
Rotavirus	60°C, 10 min	Medium	7	[60]
Rotavirus	60°C, 120 min	HBSS with 10% sorbitol	Undetectable	[61]
HAV	60°C, 10 min	Medium	>4.6	[62]
MNV-1	63°C, 0.44 min	Medium	1	[63]
MNV-1	65°C, 2 min	Medium	3	[57]
TV	65°C, 2 min	Medium	3	[57]
Human NoV GII.4	70°C, 2 min	Fecal suspension in PBS	1.62 log RNA	[64]
MNV-1, TV	70°C, 2 min	Medium	>6	[57]
Adenovirus type 5	70°C, 10 min	Stabilizing buffer	>7.5	[65]
Human NoV GII.4	73°C, 3 min	1% stool suspension	0.1 log RNA	[58]
MNV-1	80°C, 2.5 min	Medium	6.5	[66]
HAV	80°C, 3 min	Medium	>4.6	[62]
Human NoV GII.4	85°C, 2 min	Fecal suspension in PBS	2.34 log RNA	[64]
HAV	60°C, 10 min	4 mL shellfish homogenate	2	[62]
HEV	62°C, 5 min	Pork liver pate	1.19	[67]
HEV	62°C, 20 min	Pork liver pate	1.83	[67]
HEV	62°C, 120 min	Pork liver pate	2.17	[67]
HAV	63°C, 30 min	Milk	3	[68]
HAV	65°C, 10 min	Strawberry puree (5°Brix, pH 3.35)	1.5	[69]
HAV	65°C, 10 min	Raspberry puree (5°Brix, pH 3.05)	2	[69]
HAV	65°C, 10 min	Bilberry puree (5°Brix, pH 2.87)	3	[69]
MNV-1	65°C, 0.5 min	Raspberry puree (9.2°Brix)	1.86	[66]
HEV	68°C, 5, 10, 20 min	Pork liver pate	2.26–2.31	[67]
HAV	70°C, 10 min	Strawberry puree (5°Brix, pH 3.35)	2.5	[69]
HAV	70°C, 10 min	Raspberry puree (5°Brix, pH 3.05)	3	[69]
HAV	70°C, 10 min	Bilberry puree (5°Brix, pH 2.87)	3.5	[69]
HAV	71°C, 6.55 min	Skim milk (0% fat)	4	[70]
HAV	71°C, 8.31 min	Homogenized milk (3.5% fat)	4	[70]
HAV	71°C, 12.67 min	Cream (18% fat)	4	[70]
HEV	71°C, 5, 10, 20 min	Pork liver pate	2.58–2.93	[67]
Poliovirus	72°C, 0.25 min	Milk	0.56	[71]
Poliovirus	72°C, 0.5 min	Milk	>5	[71]
HAV	75°C, 10 min	Strawberry puree (5°Brix, pH 3.35)	3.5	[69]
HAV	75°C, 10 min	Raspberry puree (5°Brix, pH 3.05)	4	[69]
HAV	75°C, 10 min	Bilberry puree (5°Brix, pH 2.87)	4	[69]
MNV-1	75°C, 0.25 min	Raspberry puree (9.2°Brix)	2.81	[66]
HAV	80°C, 3 min	4 ml shellfish homogenate	2	[62]
HAV	80°C, 8.94 min	1 g strawberry mash (52°Brix)	1	[72]
HAV	85°C, 0.96 min	1 g strawberry mash (28°Brix)	1	[72]
HAV	85°C, 4.98 min	1 g strawberry mash (52°Brix)	1	[72]
HAV	90°C, 3 min	Clams	Undetectable	[54]
MNV-1	90°C, 3 min	Clams	Undetectable	[54]
Poliovirus	Steaming, 30 min	Oysters	2	[73]

reaction (RT-qPCR). Human norovirus GII.4 was added at this point in the study, as it cannot be easily grown in cell culture. Unlike the cell culture-based methods, there was less than a 1-log reduction in viral RNA due to heat treatment. This indicates that viral RNA is still intact when the viruses are rendered noninfectious by thermal treatment. Therefore, viral nucleic acid detection alone may not be applicable to determine viral thermal inactivation.

It was found at 56°C in both DMEM and 1% stool suspension, that the thermal inactivation of all viruses best fit the Weibull model. At 73°C in 1% stool suspension, the thermal inactivation of all viruses fit the monophasic reduction model. However, at 73°C in DMEM, parechovirus 1 best fit the biphasic reduction model, MNV best fit the Weibull model, and both adenovirus type 5 and poliovirus Sabin 1 best fit the monophasic model. Based on these results, it is apparent that reliable and reproducible thermal stability data are required for each foodborne virus under different experimental conditions in order to properly apply statistical models [58].

Adenovirus

It was found that the titer of infectious purified adenovirus type 5 in a proprietary stabilizing buffer could be reduced by >7.5 logs when treated at 70°C for 10 min. However, using a polymerase chain reaction (PCR)-based assay to detect the viral DNA of the same samples, only a 4.4-log reduction was observed. When deoxyribonuclease (DNase) treatment of samples was employed before PCR, it was found that a 6.1-log reduction in viral DNA was detected at the same treatment parameters [65]. This indicates that PCR-based assays alone do not accurately assess the level of thermal inactivation, because inactivity is primarily due to denaturation of the capsid and DNA may not be fully degraded.

Adenovirus has been reported to lose infectivity at temperatures greater than 56°C and it has been theorized that the initial loss of infectivity is due to the denaturation of the capsid protein [77,78]. A greater than 6-log reduction in viral DNA was reported when purified adenovirus was treated at 54°C for 3 min, using DNase treatment prior to DNA extraction followed by qPCR. Similarly, treatment at 50°C for 10 min reduced the amount of adenovirus DNA below the limit of detection of the PCR assay. Using the PCR assay, the thermal inactivation of adenovirus was fairly linear; however, using the infectivity assay the thermal inactivation was biphasic, with a second slower inactivation curve observed [65]. These phenomena of shouldering and tailing will require longer treatment times at each temperature to inactivate adenovirus and to compensate for the more thermally stable fractions of the virus population.

Rotavirus

Purified, triple-layered rotavirus YK-1 strain (simian P3, G3) suspended in Hank's Balanced Salt Solution (HBSS) supplemented with 10% sorbitol was heat treated at 60°C for 2 h [61]. The heat-killed rotavirus was passaged twice in cell culture, then, to ensure the loss of infectivity, the cells were tested for rotavirus using a commercial Environmental Impact Assessment (EIA) kit. Following treatment, the purified rotavirus was examined using EM and particles were found to maintain their biophysical structure. Additionally, the viral proteins VP1, VP2, VP4, VP6, and VP7 remained intact and antigenic as demonstrated by sodium dodecyl sulfate-polyacrylamide gel electrophoresis (SDS-PAGE) and Western blotting. In the heat-treated samples, high molecular weight bands (>125 kDa) were observed, which were absent in the untreated samples. Also, a 60 kDa band was present in the untreated control sample, but not in the heat-treated sample. This may indicate that the heat-induced cross-linking and aggregation of the viral proteins rendered the virus noninfectious. The viral double-stranded ribonucleic acid (dsRNA) was also unaffected by thermal treatment [61].

Human Norovirus Surrogates

Human norovirus surrogates, such as MNV, FCV, and TV, have long been employed to estimate the thermal stability of human norovirus. In 2006, Cannon et al. conducted a comparison study of the thermal stability of MNV and FCV. The time required to reduce the titer of each virus by 1 log was evaluated at temperatures of 56°C, 63°C, and 72°C, which correlate to the lower holding temperature limit for food on "hot bars"; low temperature, long holding time pasteurization; and high temperature, short time pasteurization, respectively [79]. At 56°C, MNV was reduced by 1 log in 3.5 min, and a 1-log reduction in FCV was achieved in 6.7 min. Both viruses were reduced by 1 log at 63°C in <25 s and at 72°C in <10 s [79].

More recently, the thermal inactivation of MNV and FCV at temperatures ranging from 50°C to 72°C was evaluated and it was found at all temperatures that inactivation could be achieved by increasing treatment times [80]. D-values were assigned at each temperature fitting the data to a first-order kinetics model. For MNV, the D value at 72°C ($D_{72°C}$) was 0.15 min and for FCV it was 0.11 min, which was similar to reports from previous studies [79,80]. However, when fitting the data from lower treatment temperatures to the Weibull model, the shape factors (β) indicated that both viruses had a monotonic upward concave (tailing) and downward concave (shouldering) curve behavior, which was dependent on the temperature [80]. This study again indicates that at lower treatment temperatures, there may be thermally resistant and thermally sensitive fractions of virus suspensions leading to nonlinear thermal inactivation kinetics.

MNV and TV thermal stability were compared using 2 min treatments at 5°C increments from 50°C to 75°C [57]. Both viruses were inactivated beyond the detection limit (an approximate 6-log reduction) at temperatures of 70°C and 75°C in cell culture medium [57]. Minimal inactivation was observed for both viruses at the other treatment temperatures (50°C–70°C), with an approximate 3-log reduction at 60°C and 65°C and <2-log reductions at 50°C and 55°C [57].

Human Norovirus

Human norovirus GII.4 thermal stability was tested at temperatures of 70°C and 85°C, with a 2 min treatment time [64]. A method that utilized human norovirus receptor binding and ribonuclease (RNase) pretreatment was used prior to RT-qPCR to ensure that the virus was still able to bind to its cellular receptors, and that the RNA detected was not from damaged/noninfectious particles after thermal treatment [64,81,82]. At 70°C for 2 min, the amount of detectable human norovirus GII.4 RNA using this method was reduced to 3.46 log RNA copies/mL compared with the initial titer of 5.08 log RNA copies/mL. Treatment at 85°C for 2 min reduced the amount of RNA in two of three replicates below the detection limit, and one positive sample had a titer of 2.74 log RNA copies/mL [64].

Compared with results using RT-qPCR alone, the combination of human norovirus cell binding and RNase pretreatment significantly reduced the amount of human norovirus RNA detectable after thermal treatment. This suggests that perhaps portions of the capsid, such as the protruding (P) domain, which is responsible for human blood group antigen (HBGA) binding, are denatured leading to loss of infectivity without degradation of viral nucleic acids. Therefore, human norovirus virus-like particles (VLPs) and P domain particles were heat-treated and their binding ability was tested using a saliva-binding, enzyme-linked immunosorbent assay (ELISA) (with samples containing soluble HBGAs). The binding of human norovirus GII.4 VLPs was significantly reduced to nearly zero at 70°C for 10 min or 85°C for 2 min and the GII.4 P domain particles binding ability was also impaired at these temperatures [64]. Additional P domains from other genotypes of human norovirus were also tested. It was found that GI.1 P domain particles were the most sensitive to thermal treatment, GII.9 P domain particles were the most resistant to higher treatment temperatures, and GI.4 behaved similarly to GII.4 [64]. These data indicate that different human norovirus genotypes may have differing thermal stabilities.

Thermal Inactivation of Viruses in the Food Matrix

HAV has been implicated in outbreaks in a wide array of foods, including shellfish, vegetables, fruits, and dairy products [83–85]. The inactivation of HAV by thermal processing is significantly impacted by the type and composition of the food product that it is contaminating. In dairy products, it was found that increasing the amount of fat in the product imparted a protective effect to HAV and increased its thermal stability [70]. Skimmed milk (0% fat), homogenized milk (3.5% fat), and table cream (18% fat) were inoculated with HAV to a final concentration of 2×10^6 plaque-forming units (pfu)/mL and treated at a range of temperatures from 65°C to 85°C. At 85°C, a 5-log reduction in HAV titer in all dairy products was achieved in <0.5 min. At all other temperatures, it was observed that a longer exposure time was required in the higher fat content dairy products to achieve a 5-log reduction in HAV titer [70].

In strawberry puree, both sucrose concentration and pH impacted the thermal stability of HAV [72]. The $D_{85°C}$ for 52° Brix strawberry puree at pH 3.8 was found to be 4.98 min. When the sucrose concentration was reduced to 28° Brix at the same pH, the $D_{85°C}$ was 0.96 min. Thus, a high sucrose concentration in strawberry puree increased the thermal resistance of HAV. Additionally, the $D_{85°C}$ for 40° Brix solutions with pH 3.3, 3.8, and 4.3 was determined to be 1.04, 2.37, and 2.78 min, respectively. Therefore, increasing the pH was found to increase the thermal stability of HAV. Sucrose concentration and pH were also found to have a synergistic effect, increasing the thermal stability of HAV in strawberry puree when a high sucrose concentration was paired with a high pH [72]. To further investigate the effect of pH on HAV thermal inactivation in red berries, strawberries (pH 3.35), raspberries (pH 3.05), and bilberries (pH 2.87) were pureed without the addition of sucrose (average 5° Brix), inoculated with HAV, and then treated at 65°C, 70°C, and 75°C [69]. Similarly, the time required at each temperature to inactivate HAV was decreased, in correlation with a decrease in pH [69].

Shellfish are a high-risk food for both HAV and human norovirus contamination. These bivalves can bioaccumulate the viruses within their tissues during filter feeding if the viruses are present in shellfish growing waters. The digestive glands are a major site for bioaccumulation and these tissues have been shown to express moieties that mimic viral HBGA binding factors, which may contribute to the high bioaccumulation rates [86]. Soft-shell clams were artificially contaminated with either HAV or MNV and the efficacy of thermal treatment on viral inactivation was evaluated. It was found that in shucked clams, a treatment at 90°C for 180 seconds was required to completely inactivate the viruses [54].

HEV, in addition to being a common food- and waterborne virus, is also a zoonotic virus that is known to infect pigs. Thus, pork products can harbor HEV and, if undercooked, can transmit the virus to humans and pigs upon ingestion. Recently, Barnaud et al. [67], used HEV-infected pig livers to prepare a pork liver pâté product that was treated at 62°C, 68°C, and 71°C for either 5, 10, 20, or 120 min. Treating the pâté at 71°C or 68°C for 5, 10, or 20 min resulted in a <3-log reduction in HEV RNA. Treating at 62°C for 5 or 20 min resulted in a <2-log reduction and treating for 120 min gave an approximate 2-log reduction in viral RNA. The HEV-infected pork liver pâté was then injected intravenously into the ears of specific pathogen-free (SPF) pigs and HEV viral shedding was monitored in the pigs' stools for 35 days postinoculation (dpi). All positive control pigs receiving pâté with no thermal treatment excreted HEV in their stools from 2 dpi to the end of the experiment. Pigs receiving pâté treated at 62°C all excreted HEV starting between 7 and 9 dpi and lasting until 22 or 35 dpi. Approximately half (6/13) of the pigs that received pâté treated at 68°C shed HEV in their stools, starting at 9–16 dpi and subsiding at 16–22 dpi. The amount of fecal HEV shedding was reduced in pigs receiving pâté treated at 71°C. There was no viral shedding in pigs that received pâté treated for 20 min at 71°C. However, two of three pigs in each group given pâté treated for 5 or 10 min at 71°C shed HEV starting at 11 dpi and lasting until 16 dpi or 18 dpi. Using logistic regression modeling it was determined that the effect of the treatment temperature of the pâté was significantly related to the probability of infection in pigs, but was not dependent on the treatment time at each temperature. Pigs receiving pâté treated at 62°C had significantly higher odds of HEV infection compared with pigs receiving pâté treated at 71°C [67].

The incidence of HEV in pork livers is problematic because the virus is located within the tissue itself and cannot be removed by surface decontamination. Products utilizing pork livers, such as pork liver sausage, can consist of up to 30% fat and contain other components such as spices, nitrite salt, dextrose, and other additives [67]. The complex nature of these pork products can provide protection to HEV during thermal treatment. These results suggest that HEV in complex food matrices is highly resistant to thermal treatment; however, the direct impact on human health due to HEV-infected pork liver has not been elucidated. The efficiency of thermal inactivation treatments required for various food- and waterborne viruses is shown in Table 5.2.

CHEMICAL PROCESSES AGAINST FOODBORNE VIRUSES

INTRODUCTION

Most of the methods currently utilized for controlling bacterial growth in foods, such as lowering pH, high-temperature processing, and decreasing water activity, are not effective against foodborne viruses [2]. The fact that foodborne viruses are nonenveloped makes them resistant to environmental stress. The protein capsid remains intact after exposure to low pH, as these viruses need to pass through the digestive tract in order to cause infection in their hosts. The use of chemical disinfectants to inactivate viruses has been a major research focus, as they could easily be integrated into the food processing chain. Currently in the fresh produce industry the spraying, washing, or dipping of fruits and vegetables in water is a common practice to reduce pathogens, prevent cross-contamination, and remove debris [87]. Chemicals including chlorine, chlorine dioxide, quaternary amines, various acids, surfactants, ozone (both gaseous and ozonated water), and electrolyzed water are either currently used in the food industry or have been investigated for their efficacy against foodborne pathogens.

CHLORINE

Currently, chlorine is the most widely used and widely researched sanitizer in the food industry. Three forms of chlorine have been approved for use: chlorine gas, calcium hypochlorite, and sodium hypochlorite. Unfortunately, a direct comparison of research studies using chlorine is difficult due to variations in the environmental factors that impact disinfection kinetics. In addition, several studies have shown that different viruses vary in their resistance/sensitivity to chlorine treatment. Therefore, chlorine inactivation data must be generated for each foodborne virus independently. In the fresh produce industry, chlorine is used at a concentration of 50–200 parts per million (ppm) at a neutral pH with a contact time of 1–2 min. This chlorine concentration has little effect on foodborne viruses and increasing the contact time to greater than 10 min has not been shown to significantly increase the antiviral and sanitization activity [88].

Several studies have investigated chlorine inactivation of human norovirus surrogates and only a handful of studies have used human norovirus, because human norovirus cannot easily be grown in cell culture, hampering its use. One human volunteer study was conducted in which the human Norwalk virus (GI.1) was treated using different concentrations of chlorine and then administered to eight human subjects. Treatment with 3.75–6.25 mg/L of chlorine, a typical water treatment concentration, was not effective at inactivating the virus, as five of the eight subjects became ill and developed an antibody response against the GI.1 human norovirus. When the chlorine concentration was increased to 10 mg/L, none of the eight volunteers developed any illness or had seroconversion [89]. These results indicate that the current routine water treatment with chlorine is not sufficient to inactivate human norovirus [89]. Another study investigated the disinfection of human norovirus GII.4, MNV, and poliovirus 1 in water by chlorine. Evaluating the amount of viral RNA reduction by RT-qPCR, it was found that treatment for 30 min with 0.5 mg/L of chlorine gave 3.64, 2.88-, and 3.21-log reductions in the RNA copies of the GII.4 human norovirus, MNV, and poliovirus 1, respectively. When the viral reduction of MNV and poliovirus 1 was determined using a viral plaque assay, it was shown that the same conditions gave >4.41- and >4.95-log reductions, respectively. While most research suggests that human norovirus is more resistant to chemical disinfectants than its surrogates, these results showed no significant difference in the viral RNA reduction between MNV and human norovirus [90].

A more recent study investigated the effect of chlorine treatment on two human norovirus surrogates present on food contact surfaces (stainless steel coupons). It was found that treatment with 5000 ppm of chlorine and a contact time of 5 min was needed to yield a 5-log $TCID_{50}$ reduction of both MNV and FCV [91]. This treatment exceeds the permissible limit for the chlorine treatment of food contact surfaces, which is 200 ppm as mandated by the FDA [92]. Another recent study integrated chlorine into a standard warewashing protocol. The addition of 200 ppm chlorine to the control sanitizer did not increase the removal of MNV during mechanical and manual warewashing compared with the use of the sanitizer alone [93]. The panel of sanitizers tested was effective for both *E. coli* and *Listeria*, indicating that MNV was more resistant to chemical disinfectants than the bacteria. Similar studies have also shown an extreme resistance of viruses to chlorine treatment (Table 5.3). In summary, chlorine does have the potential to inactivate foodborne viruses, but typically the levels of chlorine required to do so are much higher than allowed by governmental safety regulations.

Chlorine Dioxide

Chlorine dioxide (ClO_2) is an oxidizing agent that is extremely soluble in water. Common uses include the disinfection of drinking water, the sterilization of both equipment and surfaces in the pharmaceutical and food industries, and the bleaching of flour and paper [107]. Chlorine dioxide is as much as five times more soluble than chlorine and it does not react with ammonia or other organic compounds to generate trihalomethanes [108,109]. However, there are several drawbacks and limitations to using chlorine dioxide as a disinfectant. Chlorine dioxide must be generated on site, it is highly unstable, and it has the potential to be explosive. Chlorine dioxide was originally discovered in 1814, through a reaction between potassium chlorate and sulfuric acid. Today, many chemical pathways can be utilized to generate chlorine dioxide. Currently, three of these methods are used to produce ClO_2 for use as a disinfectant: (1) the oxidation reaction between sodium chlorate and chlorine to produce chlorine dioxide and sodium chloride; (2) the reaction between sodium chlorite, sodium hypochlorite, and hydrochloric acid to produce ClO_2, sodium chloride, and water; and (3) method (2) but without the sodium chlorite to produce a higher yield of ClO_2 [110].

According to Title 21 in the Code of Federal Regulations (CFR) as written by the FDA, ClO_2 is considered a secondary direct food additive and its use is only permitted for certain applications using specific methods of generation [92]. ClO_2 may be used as an antimicrobial agent in water intended only for poultry processing or produce washing at a concentration of less than 3 ppm [92]. Because it is approved for use on food, ClO_2 has been extensively investigated as a means to inactivate bacterial pathogens, yet its effectiveness against food- and waterborne viruses is still poorly understood.

In a study conducted by Zoni et al. [111], it was found that 2.5 min of exposure to 0.8 mg/L of aqueous ClO_2 was able to achieve a 4-log reduction in FCV and HAV titers. Another study done by Butot et al. [97] found that 10 ppm of ClO_2 gave less than 2 logs of reduction in MNV viral titer. Unfortunately, 10 ppm is more than three times the upper limit allowed by the FDA [92]. Xue et al. [99] performed a study that investigated the amount of time it would take to achieve a 4-log reduction in rotavirus titer by treatment with 0.1 and 0.2 mg/L ClO_2 in solution. It was found that at 0.1 mg/L, it took 1.26–2.50 min and at 0.2 mg/L, it took 0.78–1.51 min to achieve a 4-log reduction. ClO_2 gas has been found to be more effective than its aqueous counterpart in the inactivation of viruses [99]. For example, ClO_2 gas at a concentration of 0.08 ppm with a treatment time of 6 h was able to achieve a 4-log reduction in FCV, while ClO_2 in solution at 0.8 mg/L for 2 min will achieve the same inactivation [111,112]. This result is consistent with the Ct values for a 4-log reduction in FCV; the aqueous phase (Ct = 1.6 mg/L min) is far less effective than the gaseous phase (Ct = 0.072 mg/L min) [113].

TABLE 5.3
Effectiveness of Commonly Used Sanitizers against Food- and Waterborne Viruses

Sanitizer	Virus	Treatment	Matrix	Log_{10} Reduction	Reference
Chlorine	MNV-1	5000 ppm, 1.9 min	Fecally soiled stainless steel surface	3	[94]
	FCV	5000 ppm, 3.2 min	Fecally soiled stainless steel surface	3	[94]
	Human NoV GII.4	5000 ppm, 4 min	Fecally soiled stainless steel surface	1.4[a]	[94]
	MNV-1	200 ppm, warewashing	Plates	3.2	[93]
	MNV-1	200 ppm, warewashing	Forks and drinking glasses	1.4–1.5	[93]
	MNV-1	10%, 30 s	Virus suspension	2.52	[95]
	FCV, MNV-1	10%, 1 min	Virus suspension	5.14–5.16	[95]
	Human NoV	3%, 5 min	Stainless steel coupon	2[a]	[96]
	Human NoV	3%, 10 min	Stainless steel coupon	Undetectable[a]	[96]
	MNV-1	3%, 5–10 min	Stainless steel coupon	Undetectable	[96]
	HAV	200 ppm	Blueberries	2.4	[97]
	HAV	10 ppm	Parsley, raspberries	<2	[97]
	MNV-1	200 ppm, 2 min	Lettuce, cabbage, berries	<1.5	[98]
Chlorine dioxide	Rotavirus	0.1 mg/L, 1.26–2.50 min	Virus suspension	4	[99]
	Rotavirus	0.2 mg/L, 0.78–1.51 min	Virus suspension	4	[99]
	MNV-1	2.5 mg/L, 1 min	Stainless steel coupon	2.7	[100]
	MNV-1	4 mg/L, 1 min	Stainless steel coupon	>5	[100]
Quaternary amines	MNV-1	Alpet D2	Stainless steel coupon	2.23	[101]
	MNV-1, FCV	0.05 mg/mL benzalkonium chloride	Virus suspension	2.75, 3.25	[102]
	MNV-1	OASIS 146 Multi-Quat, warewashing	Forks, plates, drinking glasses	1.4–2.7	[93]
	MNV-1	N-alkyl benzyl ammonium chloride, 5 min	Stainless steel coupon	0.5	[96]
	MNV-1	N-alkyl benzyl ammonium chloride, 10 min	Stainless steel coupon	1	[96]
Acid/bases	FCV	0.4 mg/mL gallic acid	Virus suspension	0.86	[102]
	MNV-1	3% levulinic acid	Stainless steel coupon	No reduction	[63]
	FCV	3% levulinic acid	Stainless steel coupon	0.43	[63]
	HAV	100 ppm, 2 min	Lettuce	0.66	[103]
	MNV-1	80 mg/L peroxyacetic acid	Iceberg lettuce	0.77	[104]
	MNV-1	250 mg/L peroxyacetic acid	Iceberg lettuce	2.2	[104]
Ozone	HAV	>1 mg/L, 5 min	Virus suspension	5	[105]
	MNV-1	Tap water+2% SDS, spray	Stainless steel coupon	0.85	[101]
	FCV, MNV-1	6.25 ppm ozone, 5 min	Water	>6	[106]
	FCV, MNV-1	6.25 ppm ozone, 5 min	Green onions, lettuce	2	[106]

TABLE 5.3 (CONTINUED)
Effectiveness of Commonly Used Sanitizers against Food- and Waterborne Viruses

Sanitizer	Virus	Treatment	Matrix	Log_{10} Reduction	Reference
Surfactants	MNV-1	5% levulinic acid + 2% SDS, spray	Stainless steel coupon	2.71	[101]
	FCV, MNV-1	2% SDS	Stainless steel coupon	No reduction	[63]
	MNV-1	5% levulinic acid + 2% SDS, liquid, 1 min	Stainless steel coupon	3.35	[63]
	MNV-1	5% levulinic acid + 2% SDS, liquid, 5 min	Stainless steel coupon	3.38	[63]
	MNV-1	50 ppm SDS, 2 min	Lettuce, cabbage, berries	2.5–3.14	[98]
	MNV-1	50 ppm SDS + 200 ppm chlorine, 2 min	Lettuce, cabbage, berries	3.0–3.36	[98]

[a] Indicates log RNA copies reduction.

Quaternary Amines

Quaternary ammonium compounds (QACs) are a group of chemical compounds, which have been shown to exhibit strong antimicrobial properties and are widely used for cleaning and disinfecting purposes in the food industry. QACs are composed of the general structure $N^+R_1R_2R_3R_4X^-$, where the Rs represent alkyl groups, or alkyl groups with substituted functional groups, and the X represents an anion. They have been shown to be active against bacteria, fungi, yeasts, protozoa, and some viruses, and they may provide effective treatment for foods, such as fresh produce, without affecting the food quality [114,115]. One of the most commonly used QACs is benzalkonium chlorine (BAC), but other less common QACs have also been evaluated specifically for their antiviral efficacy [116]. BAC is currently used in formulations for cleaners and sanitizers for the dairy industry, the catering industry, food storage tanks, and fisheries. It is also a very popular ingredient in rinse-free hand sanitizers to prevent any cross-contamination in food service establishments and health-care settings [117].

One study found that three different QACs (Microquat, Oasis 144, and UMQ) tested alone were unable to inactivate FCV on stainless steel coupons, but the combination of each of these QACs with sodium carbonate resulted in more than a 3-log viral reduction. However, this reduction was achieved when the QACs were used at twice the concentration recommended by the manufacturer [88]. Incorporating a different QAC (Oasis 146 Micro-Quat) into a mechanical and manual warewashing protocol did not significantly increase the reduction of MNV compared with a control sanitizer or chlorine treatments alone [93]. However, in a more recent study, Su and D'Souza [102] found that 0.05 mg/mL of the QAC benalkonium chloride was able to give 2.0-, 2.75-, and 3.25-log reductions of the human norovirus surrogates bacteriophage MS2, FCV, and MNV, respectively. Therefore, while QACs have been shown to be effective against other pathogens, their effectiveness against foodborne viruses has still not been consistently proved.

Acids

Two of the most common acids considered for the inactivation of foodborne viruses are hydrogen peroxide (HP) and peroxyacetic acid (PAA). HP is a widely used commercial sterilizing agent that has been approved for use by the FDA [118]. It is highly stable in pure form, and it is even naturally present in such foods as milk and honey, where it helps prevent spoilage [119]. It has shown to be active against a wide range of organisms, but only limited effectiveness against foodborne viruses has been established. HP can be used in both liquid and vapor form. PAA is similar to HP in that it has sterilization capabilities and is effective against a wide range of microorganisms, but it is much more potent and does not create toxic residues. It can also be useful in both liquid and vapor form [119]. The reactivity of PAA is not affected by temperature or by organic matter and it is not corrosive at permissible levels. However, PAA has not been shown to be particularly effective against foodborne viruses at the levels it is permitted and it cannot be used for organic foods.

One study performed by Bentley et al. [120] investigated the use of 30% w/w of HP to sanitize five different types of surfaces inoculated with FCV (stainless steel, glass, vinyl flooring, plastic, and ceramic tile). They found that after 10 min, the HP's effect on the different surfaces had given 2.4–4.1 log reductions and after 20 min, all the surfaces had more than 4.9 logs of reduction [120]. While these results may seem promising, 30% hydrogen peroxide is very concentrated, and FCV has shown to be much more sensitive to chemical treatments than human norovirus. In a different study, a commercially available disinfectant containing both 11% HP and 15% PAA was tested to determine its effectiveness in inactivating FCV on stainless steel coupons. The use of a concentration of the disinfectant four times higher than the manufacturer's recommended level, with a contact time of 10 min, was able to completely inactivate FCV [88]. This same concentration was shown to yield a 3-log reduction in FCV when the virus was inoculated on lettuce and strawberries. However, using four times the suggested concentration is not only potentially unsafe to consumers, but it may also negatively affect the taste and quality of the food [88]. Another study found that 100 ppm of a PAA-based biocide gave 3.2- and 2.3-log reductions in FCV and MNV, respectively, and only a 0.7-log reduction of HAV [103]. Baert et al. found that the treatment of shredded iceberg lettuce with 80 and 250 mg/L of PAA gave an additional 0.77- and 1.43-log reduction, respectively, in MNV titer compared with simply washing with tap water [121]. Other acids such as tannic acid [102], gallic acid [102], and levulinic acid [63], have also shown to be relatively ineffective against FCV. Overall, acids have shown to not only have varied effectiveness on foodborne viruses, but their use may also impart negative flavors, odors, or textures to the processed foods.

Surfactants

Surfactants are "surface-acting" compounds that can essentially reduce the surface tension of a liquid. They contain both a hydrophilic and a hydrophobic group, which interacts with the substance that they are mixed with, altering the surface properties (normally lowering the interfacial free energy) of the liquid either at the water and air or the water and solid interface. Surfactants are widely used in many products ranging from toothpastes, soaps, shampoos, and detergents, to laxatives, cosmetics, herbicides, and insecticides, as well as detergents, wetting agents, emulsifiers, foaming agents, and dispersants.

It has been established that surfactants interact with viral proteins and can influence protein folding/refolding, denaturation, and aggregation, thereby inhibiting or potentially inactivating viruses [122–125]. Surfactants may even directly disrupt the viral envelope. Historically, more research has been successfully performed with nonfoodborne viruses, especially sexually transmitted viruses such as herpes virus and human immunodeficiency virus, but more research on foodborne viruses using surfactants is emerging.

One study done by Predmore and Li in 2011 [98] found that a combination of chlorine and each one of four different surfactants (sodium dodecyl sulfate [SDS], NP-40, Triton X-100, and polysorbates) were able to give 2–3 logs of MNV reduction in produce, significantly higher reductions than with tap water or chlorine alone [98]. Similarly, another study done by Cannon et al. investigated the synergistic effects of surfactants and sanitizers. MNV and FCV were inoculated on stainless steel coupons to mimic food contact surfaces [63]. It was found that neither SDS nor levulinic acid separately were able to inactivate MNV or FCV (<0.51-log reduction), but treatment with a combination of the two for 1 min was able to give 3.0 and 4.21 logs of reduction of FCV and

MNV, respectively. While these results are promising, the use of surfactants may have effects on the organoleptic properties of produce and many surfactants have yet to be approved for use by the FDA [126].

Ozone/Ozonated Water

Ozone, or triatomic oxygen, is a natural form of oxygen that is pale bluish in color as a gas. It is also soluble in water and has the ability to kill a wide range of microorganisms by disrupting or oxidizing cell membranes [127]. Ozone was first used as a means to disinfect municipal water in France in 1906, and in 1982 it was given generally recognized as safe (GRAS) status by the FDA for the treatment of bottled water [128]. In 1997, the Electric Power Research Institute (EPRI) also gave ozone GRAS status for the sanitization of food [129]. In the United States, ozone continues to be used for the disinfection of bottled water and it has the potential to be exploited in many other food-processing applications. Ozone treatment of foods is gaining increasing interest within the industry because ozone is one of the most potent sanitizers found to date, residual ozone naturally decomposes into O_2 with no residue deposited on foods, and it is effective against a wide range of microorganisms [130]. Ozonated water could also be employed as a sanitizer in the food industry, replacing chlorine, because it retains the sensory quality of produce and controls browning without diminishing the antioxidants present in the food [131].

Ozone was first shown to inactivate viruses in 1943, when Kessel et al. investigated its activity against poliovirus type 1 [132]. More recently, research has shown that many viruses are readily inactivated by ozone treatment [133]. Contrary to other viral inactivation methods, it has been suggested that the absence of a viral envelope may actually provide more access to the core nucleic acid and increase the efficacy of ozone disinfection [130]. Most foodborne viruses lack an envelope and it has been theorized that they will be more sensitive to ozone treatment. The potential mechanisms of viral inactivation by ozone include the degradation of the viral capsid proteins and/or viral nucleic acids, or the disruption of the antigenic sites on the protein capsid responsible for host cell attachment [134].

Both MNV and FCV have been shown to be susceptible to ozone treatment. Hudson et al. [135] showed that 25 ppm of gaseous ozone treatment held for 20 min was able to give a 3-log reduction in FCV titer on various surfaces in an office, cruise liner cabin, and a hotel room [135]. Additionally, Lim et al. [136] found a 2-log reduction in MNV RNA copies during treatment with 1.0 mg/L ozone for 2 min [136]. In fresh produce, Hirneisen et al. (2011) found that treatment with 6.25 ppm ozonated water gave a 2-log reduction in both FCV and MNV titers on green onions and lettuce after 5 and 1 min of treatment, respectively [106]. In water, they also found that the same conditions gave >6 logs of viral reduction for both FCV and MNV [106]. Ozone is very effective at inactivating norovirus surrogates, but there are still drawbacks to its use. More research needs to be done to evaluate the effects of ozone on the properties of food and also the sensory effects after treatment. Consumers may not approve of the use of ozone in their food products.

Electrolyzed Water

Electrolyzed water is produced by adding a dilute amount of sodium chloride to tap water and then passing it through an electrolysis chamber that has anode and cathode electrodes. The anode and cathode are on opposite sides of a bipolar membrane. The negatively charged ions present, such as chloride and hydroxide, pass through the dilute salt solution, give up their electrons and become gaseous oxygen and chlorine (O_2 and Cl_2), and hypochlorous acid. This imparts strong oxidation properties to the electrolyzed water, which then has the ability to disrupt cell membranes and metabolic processes [137]. The main advantage of using electrolyzed oxidizing water (EOW) is its safety. It is a strong acid, but it is much safer than other acids such as sulfuric or hydrochloric because it is not corrosive to the skin or mucous membranes [137].

Compared with other disinfecting methods, EOW disinfects quickly, is easy to work with, has few harmful health side effects and is relatively cheap [138]. While the effectiveness of electrolyzed water against bacterial pathogens has been extensively studied, its efficacy against foodborne viruses is very poorly understood. Very few studies have been carried out on the effectiveness of electrolyzed water on foodborne pathogens, specifically. Very recently, it was shown that EOW was not effective in removing human norovirus from produce [139]. Additionally, there was a significant decrease in the removal of human norovirus from contaminated produce after rinsing with EOW in comparison with other rinses (tap water, neutral electrolyzed water) [139]. Because there is so little research in this area, more research is needed to gain a better understanding of the kinetics of the inactivation of foodborne viruses by electrolyzed water.

PREVENTIVE MEASURES TO PROTECT AGAINST VIRAL CONTAMINATION IN FOODS

As with all foodborne disease transmission, handwashing and personal hygiene play a key role in reducing the introduction of viruses onto foods by food handlers. Washing hands with soap and water after using the toilet or changing diapers/nappies as well as before food preparation, handling, or eating is recommended by the Centers for Disease Control and Prevention (CDC) [140]. Experiments in which fecally suspended HAV was inoculated onto bare hands found that the infectious virus remains present for more than 4 h after inoculation [141]. Additionally, laboratory investigations have shown that by simply washing HAV-contaminated hands with water, viral transfer to handled lettuce was reduced by 1–2 logs [142]. Studies using human norovirus surrogates contaminated on hands have shown less than 3-log reductions in viral titers using alcohol-based sanitizers to decontaminate hands [143]. Therefore, the CDC recommends that alcohol hand sanitizers be used in conjunction with regular soap and water handwashing, but not as a stand-alone foodborne virus disinfection procedure [140].

One of the most important and simplest ways to reduce viral contamination through food handling is to restrict sick employees from working. The CDC recommends that personnel who have a norovirus infection stay home for 2–3 days after symptoms have subsided [140]. In the case of HAV, activities involving food contact should be avoided for 1–2 weeks after the onset of jaundice [144]. Many of the foodborne viruses are shed at high levels after clinical symptoms have been resolved, so it is imperative to restrict the work activities of food handlers who exhibited the symptoms of vomiting or diarrhea even after they have fully recovered.

Another option for controlling HAV contamination of foods is to administer postexposure prophylaxis to food handlers who have been infected with or exposed to individuals infected with HAV. If administered within 2 weeks of HAV exposure, immunoglobulin is more than 85% effective in preventing HAV infection [145]. However, the use of immunoglobulin prophylaxis is expensive and many food operations may not be fiscally capable of utilizing this option. Additionally, postexposure prophylaxis requires that the infected food handler receives a diagnosis of HAV infection in a timely fashion to ensure that the treatment will be effective [144]. An effective alternative is to vaccinate all food handlers against HAV infection and in recent times more impetus for this route of disease prevention has emerged, particularly in the United States.

Most of the major foodborne viruses have high stability in the environment and are resistant to many common disinfectants. HAV has been shown to remain infectious after a 1 month incubation period at an ambient temperature on environmental surfaces [146,147]. The environmental stability of HAV requires the use of appropriate disinfection techniques for eliminating residual viruses after vomiting or diarrheal episodes in food manufacturing, processing, or preparation areas. Using a 1:100 dilution of household bleach, QACs, and/or hydrochloric acid (HCl) cleaning solutions are all effective ways to inactivate HAV. However, several organic solvents, as well as solutions with pH values above 3.0, do not have efficacy against HAV [148]. Similar to HAV, human norovirus is also very resilient and remains stable in the environment for long periods of time. The CDC has published a fact sheet on human norovirus, specifically aimed at educating food handlers on ways to prevent human norovirus foodborne outbreaks [140]. They recommend cleanup procedures similar to those for HAV, using 1000–5000 ppm bleach for disinfecting any area that has become contaminated with vomit or diarrhea.

Properly implementing and regulating food handler hygiene has the potential to significantly decrease the occurrence of viral foodborne diseases. As described in the previous sections, most food processing technologies and treatments are ineffective against viruses. Therefore, preventing the viral contamination of food is a more effective means of controlling foodborne outbreaks.

REGULATORY EFFORTS TO PREVENT FOOD- AND WATERBORNE VIRAL DISEASES

Some of the highest risk foods for viral contamination include shellfish, fresh produce, and ready-to-eat foods. The FDA is working to regulate the production, processing, storage, and handling of these commodities with an aim to reduce the outbreaks associated with them. The National Shellfish Sanitation Program (NSSP) is a guide published by the FDA, which outlines the criteria for harvesting molluscan shellfish. The NSSP focuses on the water quality of shellfish growing areas, as well as general harvesting, processing, and transportation standards. The water quality of shellfish production areas is evaluated based on a sanitary survey and a bacteriological standard. The sanitary survey is an evaluation of all the environmental factors that affect the growing waters, including all actual and potential sources of pollution. The bacteriological standard mandates that shellfish harvesting waters contain less than 14 per 100 mL of total or fecal coliforms [149].

Based on the sanitary survey and the bacteriological standard, growing waters can be classified in one of five groups: approved, conditionally approved, restricted, conditionally restricted, or prohibited. Shellfish can be readily harvested from approved waters because both the water quality and the sanitary survey meet the NSSP requirements. Restricted waters have limited pollution, which can be removed by relaying, depuration, or low-acid canning processes. Prohibited waters have levels of pollution that are well above the water quality standards and no permits to harvest shellfish are issued for prohibited waters [149].

In terms of biological hazards, the only fecal indicators used are total or fecal coliforms, which have been shown to be poor indicators for viral fecal contaminates [150–153]. Additionally, the standards set forth for the removal of pathogens postharvest through processing are also aimed at bacterial and chemical contamination. Standard depuration processes have not been effective in the removal of HAV or human norovirus contamination in oyster tissues [154–156].

The Hazard Analysis and Critical Control Point (HACCP) system is designed to improve food safety by requiring the industry to identify potential biological, chemical, and physical hazards within the food production process. There are seven basic principles associated with the HACCP system, which include hazard analysis, the determination of critical control points, the establishment of critical limits, monitoring procedures, corrective actions, verification procedures, and proper documentation procedures. The FDA implemented the HACCP system for seafood, dairy, and juice safety. The United States Department of Agriculture (USDA) has also adopted the HACCP system for the production of meat and poultry products. In 2008, the FDA published a manual for the voluntary use of HACCP principles for use in food service and retail establishments [157]. If a bacterial contaminant issue was identified during the hazard analysis, then a 5D process (the time required for a five decimal reduction) would be required to target the microorganism and ensure that the food produced is free from contamination. Unfortunately, D-values have not been established for most foodborne viruses, which makes it difficult to ensure that foods are free of viral contamination using traditional HACCP systems.

In 2013, a proposed rule regulating fresh produce production was entered into the federal registrar under the Food Safety Modernization Act [158]. As of November 2, 2015, the FDA submitted the final proposal to the Federal Register for publication, as required by the court. The proposed rule will set standards for growing, harvesting, packing, and handling of produce for human consumption. Five routes of microbial contamination have been identified and targeted under the proposed rule. The first is agricultural water and the rule would require generic *E. coli* levels to be less than 235 colony-forming units (cfu)/mL in water used for irrigation and other purposes. Next, the rule would set standards for the use of biological soil amendments of animal origin, including the allowable microbial load, requirements for application, and minimum times between application and produce harvest. Another target area will be employee health and hygiene, mandating that employees will receive proper training for safe produce handing, be trained in handwashing procedures and glove use, and also ill workers will be restricted from performing job duties. Requirements will also be set for the maintenance and cleaning of facilities and equipment. Finally, issues involving domesticated and wild animals will be addressed such as the time between grazing activities and produce harvest and also monitoring to ensure that there are no visible signs of animal contamination on produce prior to harvesting [158]. Overall, the improvement in practices, such as personal hygiene and facility maintenance, could potentially minimize viral contamination in fresh produce. However, as stated previously, *E. coli* and other fecal coliforms are not good indicators of viral contamination. Hence, agricultural water that is free from *E. coli* contamination may still harbor viruses.

It remains a challenge to implement regulations in the food industry to specifically target viruses, as the viruses are difficult to detect and eliminate once contamination has occurred. In general, any process that requires enhanced sanitation and personal hygiene should increase the viral safety of the food product.

VACCINES AGAINST FOOD- AND WATERBORNE VIRAL DISEASES

Introduction

Vaccination is the most effective strategy to protect humans from infectious diseases. The increasing clinical significance of food- and waterborne virus infections suggests that there is an urgent need for an efficacious vaccine for them, particularly for the populations at high risk such as food handlers, pregnant women, military personnel, elderly, infants, children, blood donors, organ transplantation receptors, and immunocompromised individuals. To prevent viral hepatitis, it is particularly important to vaccinate individuals at high risk of infection such as travelers to countries with high or intermediate endemicity, men who have sex with men, injecting-drug users, patients with clotting factor disorders, and patients with chronic liver disease. An effective vaccine would not only prevent mortality and morbidity caused by the virus, but would also block transmission routes and thus improve foodborne illnesses, food safety, public health, and biodefense.

Commercialized Food- and Waterborne Vaccines

Current commercially available vaccines against food- and waterborne viruses include poliovirus, HAV, and HuRV, all of which are highly successful in disease prevention [159]. There are two types of vaccines that protect against the poliovirus: the inactivated polio vaccine and the oral attenuated polio vaccine. In 1953, Salk and his colleagues developed an inactivated polio vaccine [160]. In the following year, nearly two million children participated in the field trials. From 1955 to 1957, the incidence of polio in the United States fell by 85%–90% [161]. In the years 1957–1959, mass clinical trials of Albert Sabin's live, attenuated polio vaccine were performed in Russia [162]. In 1962, the Salk vaccine was replaced by the Sabin vaccine for most purposes because it was easier to administer and less expensive. Both poliovirus vaccines are highly successful. According to CDC, the last case of polio caused by the "wild" virus in the United States was in 1979. However, due to safety concerns, the inactivated polio vaccine replaced the oral polio vaccine as the recommended method of polio immunization in the United States in 1999 [163].

The World Health Organization (WHO) has also issued recommendations for the use of the HAV vaccine [159,164]. Two types of monovalent HAV vaccines are currently used worldwide. Inactivated HAV vaccines are used in most countries (usually given in two intramuscular injections at 6 and 36 month intervals); live attenuated vaccines are also manufactured and mainly used in China and sporadically in India (one injection only). The protective efficacy of the HAV vaccines against clinical disease in adults, children, and adolescents was determined to be 94%–100% following two doses of the vaccine given 1 month apart [165].

In 2009, the WHO recommended the inclusion of a rotavirus vaccine in national immunization programs and several countries have adopted rotavirus vaccine policies [166]. Two different rotavirus vaccines, RotaTeq® (RV5) and Rotarix® (RV1), are currently licensed for use for infants in the United States. These vaccines are given in a two- or three-dose series. Both vaccines were tested in clinical trials and have been shown to be safe and highly effective [166–168]. In these studies, during approximately the first year of an infant's life, the rotavirus vaccine was found to prevent almost all (85%–98%) severe rotavirus illness episodes and prevented 74%–87% of all rotavirus illness episodes.

Vaccine Candidates That Are in Human Clinical Trials

Vaccine Clinical Trials of Two Noncultivable Foodborne Viruses: Human Norovirus and HEV

Since human norovirus is not robustly cultivable, most vaccine studies have been focused on a subunit vaccine using the viral capsid protein (VP1) as the antigen. Expression of the VP1 gene in insect cells can lead to the subsequent self-assembly of VLPs that are structurally and antigenically similar to native virions [169]. These VLPs are noninfectious, since they lack viral RNA. However, VLPs contain optimal epitopes that can trigger human norovirus–specific immune responses in hosts including mice, gnotobiotic pigs, and nonhuman primates [169]. Currently, the VLPs-based vaccine candidate has been tested in human clinical trials. In 1999, Ball and colleagues performed the first clinical study to demonstrate that human norovirus VLPs were safe and immunogenic (such as serum immunoglobulin A [IgA] response, intestinal IgA response, and T cell immune response) in humans when administered orally [170]. A similar human volunteer study was performed by Tacket et al. [171] using the Norwalk VLPs as antigens. Recently, a human subject study was conducted in healthy adults to assess the protection efficacy of a VLP vaccine candidate (with chitosan and monophosphoryllipid A as adjuvants) to prevent acute viral gastroenteritis after challenge with a homologous viral strain, the Norwalk virus (GI.1) [172]. Between 98 human subjects, 50 participants received the VLP vaccine, 48 participants received the placebo, and 90 received both doses (47 participants in the vaccine group and 43 in the placebo group). It was found that the Norwalk virus–specific IgA antibody was detected in 70% of the vaccine recipients. After challenge with the Norwalk virus, it was found that vaccination significantly reduced the frequency of Norwalk virus gastroenteritis. Of the placebo participants, 67% developed gastroenteritis whereas only 37% of vaccine recipients developed symptoms. In addition, 82% of placebo recipients were infected with the Norwalk virus compared with 61% of vaccine recipients who developed an infection. It was concluded that the VLP vaccine candidate provided protection against illness and infection after challenge with a homologous virus.

It has also been technically challenging to develop a vaccine for HEV as it grows poorly in cell culture. The poor growth and low virus yield prevent the development of a live or inactivated vaccine for HEV [173]. Similar to human norovirus, the expression of a capsid gene of HEV in *E. coli* and insect cells led to the formation of VLPs, which can be used as a subunit vaccine [174–176]. This vaccine candidate was shown to be safe and immunogenic in animal models. In China, the VLP-based HEV vaccine was approved for phase I and II trials in 2004 [176]. Upon completion of phase III clinical trials in 2009, the vaccine was named Hecolin® (hepatitis E vaccine made in *E. coli*). In December 2011, China approved the hepatitis

E vaccine, Hecolin®, for use in subjects who are over 16 years old [176]. Currently, commercial production of Hecolin® is ongoing. A follow-up study on the long-term efficacy of Hecolin® within the Chinese population was published in 2015 and was found to provide immunogenicity for up to 4.5 years in almost 90% of those vaccinated [177]; however, this vaccine is currently only available in China.

Vaccine Clinical Trials of Enterovirus 71 (EV71)

EV71 is an important cause of hand, foot, and mouth disease (HFMD), and is associated with more severe diseases in young children (aged under 5 years), including aseptic meningitis and encephalitis. Both EV71 and poliovirus belong to the enterovirus genus in the *Picornaviridae* family, and share many similarities in virological and clinical aspects. Poliovirus vaccine technology, both live attenuated and inactivated virus vaccines, can potentially be adapted to control EV71 infection. Recently, Feng-Cai Zhu and colleagues reported the results of a randomized, double-blind, placebo-controlled, phase III clinical trial of an inactivated EV71 vaccine in China [178]. A total of 10,245 participants (aged from 6 to 35 months) were enrolled in this trial and assigned to two groups: 5120 to vaccine versus 5125 to placebo. Excitingly, this vaccine achieved an efficacy of 90.0% against EV71-associated HFMD and 80.4% against EV71-associated disease (including herpangina and neurological complications). Similar to the inactivated poliovirus vaccine, the EV71 vaccine provides high efficacy, satisfactory safety, and sustained immunogenicity in children [179]. It is highly promising that this inactivated vaccine can be implemented in vaccination schedules in the future.

CONCLUDING REMARKS

Despite tremendous research, extension, and education efforts, food- and waterborne viruses are still major public health issues. To combat these important pathogens, future research efforts should focus on eight major areas:

1. Developing an efficient cell culture system for uncultivable food- and waterborne viruses such as norovirus and HEV is imperative. Without a tissue culture system, many biological and virological questions such as survival and viral replication cycles cannot be answered.
2. A robust, small animal model must be developed for these viruses. Without a small animal model, many immunological and pathological questions cannot be answered.
3. The development of a highly efficacious vaccine. To date, there is no licensed vaccine for human norovirus, sapovirus, HEV (outside China), or EV71, although recent human clinical trials have yielded highly promising results. New vaccine strategies (such as live-vectored vaccines), adjuvants, and delivery strategies are needed.
4. The establishment of rapid, sensitive, and cost-effective detection methods for food- and waterborne viruses so that viral contamination can be monitored in a real-time manner.
5. A focus on the optimization and commercialization of promising thermal and nonthermal processing technologies is urgently needed. To enhance the effectiveness of viral inactivation, new norovirus technologies as well as the use of hurdle technologies or combined technologies are needed.
6. The development of sanitizers to specifically target common viruses, such as noroviruses, is required, as many studies have found that routine sanitizers that are effective for bacterial pathogens are not as effective for food- and waterborne viruses.
7. The development of antiviral drugs is required, as to date, no specific, approved drug is available to control food- and waterborne viruses.
8. Building multidisciplinary and multi-institutional teams is needed to combat food- and waterborne viruses.

ACKNOWLEDGMENTS

This study was supported by a special emphasis grant (2010-51110-21080) from the National Integrated Food Safety Initiative (NIFSI) of the USDA, a food safety challenge grant (2011-68003-30005) from the Agriculture and Food Research Initiative (AFRI) of the USDA National Institute of Food and Agriculture (NIFA), and NoroCORE (2011-68003-30395), the USDA-NIFA Food Virology Collaborative.

REFERENCES

1. Mead PS, Slutsker L, Dietz V, McCaig LF, Bresee JS, Shapiro C, Griffin PM, Tauxe RV. 1999. Food-related illness and death in the United States. *Emerg Infect Dis* 5:607–625.
2. Grove SF, Lee A, Lewis T, Stewart CM, Chen H, Hoover DG. 2006. Inactivation of foodborne viruses of significance by high pressure and other processes. *J Food Prot* 69:957–968.
3. Rastogi NK, Raghavarao KS, Balasubramaniam VM, Niranjan K, Knorr D. 2007. Opportunities and challenges in high pressure processing of foods. *Crit Rev Food Sci Nutr* 47:69–112.
4. Buckow R, Isbarn S, Knorr D, Heinz V, Lehmacher A. 2008. Predictive model for inactivation of feline calicivirus, a norovirus surrogate, by heat and high hydrostatic pressure. *Appl Environ Microbiol* 74:1030–1038.
5. Kingsley DH, Guan D, Hoover DG, Chen H. 2006. Inactivation of hepatitis A virus by high-pressure processing: The role of temperature and pressure oscillation. *J Food Prot* 69:2454–2459.
6. Lou F, Neetoo H, Li J, Chen H, Li J. 2011. Lack of correlation between virus barosensitivity and the presence of a viral envelope during inactivation of human rotavirus, vesicular stomatitis virus, and avian metapneumovirus by high-pressure processing. *Appl Environ Microbiol* 77:8538–8547.
7. Lou F, Neetoo H, Chen H, Li J. 2011. Inactivation of a human norovirus surrogate by high-pressure processing: Effectiveness, mechanism, and potential application in the fresh produce industry. *Appl Environ Microbiol* 77:1862–1871.
8. Wilkinson N, Kurdziel AS, Langton S, Needs E, Cook N. 2001. Resistance of poliovirus to inactivation by high hydrostatic pressures. *Innov Food Sci Emerg* 2:95–98.
9. Lou F, Huang P, Neetoo H, Gurtler JB, Niemira BA, Chen H, Jiang X, Li J. 2012. High-pressure inactivation of human norovirus virus-like particles provides evidence that the capsid of human norovirus is highly pressure resistant. *Appl Environ Microbiol* 78:5320–5327.
10. Dancho BA, Chen H, Kingsley DH. 2012. Discrimination between infectious and non-infectious human norovirus using porcine gastric mucin. *Int J Food Microbiol* 155:222–226.
11. Tang Q, Li D, Xu J, Wang J, Zhao Y, Li Z, Xue C. 2010. Mechanism of inactivation of murine norovirus-1 by high pressure processing. *Int J Food Microbiol* 137:186–189.
12. Chen H, Hoover DG, Kingsley DH. 2005. Temperature and treatment time influence high hydrostatic pressure inactivation of feline calicivirus, a norovirus surrogate. *J Food Prot* 68:2389–2394.
13. Grove SF, Lewis T, Ross T, Forsyth S, Wan J, Coventry J, Lee A, Cole M, Stewart CM. 2008. Inactivation of hepatitis A virus, poliovirus and a norovirus surrogate by high pressure processing. *Innov Food Sci Emerg* 9:206–210.
14. Kingsley DH, Hoover DG, Papafragkou E, Richards GP. 2002. Inactivation of hepatitis A virus and a calicivirus by high hydrostatic pressure. *J Food Prot* 65:1605–1609.
15. Calci KR, Meade GK, Tezloff RC, Kingsley DH. 2005. High-pressure inactivation of hepatitis A virus within oysters. *Appl Environ Microbiol* 71:339–343.
16. Kingsley DH, Guan D, Hoover DG. 2005. Pressure inactivation of hepatitis A virus in strawberry puree and sliced green onions. *J Food Prot* 68:1748–1751.
17. Sharma M, Shearer AEH, Hoover DG, Liu MN, Solomon MB, Kniel KE. 2008. Comparison of hydrostatic and hydrodynamic pressure to inactivate foodborne viruses. *Innov Food Sci Emerg* 9:418–422.
18. Kingsley DH, Chen H. 2009. Influence of pH, salt, and temperature on pressure inactivation of hepatitis A virus. *Int J Food Microbiol* 130:61–64.
19. Kingsley DH, Chen H, Hoover DG. 2004. Inactivation of selected picornaviruses by high hydrostatic pressure. *Virus Res* 102:221–224.
20. Kingsley DH, Chen H. 2008. Aqueous matrix compositions and pH influence feline calicivirus inactivation by high pressure processing. *J Food Prot* 71:1598–1603.
21. Kingsley DH, Holliman DR, Calci KR, Chen H, Flick GJ. 2007. Inactivation of a norovirus by high-pressure processing. *Appl Environ Microbiol* 73:581–585.

22. Li X, Ye M, Neetoo H, Golovan S, Chen H. 2013. Pressure inactivation of Tulane virus, a candidate surrogate for human norovirus and its potential application in food industry. *Int J Food Microbiol* 162:37–42.
23. Leon JS, Kingsley DH, Montes JS, Richards GP, Lyon GM, Abdulhafid GM, Seitz SR, et al. 2011. Randomized, double-blinded clinical trial for human norovirus inactivation in oysters by high hydrostatic pressure processing. *Appl Environ Microbiol* 77:5476–5482.
24. Sanchez G, Aznar R, Martinez A, Rodrigo D. 2011. Inactivation of human and murine norovirus by high-pressure processing. *Foodborne Pathog Dis* 8:249–253.
25. Guan D, Kniel K, Calci KR, Hicks DT, Pivarnik LF, Hoover DG. 2006. Response of four types of coliphages to high hydrostatic pressure. *Food Microbiol* 23:546–551.
26. Baert L, Debevere J, Uyttendaele M. 2009. The efficacy of preservation methods to inactivate foodborne viruses. *Int J Food Microbiol* 131:83–94.
27. Balny C, Masson P, Heremans K. 2002. High pressure effects on biological macromolecules: From structural changes to alteration of cellular processes. *Biochim Biophys Acta* 1595:3–10.
28. Gross M, Jaenicke R. 1994. Proteins under pressure: The influence of high hydrostatic pressure on structure, function and assembly of proteins and protein complexes. *Eur J Biochem* 221:617–630.
29. Murchie LW, Kelly AL, Wiley M, Adair BM, Patterson M. 2007. Inactivation of a calicivirus and enterovirus in shellfish by high pressure. *Innov Food Sci Emerg* 8:213–217.
30. Rawsthorne H, Phister TG, Jaykus LA. 2009. Development of a fluorescent in situ method for visualization of enteric viruses. *Appl Environ Microbiol* 75:7822–7827.
31. Urbanucci A, Myrmel M, Berg I, von Bonsdorff CH, Maunula L. 2009. Potential internalisation of caliciviruses in lettuce. *Int J Food Microbiol* 135:175–178.
32. Wei J, Jin Y, Sims T, Kniel KE. 2010. Manure- and biosolids-resident murine norovirus 1 attachment to and internalization by Romaine lettuce. *Appl Environ Microbiol* 76:578–583.
33. Dicaprio E, Ma Y, Purgianto A, Hughes J, Li J. 2012. Internalization and dissemination of human norovirus and animal caliciviruses in hydroponically grown romaine lettuce. *Appl Environ Microbiol* 78:6143–6152.
34. Cascarino J. 2007. Inactivation of viruses by high hydrostatic pressure in ready-to-eat food products. MS thesis. University of Delaware, Newark, DE.
35. Abdel Karim P. 2011. High pressure processing as an alternative food preservation technology and its applications for fruits and vegetables. MS thesis. Kansas State University, Manhattan, KS.
36. Gogal RM, Jr., Kerr R, Kingsley DH, Granata LA, LeRoith T, Holliman SD, Dancho BA, Flick GJ, Jr. 2011. High hydrostatic pressure processing of murine norovirus 1-contaminated oysters inhibits oral infection in STAT-1(-/-)-deficient female mice. *J Food Prot* 74:209–214.
37. Cruz-Romero M, Smiddy M, Hill C, Kerry JP, Kelly AL. 2004. Effects of high pressure treatment on physicochemical characteristics of fresh oysters (*Crassostrea gigas*). *Innov Food Sci Emerg* 5:161–169.
38. He H, Adamas RM, Farkas DF, Morrissey MT. 2002. Use of high-pressure processing for oyster shucking and shelf life extension. *J Food Sci* 67:640–645.
39. Hoover DG, Metrick C, Papineau AM, Farkas DF, Knorr D. 1989. Biological effects of high hydrostatic pressure on food microorganisms. *Food Tech* 43:99.
40. Lopez-Caballero ME, Perez-Mateos M, Montero P, Borderias AJ. 2000. Oyster preservation by high-pressure treatment. *J Food Prot* 63:196–201.
41. Farkas J. 1998. Irradiation as a method for decontaminating food: A review. *Int J Food Microbiol* 44:189–204.
42. Sanglay GC, Li J, Uribe RM, Lee K. 2011. Electron-beam inactivation of a norovirus surrogate in fresh produce and model systems. *J Food Prot* 74:1155–1160.
43. Espinosa AC, Jesudhasan P, Arredondo R, Cepeda M, Mazari-Hiriart M, Mena KD, Pillai SD. 2012. Quantifying the reduction in potential health risks by determining the sensitivity of poliovirus type 1 chat strain and rotavirus SA-11 to electron beam irradiation of iceberg lettuce and spinach. *Appl Environ Microbiol* 78:988–993.
44. Feng K, Divers E, Ma Y, Li J. 2011. Inactivation of a human norovirus surrogate, human norovirus virus-like particles, and vesicular stomatitis virus by gamma irradiation. *Appl Environ Microbiol* 77:3507–3517.
45. Sommers C, Fan X. 2010. Irradiation of ground beef and fresh produce. *In* Zhang HQ, Barbosa-Canovas V, Balasubramaniam VM, Dunne CP, Farkas DF, and Yuan JTC (eds), *Nonthermal Processing Technologies for Food*. Wiley-Blackwell, Oxford, UK. doi: 10.1002/9789470958360.ch17.
46. Fino VR, Kniel KE. 2008. UV light inactivation of hepatitis A virus, Aichi virus, and feline calicivirus on strawberries, green onions, and lettuce. *J Food Prot* 71:908–913.

47. Roberts P, Hope A. 2003. Virus inactivation by high intensity broad spectrum pulsed light. *J Virol Methods* 110:61–65.
48. Lee J, Zoh K, Ko G. 2008. Inactivation and UV disinfection of murine norovirus with TiO2 under various environmental conditions. *Appl Environ Microbiol* 74:2111–2117.
49. Thurston-Enriquez JA, Haas CN, Jacangelo J, Riley K, Gerba CP. 2003. Inactivation of feline calicivirus and adenovirus type 40 by UV radiation. *Appl Environ Microbiol* 69:577–582.
50. Guerrero-Beltran JA, Barbosa-Canovas GV. 2010. Ultraviolet-C light processing of liquid food products. *In* Zhang HQ, Barbosa-Canovas V, Balasubramaniam VM, Dunne CP, Farkas DF, Yuan JTC (eds), *Nonthermal Processing Technologies for Food*. Wiley-Blackwell, Oxford, UK. doi: 10.1002/9789470958360.ch19.
51. Demirci A, Krishnamurthy K. 2011. Pulsed ultraviolet light. *In* Zhang HQ, Barbosa-Cánovas GV, Balasubramaniam VM, Dunne CP, Farkas Df, Yuan JTC (eds), *Nonthermal Processing Technologies for Food*, pp. 249–261. IFT Press & Wiley-Blackwell, Chichester, UK.
52. Lamont Y, Rzezutka A, Anderson JG, MacGregor SJ, Given MJ, Deppe C, Cook N. 2007. Pulsed UV-light inactivation of poliovirus and adenovirus. *Lett Appl Microbiol* 45:564–567.
53. Jean J, Morales-Rayas R, Anoman MN, Lamhoujeb S. 2011. Inactivation of hepatitis A virus and norovirus surrogate in suspension and on food-contact surfaces using pulsed UV light (pulsed light inactivation of food-borne viruses). *Food Microbiol* 28:568–572.
54. Sow H, Desbiens M, Morales-Rayas R, Ngazoa SE, Jean J. 2011. Heat inactivation of hepatitis A virus and a norovirus surrogate in soft-shell clams (*Mya arenaria*). *Foodborne Pathog Dis* 8:387–393.
55. Su X, Zivanovic S, D'Souza DH. 2010. Inactivation of human enteric virus surrogates by high-intensity ultrasound. *Foodborne Pathog Dis* 7:1055–1061.
56. United States, Food and Drug Administration (FDA). 2002. Guidance for industry: Exemptions from the warning label requirement for juice—recommendations for effectively achieving a 5-log pathogen reduction; Final Guidance. http://www.fda.gov/Food/GuidanceRegulation/GuidanceDocumentsRegulatoryInformation/Juice/ucm072965.htm. Accessed October 7, 2015.
57. Hirneisen KA, Kniel KE. 2013. Comparing human norovirus surrogates: Murine norovirus and Tulane virus. *J Food Prot* 76:139–143.
58. Tuladhar E, Bouwknegt M, Zwietering MH, Koopmans M, Duizer E. 2012. Thermal stability of structurally different viruses with proven or potential relevance to food safety. *J Appl Microbiol* 112:1050–1057.
59. Dolin R, Blacklow NR, DuPont H, Buscho RF, Wyatt RG, Kasel JA, Hornick R, Chanock RM. 1972. Biological properties of Norwalk agent of acute infectious nonbacterial gastroenteritis. *Proc Soc Exp Biol Med* 140:578–583.
60. O'Mahony J, O'Donoghue M, Morgan JG, Hill C. 2000. Rotavirus survival and stability in foods as determined by an optimised plaque assay procedure. *Int J Food Microbiol* 61:177–185.
61. Jiang B, Wang Y, Saluzzo JF, Bargeron K, Frachette MJ, Glass RI. 2008. Immunogenicity of a thermally inactivated rotavirus vaccine in mice. *Hum Vaccin* 4:143–147.
62. Croci L, Ciccozzi M, De Medici D, Di Pasquale S, Fiore A, Mele A, Toti L. 1999. Inactivation of hepatitis A virus in heat-treated mussels. *J Appl Microbiol* 87:884–888.
63. Cannon JL, Aydin A, Mann AN, Bolton SL, Zhao T, Doyle MP. 2012. Efficacy of a levulinic acid plus sodium dodecyl sulfate-based sanitizer on inactivation of human norovirus surrogates. *J Food Prot* 75:1532–1535.
64. Li D, Baert L, Xia M, Zhong W, Van Coillie E, Jiang X, Uyttendaele M. 2012. Evaluation of methods measuring the capsid integrity and/or functions of noroviruses by heat inactivation. *J Virol Methods* 181:1–5.
65. Maheshwari G, Jannat R, McCormick L, Hsu D. 2004. Thermal inactivation of adenovirus type 5. *J Virol Methods* 118:141–146.
66. Baert L, Uyttendaele M, Van Coillie E, Debevere J. 2008. The reduction of murine norovirus 1, *B. fragilis* HSP40 infecting phage B40-8 and E. coli after a mild thermal pasteurization process of raspberry puree. *Food Microbiol* 25:871–874.
67. Barnaud E, Rogee S, Garry P, Rose N, Pavio N. 2012. Thermal inactivation of infectious hepatitis E virus in experimentally contaminated food. *Appl Environ Microbiol* 78:5153–5159.
68. Mariam TC, Cliver DO. 2000. Small round coliphages as surrogates for human viruses in process assessment. *Dairy, Food and Environmental Sanitation* 20:684–689.
69. Deboosere N, Pinon A, Delobel A, Temmam S, Morin T, Merle G, Blaise-Boisseau S, Perelle S, Vialette M. 2010. A predictive microbiology approach for thermal inactivation of hepatitis A virus in acidified berries. *Food Microbiol* 27:962–967.

70. Bidawid S, Farber JM, Sattar SA, Hayward S. 2000. Heat inactivation of hepatitis A virus in dairy foods. *J Food Prot* 63:522–528.
71. Strazynski M, Kramer J, Becker B. 2002. Thermal inactivation of poliovirus type 1 in water, milk and yoghurt. *Int J Food Microbiol* 74:73–78.
72. Deboosere N, Legeay O, Caudrelier Y, Lange M. 2004. Modelling effect of physical and chemical parameters on heat inactivation kinetics of hepatitis A virus in a fruit model system. *Int J Food Microbiol* 93:73–85.
73. DiGirolamo R, Liston J, Matches JR. 1970. Survival of virus in chilled, frozen, and processed oysters. *Appl Microbiol* 20:58–63.
74. Chick H. 1908. An investigation of the laws of disinfection. *J Hyg (Lond)* 8:92–158.
75. de Roda Husman AM, Lodder WJ, Rutjes SA, Schijven JF, Teunis PF. 2009. Long-term inactivation study of three enteroviruses in artificial surface and groundwaters, using PCR and cell culture. *Appl Environ Microbiol* 75:1050–1057.
76. van Boekel MA. 2002. On the use of the Weibull model to describe thermal inactivation of microbial vegetative cells. *Int J Food Microbiol* 74:139–159.
77. Wallis C, Yang CS, Melnick JL. 1962. Effect of cations on thermal inactivation of vaccinia, herpes simplex, and adenoviruses. *J Immunol* 89:41–46.
78. Wigand R, Bachmann P, Bradner G. 1981. Stabilization of isometric DNA viruses against thermoinactivation by lowered ionic strength. *Arch Virol* 69:113–138.
79. Cannon JL, Papafragkou E, Park GW, Osborne J, Jaykus LA, Vinje J. 2006. Surrogates for the study of norovirus stability and inactivation in the environment: A comparison of murine norovirus and feline calicivirus. *J Food Prot* 69:2761–2765.
80. Bozkurt H, D'Souza DH, Davidson PM. 2013. Determination of the thermal inactivation kinetics of the human norovirus surrogates, murine norovirus and feline calicivirus. *J Food Prot* 76:79–84.
81. Li D, Baert L, Van Coillie E, Uyttendaele M. 2011. Critical studies on binding-based RT-PCR detection of infectious noroviruses. *J Virol Methods* 177:153–159.
82. Topping JR, Schnerr H, Haines J, Scott M, Carter MJ, Willcocks MM, Bellamy K, et al. 2009. Temperature inactivation of Feline calicivirus vaccine strain FCV F-9 in comparison with human noroviruses using an RNA exposure assay and reverse transcribed quantitative real-time polymerase chain reaction-A novel method for predicting virus infectivity. *J Virol Methods* 156:89–95.
83. United States Centers for Disease Control and Prevention (CDC). 1994. Hepatitis surveillance report. Centers for Disease Control and Prevention, Atlanta, GA.
84. Cliver DO. 1983. *Manual on Food Virology*. World Health Organization, Geneva.
85. Raska K, Helcl J, Jezek J, Kubelka Z, Litov M, Novak K, Radkovsky J, Sery V, Zejdl J, Zikmund V. 1966. A milk-borne infectious hepatitis epidemic. *J Hyg Epidemiol Microbiol Immunol* 10:413–428.
86. Le Guyader F, Loisy F, Atmar RL, Hutson AM, Estes MK, Ruvoen-Clouet N, Pommepuy M, Le Pendu J. 2006. Norwalk virus-specific binding to oyster digestive tissues. *Emerg Infect Dis* 12:931–936.
87. Nguyen-the C, Carlin F. 1994. The microbiology of minimally processed fresh fruits and vegetables. *Crit Rev Food Sci Nutr* 34:371–401.
88. Gulati BR, Allwood PB, Hedberg CW, Goyal SM. 2001. Efficacy of commonly used disinfectants for the inactivation of calicivirus on strawberry, lettuce, and a food-contact surface. *J Food Prot* 64:1430–1434.
89. Keswick BH, Satterwhite TK, Johnson PC, DuPont HL, Secor SL, Bitsura JA, Gary GW, Hoff JC. 1985. Inactivation of Norwalk virus in drinking water by chlorine. *Appl Environ Microbiol* 50:261–264.
90. Kitajima M, Tohya Y, Matsubara K, Haramoto E, Utagawa E, Katayama H. 2010. Chlorine inactivation of human norovirus, murine norovirus and poliovirus in drinking water. *Lett Appl Microbiol* 51:119–121.
91. Kim SW, Baek SB, Ha JH, Lee MH, Choi C, Ha SD. 2012. Chlorine treatment to inactivate norovirus on food contact surfaces. *J Food Prot* 75:184–188.
92. United States, Food and Drug Administration (FDA). December 16, 2014. Chapter V. Methods to Reduce/Eliminate Pathogens from Produce and Fresh-Cut Produce, *on* FDA. http://www.fda.gov/Food/FoodScienceResearch/SafePracticesforFoodProcesses/ucm091363.htm. Accessed November 10, 2015.
93. Feliciano L, Li J, Lee J, Pascall MA. 2012. Efficacies of sodium hypochlorite and quaternary ammonium sanitizers for reduction of norovirus and selected bacteria during ware-washing operations. *PLoS One* 7:e50273.
94. Park GW, Linden KG, Sobsey MD. 2011. Inactivation of murine norovirus, feline calicivirus and echovirus 12 as surrogates for human norovirus (NoV) and coliphage (F+) MS2 by ultraviolet light (254 nm) and the effect of cell association on UV inactivation. *Lett Appl Microbiol* 52:162–167.

95. D'Souza DH, Su X. 2010. Efficacy of chemical treatments against murine norovirus, feline calicivirus, and MS2 bacteriophage. *Foodborne Pathog Dis* 7:319–326.
96. Girard M. 2009. Attachement des norovirus aux surfaces inertes et évaluation de leur sensibilité aux désinfectants domestiques, Université Laval.
97. Butot S, Putallaz T, Sanchez G. 2008. Effects of sanitation, freezing and frozen storage on enteric viruses in berries and herbs. *Int J Food Microbiol* 126:30–35.
98. Predmore A, Li J. 2011. Enhanced removal of a human norovirus surrogate from fresh vegetables and fruits by a combination of surfactants and sanitizers. *Appl Environ Microbiol* 77:4829–4838.
99. Xue B, Jin M, Yang D, Guo X, Chen Z, Shen Z, Wang X, Qiu Z, Wang J, Zhang B, Li J. 2013. Effects of chlorine and chlorine dioxide on human rotavirus infectivity and genome stability. *Water Res* 47:3329–3338.
100. Yeap JW, Kaur S, Lou F, DiCaprio E, Morgan M, Linton R, Li J. 2015. Inactivation kinetics and mechanism of a human norovirus surrogate on stainless steel coupons using chlorine dioxide gas. *Appl Environ Microbiol.* 82(1):116–123.
101. Bolton SL, Kotwal G, Harrison MA, Law SE, Harrison JA, Cannon JL. 2013. Sanitizer efficacy against murine norovirus, a surrogate for human norovirus, on stainless steel surfaces when using three application methods. *Appl Environ Microbiol* 79:1368–1377.
102. Su X, D'Souza DH. 2012. Inactivation of human norovirus surrogates by benzalkonium chloride, potassium peroxymonosulfate, tannic acid, and gallic acid. *Foodborne Pathog Dis* 9:829–834.
103. Fraisse A, Temmam S, Deboosere N, Guillier L, Delobel A, Maris P, Vialette M, Morin T, Perelle S. 2011. Comparison of chlorine and peroxyacetic-based disinfectant to inactivate feline calicivirus, murine norovirus and hepatitis A virus on lettuce. *Int J Food Microbiol* 151:98–104.
104. Baert L, Vandekinderen I, Devlieghere F, Van Coillie E, Debevere J, Uyttendaele M. 2008. Inactivation of murine norovirus 1 and *Bacteroides fragilis* infecting phage B40-8 by the use of sodium hypochlorite and peroxyacetic acid as decontaminating agents for shredded iceberg lettuce. *Commun Agric Appl Biol Sci* 73:97–101.
105. Vaughn JM, Chen Y-S, Novotny JF, Strout D. 1990. Effects of ozone treatment on the infectivity of hepatitis A virus. *Can J Microbiol* 36:557–560.
106. Hirneisen KA, Markland SM, Kniel KE. 2011. Ozone inactivation of norovirus surrogates on fresh produce. *J Food Prot* 74:836–839.
107. Keskinen LA, Annous BA. 2011. Efficacy of adding detergents to sanitizer solutions for inactivation of *Escherichia coli* O157:H7 on Romaine lettuce. *Int J Food Microbiol* 147:157–161.
108. White GC (ed). 1978. *Disinfection of Wastewater and Water for Reuse.* Van Nostrand-Reinhold, New York.
109. Schnoor JL, Nitzschke JL, Lucas RD, Veenstra JN. 1979. Trihalomethane yields as a function of precursor molecular weight. *Environ Sci Technol* 13:1134–1138.
110. Vogt H, Balej J, Bennett JE, Wintzer P, Sheikh SA, Gallone P, Vasudevan S, Pelin K. 2000. Chlorine oxides and chlorine oxygen acids. *Ullmann's Encyclopedia of Industrial Chemistry.* doi:10.1002/14356007.a06_483.pub2. Wiley-VCH Verlag GmbH & Co. KGaA, 1–43.
111. Zoni R, Zanelli R, Riboldi E, Bigliardi L, Sansebastiano G. 2007. Investigation on virucidal activity of chlorine dioxide: Experimental data on feline calicivirus, HAV and Coxsackie B5. *J Prev Med Hyg* 48:91–95.
112. Morino H, Fukuda T, Miura T, Lee C, Shibata T, Sanekata T. 2009. Inactivation of feline calicivirus, a norovirus surrogate, by chlorine dioxide gas. *Biocontrol Sci* 14:147–153.
113. Morino H, Fukuda T, Miura T, Shibata T. 2011. Effect of low-concentration chlorine dioxide gas against bacteria and viruses on a glass surface in wet environments. *Lett Appl Microbiol* 53:628–634.
114. Khajavi R, Sattari M, Ashjaran A. 2007. The antimicrobial effect of benzalkonium chloride on some pathogenic microbes observed on fibers of acrylic carpet. *Pak J Biol Sci* 10:598–601.
115. Houari A, Di Martino, P. 2007. Effect of chlorhexidine and benzalkonium chloride on bacterial biofilm formation. *Lett Appl Microbiol* 45:652–656.
116. Buffet-Bataillon S, Tattevin P, Bonnaure-Mallet M, Jolivet-Gougeon A. 2012. Emergence of resistance to antibacterial agents: The role of quaternary ammonium compounds—a critical review. *Int J Antimicrob Agents* 39:381–389.
117. Moadab A, Rupley KF, Wadhams P. 2001. Effectiveness of a nonrinse, alcohol-free antiseptic hand wash. *J Am Podiatr Med Assoc* 91:288–293.
118. Demrkol O. 2009. Effects of hydrogen peroxide treatment on thiol contents in fresh-cut asparagus (*Asparagus officinalis*) spears. *Int J Food Sci Nutr* 60:80–88.
119. Block S. 1991. Peroxygen compound. *In* Block S (ed), *Disinfection, Sterilization, and Preservation*, pp. 167–181, 4th edn. Lea & Febiger, Philadelphia, PA.

120. Bentley K, Dove BK, Parks SR, Walker JT, Bennett AM. 2012. Hydrogen peroxide vapour decontamination of surfaces artificially contaminated with norovirus surrogate feline calicivirus. *J Hosp Infect* 80:116–121.
121. Baert L, Vandekinderen I, Devlieghere F, Van Coillie E, Debevere J, Uyttendaele M. 2009. Efficacy of sodium hypochlorite and peroxyacetic acid to reduce murine norovirus 1, B40-8, Listeria monocytogenes, and *Escherichia coli* O157:H7 on shredded iceberg lettuce and in residual wash water. *J Food Prot* 72:1047–1054.
122. Banat IM, Makkar RS, Cameotra SS. 2000. Potential commercial applications of microbial surfactants. *Appl Microbiol Biotechnol* 53:495–508.
123. Hartmann SU, Wigdahl B, Neely EB, Berlin CM, Schengrund CL, Lin HM, Howett MK. 2006. Biochemical analysis of human milk treated with sodium dodecyl sulfate, an alkyl sulfate microbicide that inactivates human immunodeficiency virus type 1. *J Hum Lact* 22:61–74.
124. Koopmans M, Duizer E. 2004. Foodborne viruses: An emerging problem. *Int J Food Microbiol* 90:23–41.
125. Krebs FC, Miller SR, Malamud D, Howett MK, Wigdahl B. 1999. Inactivation of human immunodeficiency virus type 1 by nonoxynol-9, C31G, or an alkyl sulfate, sodium dodecyl sulfate. *Antiviral Res* 43:157–173.
126. Adams M, Hartley AD, Cox LJ. 1989. Factors affecting the efficacy of washing procedures used in the production of prepared salads. *Food Microbiol* 6:69–77.
127. Langlais B, Reckhow DA, Brink DR. 1991. *Ozone in Water Treatment: Application and Engineering*. Lewis Publishers, Chelsea, MI.
128. Guzel-Seydim Z, Greene AK, Seydim AC. 2004. Use of ozone in the food industry. *Lebensmittel-Wissenschaft Und -Technologie* 37:453–460.
129. Graham DM. 1997. Use of ozone for food processing. *J Food Technol* 51:66–71.
130. Khadre MA, Yousef AE, Kim J-G. 2001. Microbiological aspects of ozone applications in food: A review. *J Food Sci* 66:1242–1252.
131. Wang T, Lam KS, Tsang CW, Kot SC. 2004. On the variability and correlation of surface ozone and carbon monoxide observed in hong kong using trajectory and regression analyses. *Adv Atmos Sci* 21:141–152.
132. Kessel JF, Allison DK, Moore FJ, Kaime M. 1943. Comparison of chlorine and ozone as virucidal agents of poliomyelitis virus. *Exp Biol Med* 53:71–73.
133. Khadre, Yousef AE. 2001. Sporicidal action of ozone and hydrogen peroxide: A comparative study. *Int J Food Microbiol* 71:131–138.
134. Sattar SA and Springthorpe. 1996. Transmission of viral infections through animate and inanimate surfaces and infection control through chemical disinfection. *In* Hurst CJ (ed), *Modeling Disease Transmission and its Prevention by Disinfection*, pp. 224–257. Cambridge University Press, New York NY.
135. Hudson JB, Sharma M, Petric M. 2007. Inactivation of Norovirus by ozone gas in conditions relevant to healthcare. *J Hosp Infect* 66:40–45.
136. Lim MY, Kim JM, Lee JE, Ko G. 2010. Characterization of ozone disinfection of murine norovirus. *Appl Environ Microbiol* 76:1120–1124.
137. Subrota H, Mandal S, Minz PS, Vij S, Khetra, Y, Singh BP, Yadav D. 2012. Electrolyzed oxidized water (EOW): Non-thermal approach for decontamination of food borne microorganisms in food industry. *J Nutr Food Sci* 3:760–768.
138. Tanaka N, Fujisawa T, Daimon T, Fujiwara K, Yamamoto M, Abe T. 1999. The effect of electrolyzed strong acid aqueous solution on hemodialysis equipment. *Artif Organs* 23:1055–1062.
139. Tian P, Yang D, Mandrell R. 2011. Differences in the binding of human norovirus to and from romaine lettuce and raspberries by water and electrolyzed waters. *J Food Prot* 74:1364–1369.
140. United States Centers for Disease Control and Prevention (CDC). 2013. Norovirus: Facts for food handlers. http://www.cdc.gov/norovirus/food-handlers/work-with-food.html. Accessed November 2015.
141. Mbithi JN, Springthorpe VS, Boulet JR, Sattar SA. 1992. Survival of hepatitis A virus on human hands and its transfer on contact with animate and inanimate surfaces. *J Clin Microbiol* 30:757–763.
142. Bidawid S, Farber JM, Sattar SA. 2000. Contamination of foods by food handlers: Experiments on hepatitis A virus transfer to food and its interruption. *Appl Environ Microbiol* 66:2759–2763.
143. Sattar SA, Ali M, Tetro JA. 2011. *In vivo* comparison of two human norovirus surrogates for testing ethanol-based handrubs: The mouse chasing the cat! *PLoS One* 6:e17340.
144. Fiore AE. 2004. Hepatitis A transmitted by food. *Clin Infect Dis* 38:705–715.

145. Stapleton JT, Winokur PL. 1992. Immunglobulin prophylaxsis for hepatitis A. *Clin Infect Dis* 14:580–586.
146. McCaustland KA, Bond WW, Bradley DW, Ebert JW, Maynard JE. 1982. Survival of hepatitis A virus in feces after drying and storage for 1 month. *J Clin Microbiol* 16:957–958.
147. Mbithi JN, Springthorpe VS, Sattar SA. 1991. Effect of relative humidity and air temperature on survival of hepatitis A virus on environmental surfaces. *Appl Environ Microbiol* 57:1394–1399.
148. Favero MS, Bond WW. 1998. Disinfection and sterilization. *In* Zuckerman AJ, Thomas HC (eds), *Viral Hepatitis*, pp. 627–635. Churchill Livingstone, London.
149. United States, Food and Drug Administration (FDA). 2011. National Shellfish Sanitation Program: Guide for the control of molluscan shellfish (2011 Revision).
150. Ebdon JE, Sellwood J, Shore J, Taylor HD. 2012. Phages of Bacteroides (GB-124): A novel tool for viral waterborne disease control? *Environ Sci Technol* 46:1163–1169.
151. Finance C, Brigaud M, Lucena F, Aymard M, Bosch A, Schwartzbrod L. 1982. Viral pollution of seawater at Barcelona. *Zentralbl Bakteriol Mikrobiol Hyg B* 176:530–536.
152. Skraber S, Gassilloud B, Gantzer C. 2004. Comparison of coliforms and coliphages as tools for assessment of viral contamination in river water. *Appl Environ Microbiol* 70:3644–3649.
153. McQuaig S, Griffith J, Harwood VJ. 2012. Association of fecal indicator bacteria with human viruses and microbial source tracking markers at coastal beaches impacted by nonpoint source pollution. *Appl Environ Microbiol* 78:6423–6432.
154. Corrêa Ade A, Rigotto C, Moresco V, Kleemann CR, Teixeira AL, Poli CR, Simões CM, Barardi CR. 2012. The depuration dynamics of oysters (*Crassostrea gigas*) artificially contaminated with hepatitis A virus and human adenovirus. *Mem Inst Oswaldo Cruz* 107:11–17.
155. Love DC, Lovelace GL, Sobsey MD. 2010. Removal of *Escherichia coli*, *Enterococcus fecalis*, coliphage MS2, poliovirus, and hepatitis A virus from oysters (*Crassostrea virginica*) and hard shell clams (*Mercinaria mercinaria*) by depuration. *Int J Food Microbiol* 143:211–217.
156. Ueki Y, Shoji M, Suto A, Tanabe T, Okimura Y, Kikuchi Y, Saito N, Sano D, Omura T. 2007. Persistence of caliciviruses in artificially contaminated oysters during depuration. *Appl Environ Microbiol* 73:5698–5701.
157. United States, Food and Drug Administration (FDA). 2008. Managing Food Safety: Manual for voluntary use of HACCP principles for operators of food service and retail establishments. CFSAN, FDA. College Park, MD. http://www.fda.gov/downloads/Food/GuidanceRegulation/HACCP/UCM077957.pdf.
158. United States, Food and Drug Administration (FDA). 2013. FMSA proposed rule for produce safety: Standards for growing, harvesting, packing, and holding of produce for human consumption. Food and Drug Administration. 2015 Standards for the Growing, Harvesting, Packing, and Holding of Produce for Human Consumption: Final Rule. 21 CFR Parts 11, 16, and 112. Federal Register 80(228):74353–74672. https://www.gpo.gov/fdsys/pkg/FR-2015-11-27/pdf/2015-28159.pdf.
159. Gans HA. 2013. The status of live viral vaccination in early life. *Vaccine* 31:2531–2537.
160. Salk JE. 1956. Poliomyelitis vaccine in the fall of 1955. *Am J Public Health Nations Health* 46:1–14.
161. Salk JE. 1960. Persistence of immunity after administration of formalin-treated poliovirus vaccine. *Lancet* 2:715–723.
162. Sabin AB. 1985. Oral poliovirus vaccine: History of its development and use and current challenge to eliminate poliomyelitis from the world. *J Infect Dis* 151:420–436.
163. Mateen FJ, Shinohara RT, Sutter RW. 2013. Oral and inactivated poliovirus vaccines in the newborn: A review. *Vaccine* 31:2517–2524.
164. Andre FE. 2006. Universal mass vaccination against hepatitis A. *Curr Top Microbiol Immunol* 304:95–114.
165. Lavanchy D. 2012. Viral hepatitis: Global goals for vaccination. *J Clin Virol* 55:296–302.
166. Aballea S, Millier A, Quilici S, Caroll S, Petrou S, Toumi M. 2013. A critical literature review of health economic evaluations of rotavirus vaccination. *Hum Vaccin Immunother* 9:1272–1288.
167. Lepage P, Vergison A. 2012. Impact of rotavirus vaccines on rotavirus disease. *Expert Rev Anti Infect Ther* 10:547–561.
168. Mameli C, Fabiano V, Zuccotti GV. 2012. New insights into rotavirus vaccines. *Hum Vaccin Immunother* 8:1022–1028.
169. Richardson C, Bargatze RF, Goodwin R, Mendelman PM. 2013. Norovirus virus-like particle vaccines for the prevention of acute gastroenteritis. *Expert Rev Vaccines* 12:155–167.
170. Ball JM, Graham DY, Opekun AR, Gilger MA, Guerrero RA, Estes MK. 1999. Recombinant Norwalk virus-like particles given orally to volunteers: Phase I study. *Gastroenterol J* 117:40–48.

171. Tacket CO, Sztein MB, Losonsky GA, Wasserman SS, Estes MK. 2003. Humoral, mucosal, and cellular immune responses to oral Norwalk virus-like particles in volunteers. *Clin Immunol* 108:241–247.
172. Atmar RL, Bernstein DI, Harro CD, Al-Ibrahim MS, Chen WH, Ferreira J, Estes MK, et al. 2011. Norovirus vaccine against experimental human Norwalk virus illness. *N Engl J Med* 365:2178–2187.
173. Shukla P, Nguyen HT, Torian U, Engle RE, Faulk K, Dalton HR, Bendall RP, Keane FE, Purcell RH, Emerson SU. 2011. Cross-species infections of cultured cells by hepatitis E virus and discovery of an infectious virus-host recombinant. *Proc Natl Acad Sci USA* 108:2438–2443.
174. Kamili S. 2011. Toward the development of a hepatitis E vaccine. *Virus Res* 161:93–100.
175. Wu T, Li SW, Zhang J, Ng MH, Xia NS, Zhao Q. 2012. Hepatitis E vaccine development: A 14 year odyssey. *Hum Vaccin Immunother* 8:823–827.
176. Zhu FC, Zhang J, Zhang XF, Zhou C, Wang ZZ, Huang SJ, Wang H, et al. 2010. Efficacy and safety of a recombinant hepatitis E vaccine in healthy adults: A large-scale, randomised, double-blind placebo-controlled, phase 3 trial. *Lancet* 376:895–902.
177. Zhang J, Zhang XF, Huang SJ, Wu T, Hu YM, Wang ZZ, Wang H, et al. 2015. Long-term efficacy of a hepatitis E vaccine. *N Engl J Med* 372:914–922.
178. Zhu FC, Meng FY, Li JX, Li XL, Mao QY, Tao H, Zhang YT, et al. 2013. Efficacy, safety, and immunology of an inactivated alum-adjuvant enterovirus 71 vaccine in children in China: A multicentre, randomised, double-blind, placebo-controlled, phase 3 trial. *Lancet* 381:2024–2032.
179. Liang ZL, Mao QY, Wang YP, Zhu FC, Li JX, Yao X, Gao F, Wu X, Xu M, Wang JZ. 2013. Progress on the research and development of inactivated EV71 whole-virus vaccines. *Hum Vaccin Immunother* 9:1701–1705.

Section II

Important Agents in Foodborne Viral Infections

6 Norovirus and Sapovirus

Kun Lee Lim, Natalie E. Netzler, Grant S. Hansman, Jason M. Mackenzie, and Peter A. White

CONTENTS

NOROVIRUS

Introduction

Human norovirus (NoV) is the leading cause of acute gastroenteritis (AGE) globally [1,2], and carries with it a large social and financial burden resulting from forced time off work, business closures, and hospitalizations [3]. Foodborne transmission is a key route for NoV infection, as the virus is spread rapidly via contaminated food handlers and food products. In many developed nations, NoV has been identified as the leading cause of foodborne disease, accounting for more than half of the cases of foodborne illness each year [4–7]. As the etiological agent responsible for the majority of foodborne illness (up to 80%) is never identified, estimates additionally attribute NoV with a large proportion of these unresolved cases [4].

NoV remains challenging to develop vaccinations against, with the virus continually mutating to evade herd immunity. Human NoV is also very difficult to culture in immortalized cells, severely hampering efforts to develop treatments and vaccines against the virus [8]. Without a reliable animal model to study human NoV directly, most studies rely on murine norovirus (MNV), a genogroup V NoV. MNV has been used to infer details of the infection cycle, including viral cell entry, and is frequently used to study aspects of NoV replication in both cell culture and *in vivo* mouse studies [9,10].

NoVs are classified within the *Caliciviridae* family and named after calyx (the Greek word for cup) because of the cuplike pattern on the virion surface [1]. Of the six *Caliciviridae* genera, NoV is attributed with the highest incidence of human disease [3]. Symptoms of NoV infection include vomiting and diarrhea, often accompanied by nausea, stomach cramps, malaise, and fever [11], which follow a relatively short incubation period of 24–48 h. Symptoms usually persist for 2–4 days in healthy adults [12], but infections can persist for months or even years in immunosuppressed or immunocompromised individuals [13–15]. While the main route of NoV transmission is person-to-person [16], foodborne NoV is another significant route [3]. This section of the chapter reviews NoV and its transmission via contaminated food and water.

HISTORY

NoV was first described as hyperemesis hiemis or winter vomiting disease by Zahorsky in 1929 [17,18]. However, the disease was not fully understood at that time [19]. In the 1970s, several transmissibility studies in volunteer subjects who were orally administered stool filtrates obtained from infected patients showed that they developed similar gastroenteritis symptoms [20–22]. These studies demonstrated that an unknown nonbacterial agent was the primary cause of transmissible AGE.

The original Norwalk virus was discovered in a fecal sample obtained from an AGE outbreak at an elementary school in Norwalk, Ohio, in 1968. The outbreak had a high attack rate of 50% (116 of 232) among the schoolchildren [23,24]. Norwalk virus, subsequently named after the town where it was found, is the prototype of the *Norovirus* genus. However, it took four more years before the actual physical virus particle was first visualized by Kapikian et al. in 1972 using electron microscopy (EM) [25].

Following the discovery of Norwalk virus, other similar viruses were subsequently detected and collectively termed Norwalk-like viruses (NLV) or small round structured viruses (SRSV) [26]. These viruses were typically named after the location where they were identified or by their appearance as visualized with EM [27,28]. Norwalk virus, the prototype NoV, is classified as NoV genogroup I, genotype 1 (GI.1). Complete sequencing and characterization of the Norwalk virus genome was accomplished by Jiang et al. in the early 1990s [29]. This achievement allowed further investigation into viral taxonomy using molecular techniques, including reverse transcription polymerase chain reaction (RT-PCR) and sequencing. In 2002, NLVs were collectively renamed as a single genus, *Norovirus* [30]. In summary, NoV-associated AGE with vomiting was recognized for many years before the actual virus was finally discovered. The true health burden of NoV has really only come to light in recent years.

NOROVIRUS IS THE MOST COMMON CAUSE OF GASTROENTERITIS

Rise of Norovirus

Historically, rotavirus (RoV) was reported as the major cause of diarrhea in children globally and was estimated to kill more than 440,000 children each year [31]. However, since 2006, the vaccines Rotarix (GlaxoSmithKline) and RotaTeq (Merck) have been introduced into the health-care programs of many countries. Following their inclusion, a dramatic decrease (~70%) in the incidence of RoV-associated AGE has been reported in these countries [32–35]. A similar trend has been seen in New South Wales (NSW), Australia, where the incidence of RoV infections decreased substantially following the introduction of the RoV vaccine in 2007, with the last major epidemic in 2005 (Figure 6.1).

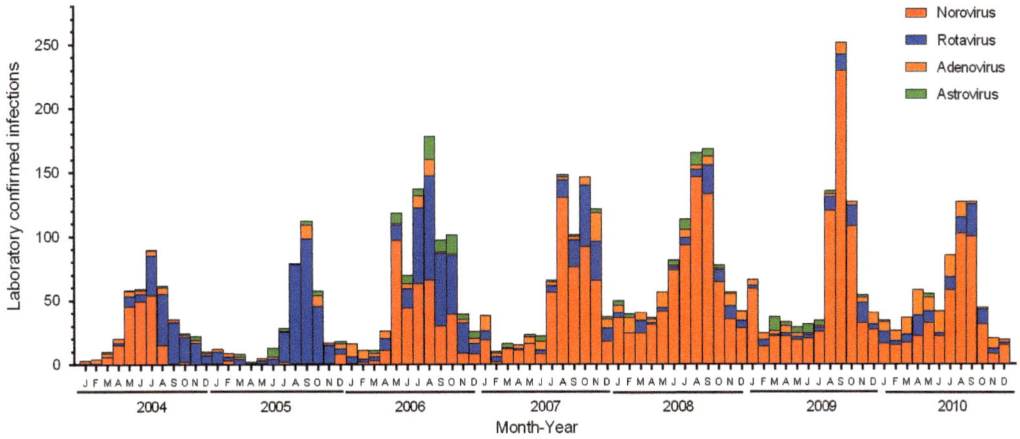

FIGURE 6.1 Detection of enteric viruses at the South Eastern Area Laboratory Services (SEALS) diagnostic facility, Prince of Wales Hospital, Sydney, Australia. The monthly total positive stool specimens for norovirus (NoV), adenovirus (AdV), astrovirus (AsV), and rotavirus (RoV) detected in cases of acute gastroenteritis were evaluated between January 2004 and December 2010. Enzyme immunoassays (EIAs) were performed to detect viral antigens in stool specimens, and viruses are color-coded as per the legend. The RoV vaccine was introduced to the NSW health-care system in 2007 and has contributed to a dramatic decrease of RoV associated infections in subsequent years. (Adapted and modified from Eden, J.S., *The Evolutionary Dynamics of Norovirus*, UNSW Australia, 2012 [36].)

Norovirus Is the Dominant Cause of AGE in NSW, Australia

In Australia, the first NoV foodborne-associated outbreak was reported in 1978 and was linked to oyster consumption [37]. The etiological agent, NoV, was confirmed by visualization of virus particles in 39% of fecal specimens by EM and in 75% of samples by immune-EM [37].

To determine the role of NoV as an etiological agent of AGE, a total of 68,154 viral diagnostic tests were analyzed between January 2004 and December 2010. The tests were performed by the South East Area Health Service (SEALS) to detect NoV, RoV, adenovirus (AdV), and astrovirus (AsV) in stool from cases of gastroenteritis. Of these, 7% (n = 4780) of samples tested positive for one of these enteric viruses. Overall, NoV was the predominant virus identified in all positive assays (61.7%), followed by RoV (24.7%), AdV (9.5%), and AsV (4.1%). By comparing their monthly totals (Figure 6.1), a shift in predominance from RoV to NoV was observed in 2007 compared with 2006, which coincided with the introduction of a free rotavirus vaccine for all newborn infants across NSW in May of that year. As a result, NoV, previously the second most prevalent etiological agent of AGE, has now become the most common in NSW. For this and other reasons, NoV has also become the leading etiological agent of AGE in developed countries [38].

NoV CLASSIFICATION AND EPIDEMIOLOGY

NoV Classification and Genetic Diversity

NoV is one of six genera within the *Caliciviridae* family, along with *Lagovirus* (host species: rabbit), *Vesivirus* (host species: feline, canine, primate, porcine, and marine species), *Sapovirus* (SaV; host species: humans, porcine, and mink), *Recovirus* (host species: primates), and *Nebovirus* (host species: bovine). Only NoV and SaV are known to infect humans, with NoV significantly more prevalent than SaV [39].

Norovirus is a genetically and antigenically diverse genus of the *Caliciviridae* family. The virus can be classified further into seven genogroups (GI—GVII) based on the open reading frame (ORF2) amino acid (aa) sequence (Figure 6.1) [40–42]. GI, GII, and GIV contain human viruses, with GII being by far the most prevalent. GI viruses infect only humans, GII NoVs infect pigs as well as humans, while GIII viruses infect ovine and bovine species. GIV NoVs have a broader host range than other genogroups and infect humans, lions, and dogs, while GV viruses infect mice [43–45]. It should be noted that GV NoV represents the best surrogate and model system to date for human NoV, as it can be cultured *in vivo* and *in vitro* [42]. The last two genogroups, VI and VII, were identified in canine species [41,42].

Analysis of full-length VP1 aa sequences reveals that the divergence between different genogroups is approximately 45%–60%, while about 15% aa difference in the capsid distinguishes the various genotypes [40]. Each genogroup can be further classified into genotypes. Over 38 genotypes have now been reported, with the majority belonging to GII [46]. Genogroup II, genotype 4 (termed GII.4) is by far the most prevalent and clinically significant NoV genotype; it causes pandemics of gastroenteritis about every 3 years and is responsible for approximately 70% of all NoV-associated cases of gastroenteritis worldwide [47,48]. The term *variant* is assigned to phylogenetic clusters within the pandemic GII.4 lineage [49].

Molecular Epidemiology of NoV

The understanding of NoV molecular epidemiology has progressed significantly following the deduction of the genome sequence by Jiang et al., in 1993 [29]. This led to the development of advanced and sensitive detection techniques including RT-PCR and EIA-based diagnostic assays. It was not until the late 1990s that NoV was recognized as a leading cause of epidemic gastroenteritis [50]. NoV is now the predominant etiological agent of gastroenteritis worldwide; it is present throughout the year and is most prevalent during the autumn and winter months in temperate climates [47,51,52].

Pandemic Noroviruses of the Genogroup II, Genotype 4 Lineage

Over the past two decades, a new NoV variant has emerged every 2–3 years from within the genogroup II, genotype 4 genetic lineage of viruses (Figure 6.1) to cause an AGE pandemic. The capsid sequences of the new GII.4 variants demonstrate aa divergence of approximately 5% compared with the outgoing variant [40], and are thought to contribute to antigenic escape from herd immunity. Contemporary GII.4 variants such as New Orleans 2009 and Sydney 2012 also demonstrate intra-genotypic recombination at the ORF1/ORF2 overlap [53], suggesting that antigenic shift could also play a role in the emergence of GII.4 pandemic NoVs. Although the reasons why GII.4 strains evolve to cause pandemics of AGE are not fully elucidated [50], the fact that they do suggests that GII.4 has acquired a higher epidemiological fitness compared with other genotypes [50]. Generally, the NoV GII.4 variants identified are named after the location (city) and year in which they were first identified. For example, the recent pandemic strain GII.4 Sydney 2012 was first identified in Sydney, Australia during March 2012 [54,55].

Over the past two decades, six GII.4 variants have been associated with global pandemics of gastroenteritis (Figure 6.2). The NoV variant termed GII.4 US 1995/96 was identified as the cause of a global increase of NoV outbreaks in the late 1990s in many countries including the United States, Europe, and Australia [56–59]. This variant was the first recognized pandemic NoV and dominated for several years before it was replaced by the GII.4 Farmington Hills variant in 2002 [60]. This new variant was associated with 64% of cruise ship and 45% of land-based outbreaks in the United States [60], and was later identified as a dominant cause of AGE in Europe and Australia in 2002 [59,61]. In 2004, NoV GII.4 Hunter 2004 was identified in Australia [61]. This new variant replaced the Farmington Hills 2002 strain and continued for 2 years to cause major gastroenteritis outbreaks in Europe, Asia, and across the globe [62–66]. In 2006, two GII.4 variants, Yerseke 2006a and Den Haag 2006b, emerged at around the same time and were implicated in widespread global outbreaks of AGE [62,67]. The Yerseke 2006a variant is thought to have evolved from the previous 2004 Hunter virus [47,53]. Yerseke 2006a is classified as an epidemic variant rather than a pandemic variant as it did not cause a recognized pandemic [47]. Den Haag 2006b is thought to be a descendent of the earlier 2002 Farmington Hills virus [47]. This pandemic variant did not dominate NoV-associated gastroenteritis outbreaks of gastroenteritis until 2007, when it was associated with a rise in global NoV activity [68–71]. The GII.4 New Orleans 2009 variant replaced Den Haag 2006b in late 2009 as the pandemic NoV and was responsible for increased global NoV activity and around 80% of all NoV infections between 2010 and 2012 [47,72]. The most recent pandemic strain, Sydney 2012, was identified in March 2012 and later that year emerged to replace New Orleans 2009 [53,54]. The newly emerged GII.4 Sydney 2012 variant caused NoV-related epidemics of AGE in various countries in late 2012 and early 2013 [55], and remains the dominant NoV detected in 2016 (Figure 6.3).

Early variants (US 1995/96, Farmington Hills, Hunter 2004 and Den Haag 2006b) likely evolved from their predecessors through mutations in their capsid, through a process termed *antigenic drift* [73–76]. However, the two most recent GII.4 variants, New Orleans 2009 and Sydney 2012, underwent a more complex evolution process via intra-genotype recombination at the ORF1/ORF2 overlap region, as well as through mutations within the P2 domain [53].

Epidemic Noroviruses

Other than pandemic strains, there are several GII.4 epidemic variants that have circulated in different continents. These variants did not cause global pandemics but rather more localized epidemics within particular geographical regions. These additional epidemic GII.4 variants include Henry 2001, Japan 2001, Asia 2003, the aforementioned Yerseke 2006a, Apeldoorn 2007, and Osaka 2007 [47]. Besides these viruses, there are also non-GII.4 strains circulating across the world

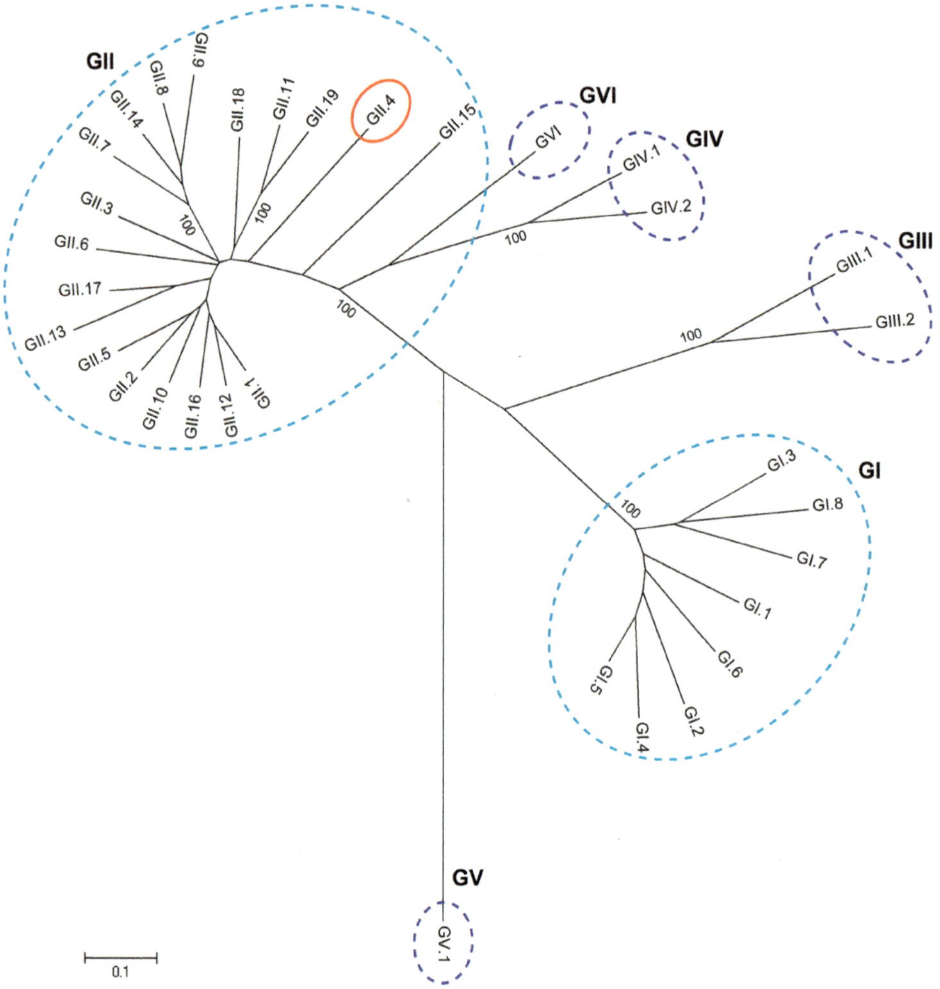

FIGURE 6.2 Phylogenetic analysis of NoV full-length amino acid VP1 sequences (approximately 540 aa). *Norovirus*, a genus within the *Caliciviridae* family, can be further classified into seven genogroups and numerous genotypes (genogroup VII is not shown). Genogroups I, II, and IV contain human viruses, with genogroup II, genotype 4 (GII.4, red circle) being the most predominant, causing more than 70% of NoV-associated human cases. GII also infects pigs (specifically, GII.11 and GII.18) while GIV.2 has been identified in lions and cats. GIII infects cows, GV infects mice, and the recently discovered GVI was identified from a canine *spp.* The viral protein (VP1) amino acid sequences used here are from representative viruses of each genotype retrieved from GenBank. The multiple alignment was performed with Clustal-W, and the neighbor-joining tree was constructed with 500 bootstraps using the Poisson correction method in MEGA 5.2. The scale bar represents the number of amino acid substitutions per site. (Adapted and modified from Glass, R.I. et al. *N Engl J Med,* 361, 1776–1785, 2009 [1].)

that have caused significant epidemics, including GII.3, GII.6, GII.7, and GII.13. For example, the recombinant GII.Pb/GII.3 NoV is often identified in sporadic infections, particularly among children [77,78]. As widespread NoV epidemics continue to occur globally, continuous epidemiological surveillance is essential to characterize circulating NoV strains and the emergence of epidemic strains. Knowledge of the changing genetic patterns of NoV will also be essential for effective vaccine design.

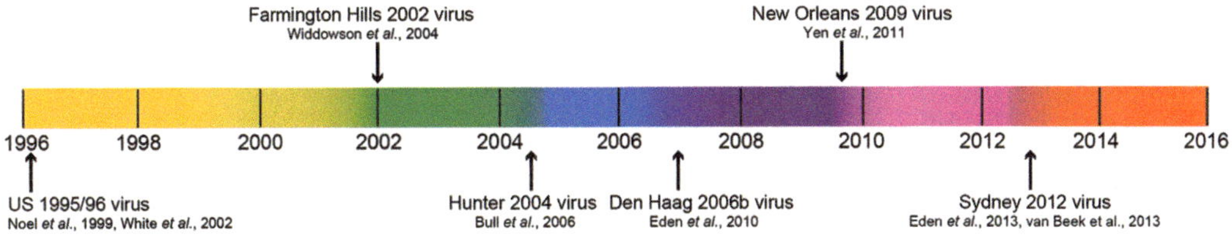

FIGURE 6.3 Time line of NoV pandemic GII.4 variants in Australia. Six pandemic NoV GII.4 variants have been identified between 1996 and 2016. The molecular epidemiology of NoV in Australia mirrors that of many countries globally. The U.S. 1995/96 variant was the first to be identified in Australia in the late 1990s (from Noel, J.S. et al. *J Infect Dis,* 179, 1334–1344, 1999; White, P.A. et al. *J Med Virol,* 68, 113–118, 2002) and was later displaced by another GII.4 variant, Farmington Hills around 2002 (from Widdowson, M.A. et al. *J Infect Dis,* 190, 27–36, 2004). The emergence of novel pandemic NoV variants has subsequently occurred every 2–3 years. The Hunter 2004 (from Bull, R.A. et al. *J Clin Microbiol,* 44, 327–333, 2006) and Den Haag 2006b variants (from Tu, E.T. et al. *Clin Infect Dis,* 46, 413–420, 2008) were detected in Australia in 2004 and 2007, respectively (from Eden, J.S. et al. *J Clin Virol,* 49, 265–271, 2010). New Orleans 2009 was detected in late 2009 (from Yen, C. et al. *Clin Infect Dis,* 53, 568–571, 2011) and replaced the Den Haag 2006b variant (from Vega, E. et al. *J Clin Microbiol,* 52, 147–155, 2014). The current circulating pandemic GII.4, the Sydney 2012 variant that rapidly replaced New Orleans 2009, was first identified in Australia in March 2012 (from Eden, J.S. et al. *J Virol,* 87, 6270–6282, 2013; van Beek, J. et al. *Euro Surveill,* 18, 8–9, 2013). The black arrows depict the time that the NoV variant was initially identified.

NOROVIRUS STRUCTURE AND GENOME ORGANIZATION

NoV has a small, round, nonenveloped icosahedral virion measuring 27–35 nm in diameter, enclosing a single-stranded, positive-sense polyadenylated RNA genome of approximately 7.5 kb [79] (Figure 6.4). The NoV genome contains three ORFs, ORF1–ORF3. ORF1 is approximately 5.5 kb and encodes a 200 kDa polyprotein that is post-translationally cleaved by a 3C-like protease ($3CL^{pro}$) into six or seven nonstructural (NS) proteins [80]. These proteins include N-terminal protein, NTPase, 3A-like, VPg, protease, and a viral RNA-dependent RNA polymerase (RdRp) (Table 6.1, Figure 6.4). The ORF1 proteins have an adapted nomenclature of NS1-2, NS3, NS4, NS5, NS6, and NS7 (Table 6.1). ORF2 and ORF3 encode structural proteins VP1 and VP2, respectively [29] (Figure 6.4).

Another characteristic of the calicivirus genome, including that of NoV, is the highly conserved 5′ terminal sequence (approximately 17–20 nucleotides long; the black box in Figure 6.4), which has over 95% homology across each NoV genogroup. This conserved region contains a stem loop and is repeated internally at the overlapping 3′ end of ORF1 as well as the 5′ end of the ORF2. From the internal position, the ORF2/3 region is transcribed as a subgenomic RNA to generate VP1 and VP2 [81,82]. Therefore, it is likely that these conserved regions play an important role in recruiting the viral RdRp for transcription of both genomic and subgenomic RNA [83].

NoV Nonstructural Proteins Involved in Genome Replication

Our understanding of individual NoV NS proteins encoded by ORF1 varies. The exact functions of NoV NS1–NS4 are not well understood, while NS5–NS7 are relatively well studied. Hyde and Mackenzie have characterized the subcellular localization of MNV ORF1 proteins in the host cell and shown that NS1-2 protein was associated with the endoplasmic reticulum (Table 6.1) [84]. This suggests that NS1-2 may recruit host membranes to the MNV replication complex (RC) during the replication process. Another study by Baker et al. showed that NS1-2 lacks significant sequence identity to other viral or cellular proteins [86]. The protein has an inherent disorder that has been suggested to provide structural flexibility to bind to numerous targets [86], allowing the NS1-2 protein to have a multifunctional role.

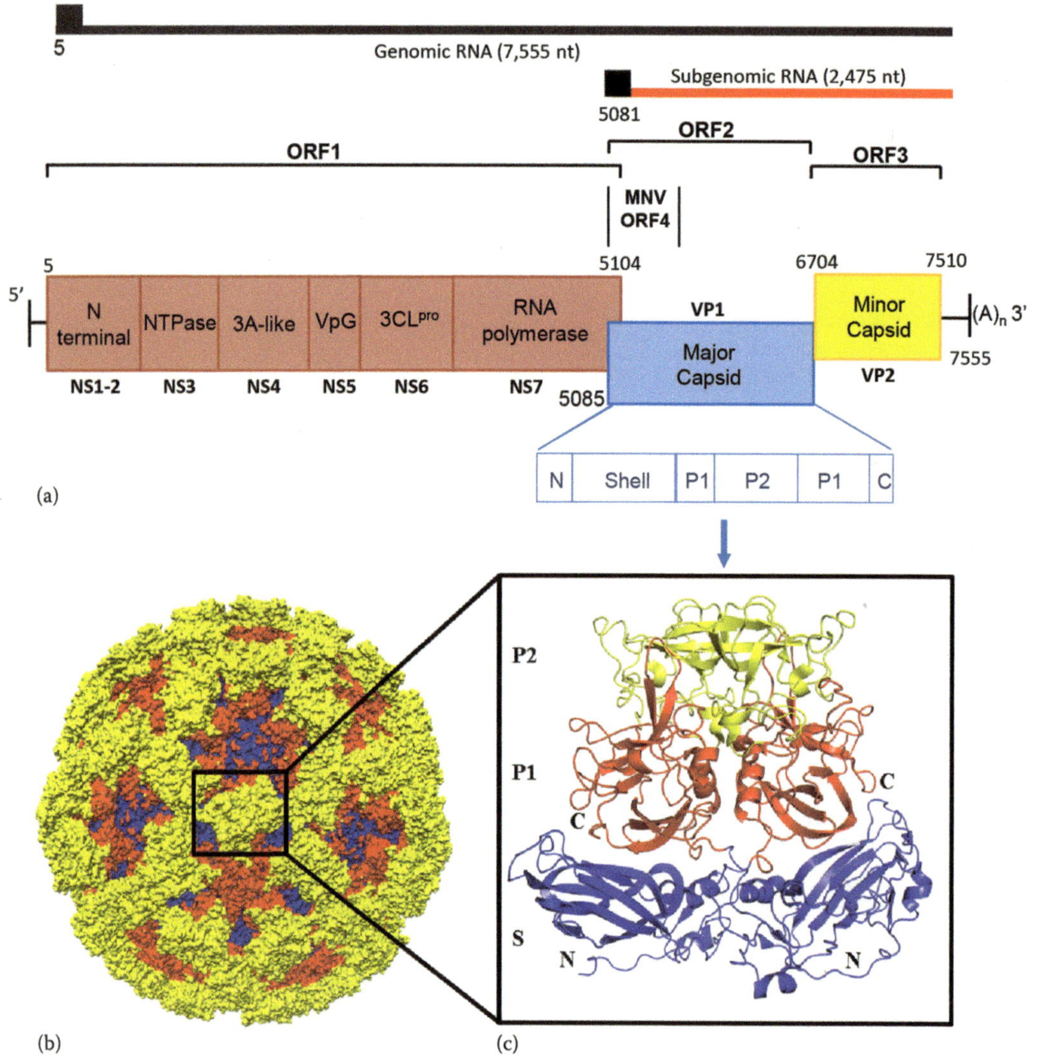

FIGURE 6.4 NoV genome organization and capsid structure of NoV. (a) The NoV genome is organized into three ORFs (ORF1–ORF3). ORF1 encodes a large polyprotein which undergoes proteolytic cleavage by the virus-encoded protease (NS6 or 3CL^{pro}) to produce mature nonstructural viral proteins (NS1–7). ORF2 partially overlaps ORF1 and encodes the major capsid protein (VP1). A minor capsid protein, VP2, is encoded by ORF3. A novel ORF4 was identified in the MNV genome and is represented by vertical dotted lines. The VPg protein is covalently attached to the 5′ end of genomic and subgenomic RNAs, the initiation sequence of which is represented here by a black square box. Nucleotide positions are with reference to the sequence of the NoV GII.4 Lordsdale virus (GenBank accession number X86557). (b) The T = 3 capsid structure of Norwalk virus is generated from 90 dimers of VP1 subunits (180 VP1 molecules). The shell (S), P1, and P2 domains of VP1 are shown in blue, red, and yellow, respectively [85]. (c) A ribbon representation of the Norwalk virus (pdb: 1IHM) VP1 dimer was modeled from the viral capsid with S, P1, and P2 domains colored as in (b). The N-terminal (N) and shell domains (S) form the interior shell of the viral particle. P1 and P2 domains are exposed by protruding out from the viral capsid structure. (Picture modified and adapted from Choi, J.M. et al. *Proc Natl Acad Sci USA,* 105, 9175–9180, 2008.)

TABLE 6.1
Description of ORF1 Encoded Nonstructural Proteins (NS1–7)

ORF1 Encoding Region	Coding Protein[a]	Proposed Function	Cellular Localization[b]
NS1-2	N-terminal (p48)	Replication complex formation	Endoplasmic reticulum
NS3	NTPase (2C-like)	Nucleotide triphosphate activity; RNA helicase	Discrete foci in cytoplasm
NS4	p22 or p20 (3A-like)	Replication complex formation	Golgi; endosome
NS5	VPg	Recruitment of translation machinery; protein primer for replication	Unknown
NS6	Protease (3C-like)	Cleavage of viral polyprotein; inhibits the translation of host proteins	Cytoplasm; mitochondria
NS7	RNA Polymerase	Viral genome replication and transcription	Cytoplasm; nucleus

[a] Alternative names for the nonstructural proteins are stated in parentheses.
[b] Based on a study by Hyde, J.L. and Mackenzie, J.M. *Virology*, 406, 138–148, 2010 [84].

The NoV NS3 protein, also known as p41 or 2C-like protein, is an NTPase protein. The study by Hyde and Mackenzie showed that NS3 localized to discrete foci within the cytoplasm [84]. Another study using the GI.2 Southampton virus, revealed that purified NS3 protein exhibited nucleoside triphosphate (NTP)-binding and NTP hydrolysis activities [87].

The NS4 protein (designated p22 for NoV GI or p20 for NoV GII) is equivalent to the 3A protein of picornaviruses [88], a protein known to interfere with host cell secretions [89]. This function has also been proposed for the Norwalk virus NS4 protein, as it inhibits cellular protein secretion by acting as an antagonist of endoplasmic reticulum or Golgi trafficking [90]. Hyde and Mackenzie proposed that NS4 is also involved in the recruitment of host membranes during the replication process [84]. However, the function of NoV NS4 remains largely unstudied.

Similar to feline calicivirus (FCV), the uncapped NoV genome does not form an internal ribosome entry site (IRES). Instead the 15 kDa viral protein, known as VPg (NS5), is covalently bound to genomic and subgenomic RNA (Figure 6.1), (reviewed in [91]). VPg is thought to initiate the transcription and translation of the positive-sense NoV RNA [79]. The translation process is accomplished via the interaction of VPg with host proteins eIF3 and eIF4, which recruit the host cellular machinery [92,93]. Therefore, VPg plays an important role in initiating NoV replication.

The 19 kDa NoV protease (NS6 or $3CL^{pro}$) [94,95] is similar to the well-characterized picornavirus 3C protease [96]. The complete nucleotide sequence of the GI.4 Chiba virus genome and functional expression of the 3C-like protease ($3CL^{pro}$) in *Escherichia coli* has been studied [94]. The NS6 or viral protease plays a role in cleaving the large 200 kDa polyprotein, encoded by ORF1, into individual NS proteins. It has also been suggested that the protease cleaves the host-encoded polyA-binding protein to inhibit cellular translation [97], an important viral mechanism to modulate host cell expression during replication.

The C-terminal NS7 or the NoV RdRp is involved in viral replication and transcription. Several studies have purified recombinant RdRp using *E. coli* expression systems [73,80,98,99]. These studies have elucidated many of the biochemical properties of the NoV polymerase using *in vitro* assays. Studies have also shown that the recombinant RdRp can be used to initiate both *de novo* and primed RNA synthesis [73,99]. Similar to other RNA viruses, the NoV RdRp lacks a proof-reading mechanism to correct any mutations introduced during replication [100,101] and, as a result, the RdRp has a high mutation rate [73].

NoV Structural Proteins

ORF2 encodes a 60 kDa major capsid protein, VP1, which consists of a shell (S) domain and two protruding domains, P1 and P2 (reviewed in [102] and shown in Figure 6.4). The S domain is linked to the P domain by a flexible hinge [103]. The capsid comprises 180 VP1 molecules that are organized into 90 dimers, with T = 3 icosahedral symmetry [103]. The S domain comprises the first 225 residues of the capsid protein and is assembled to form the shell surface of VP1 (Figure 6.4). The P domain extends from the surface domain, with the P1 domain enhancing the stability of the viral particle. The P2 domain is the most exposed domain on the surface, protruding out from the viral particle (Figure 6.4) [106]. This hypervariable P2 domain is involved in the interaction with histo-blood group antigens (HBGAs), which are viral attachment factors and are thought to help bind the virus to permissive cells [104–108].

The third ORF, ORF3, encodes a 25 kDa minor structural protein, VP2, which has a poorly defined function [29], although one study has proposed that VP2 enhances VP1 production and stability [109]. The first complete genetic map of a calicivirus to be described was the rabbit hemorrhagic disease virus (RHDV); VP2 was found to be present in one to two copies per virion [110]. This minor structural protein was later described in NoV and other caliciviruses [111,112].

In GV murine NoV (MNV) genomes, an additional ORF (ORF4) has been identified, which is encoded from a different reading frame within ORF2 (Figure 6.4) [113]. This additional ORF encodes a protein, termed *virulence factor 1* (VF1) [113]. VF1 was shown to localize to mitochondria and, the authors suggested, in that location it acts as an antagonist of the host antiviral response by stimulating apoptosis [113]. In summary, VP2 is not essential for the formation of the viral particle, but it may be critical for the specificity and conformation of VP1 [111].

REPLICATION AND LIFE CYCLE OF NoV

NoV Binding and Cell Entry

While the cellular receptor for NoV entry is unknown, many studies show that HBGAs are able to bind the NoV capsid and therefore could be involved in viral entry [104,108,114–116]. HBGAs are complex carbohydrates (CHO) expressed on the cellular surface of mucosal epithelial and red blood cells. They are also present in some secretions, such as saliva and breast milk [117]. The role of HBGAs in NoV cell entry and host susceptibility is generally not well understood. Heparan sulfate (HS) is also considered an attachment factor for NoV [106], although studies on this long, polyanionic carbohydrate are significantly less extensive than those on HBGAs. HS is constitutively expressed on the cell surface as well as in extracellular matrices [118,119]. Tamura and colleagues [120] demonstrated that HS is required for the binding of NoV, with GII viruses showing more than 10 times the affinity for HS than GI viruses. In general, cell surface carbohydrates, including HBGAs and HS, are considered initial cell attachment sites but are not necessarily the cell receptor for NoV entry.

The processes following internalization of human NoV are not well understood. As human NoV studies have shown limited success in the establishment of a permissive cell culture system, MNV is commonly used as a cell culture model to study NoV replication [9]. The mechanism of cell entry by the surrogate MNV is also not fully elucidated. However, studies with MNV suggest that cholesterol and a dynamin-dependent pathway are required for the endocytosis of MNV into host cells to establish infection [121–123]. It is likely that additional molecules, such as membrane proteins, may also serve as receptors or co-receptors to facilitate cell penetration of the NoV (reviewed in [124]). In view of this, cell receptor attachment or the entry process may be contributing factors that limit the success of permissive cell culture of human NoVs.

Translation of Viral Proteins

Upon cell entry, the viral genome is released into the cytoplasm and viral proteins are initially translated directly from the positive-sense RNA genome [125]. Small, single-stranded, positive-sense RNA (+ssRNA) viruses have limited coding capacity in their genomes so they rely heavily on host cell nucleic acid binding proteins to facilitate genome translation and replication (reviewed in [126]). The viral RNA is translated into viral proteins by the host cellular translational apparatus (reviewed in [127]) through the recruitment of eukaryotic translation initiation factors to the virus-encoded VPg (NS5) protein. VPg is covalently linked to the 5′ end of the viral genome, and the interaction between VPg and the host translation factors eIF3 and/or eIF4E has been demonstrated to initiate translation [92,93]. Meanwhile, the translation of subgenomic RNA enables the expression of the viral capsid protein. Synthesized NoV genomic RNA is packaged into the viral capsid to generate infectious viral particles. The lifecycle of NoV is depicted in Figure 6.5.

Like other +ssRNA viruses, MNV replication is associated with intracellular host membranes. Indeed, the N-terminal (p48) protein encoded by NS1/2 displays a vesicular localization pattern in transfected cells [128]. Studies with MNV suggest that viral replication occurs in the perinuclear region in association with membranes derived from the endoplasmic reticulum, trans-Golgi apparatus, and endosomes [84]. During the replication process, the MNV replication complex localizes to the virus-induced double membrane vesicle clusters that form in the cytoplasm. These clusters originate from membranes derived from the secretory pathway [129]. MNV also utilizes microtubules to position the replication complex adjacent to the microtubule organizing center, as shown by impaired production of viral proteins and infectious virus following chemical disruption of the microtubule network [130]. Clearly, the MNV model system has provided useful insights into host–viral interactions and generated questions pertaining to viral replication and the role of host immunity in controlling viral infection.

NOROVIRUS EVOLUTION

Antigenic drift and shift are thought to drive the evolution of NoV [50]. Antigenic drift occurs when the virus accumulates mutations within genes encoding for antigenic properties such as epitope or putative epitope binding sites. Capsid aa variation within the P2 domain of NoV GII.4 variants enables the virus to evade herd immunity to cause pandemics [132]. Antigenic shift or recombination is another major mechanism of NoV evolution, and is similar to reassortment in influenza [133]. Recombination plays an important role in the genetic diversification of NoV and thus increases selective advantages.

Antigenic Variation NoV Evolution

Antigenic variation is one of the factors attributed to the emergence of novel NoV strains. This process facilitates continuous NoV infections in the population. Much of the NoV GII.4 aa variation observed is localized to five evolving blockade epitopes (A–E), which are potential sites of virus neutralization; hence antibodies directed to these sites confer protection against clinical disease [74,134,135]. There are also motifs within the hypervariable P2 domain of VP1 that are involved in host cell binding [136,137]. Epitopes A, D, and E are the three most important antigenic motifs of the P domain in GII.4 variants. Amino acid variation occurring within these epitopes enables the virus to evade the hosts' neutralizing antibodies. This suggests that the P2 domain is under the greatest selective pressure [76,138]. This selection can lead to the emergence of new GII.4 pandemic variants [132,138,139]. Antigenically novel strains are capable of causing epidemics of gastroenteritis across the world because they have the ability to escape from herd immunity [76,140]. Several studies have demonstrated that the evolution of GII.4 viruses has generated novel antigenic variation in the P2 domain [69,106,135,140,141]. These changes have led to differential receptor binding

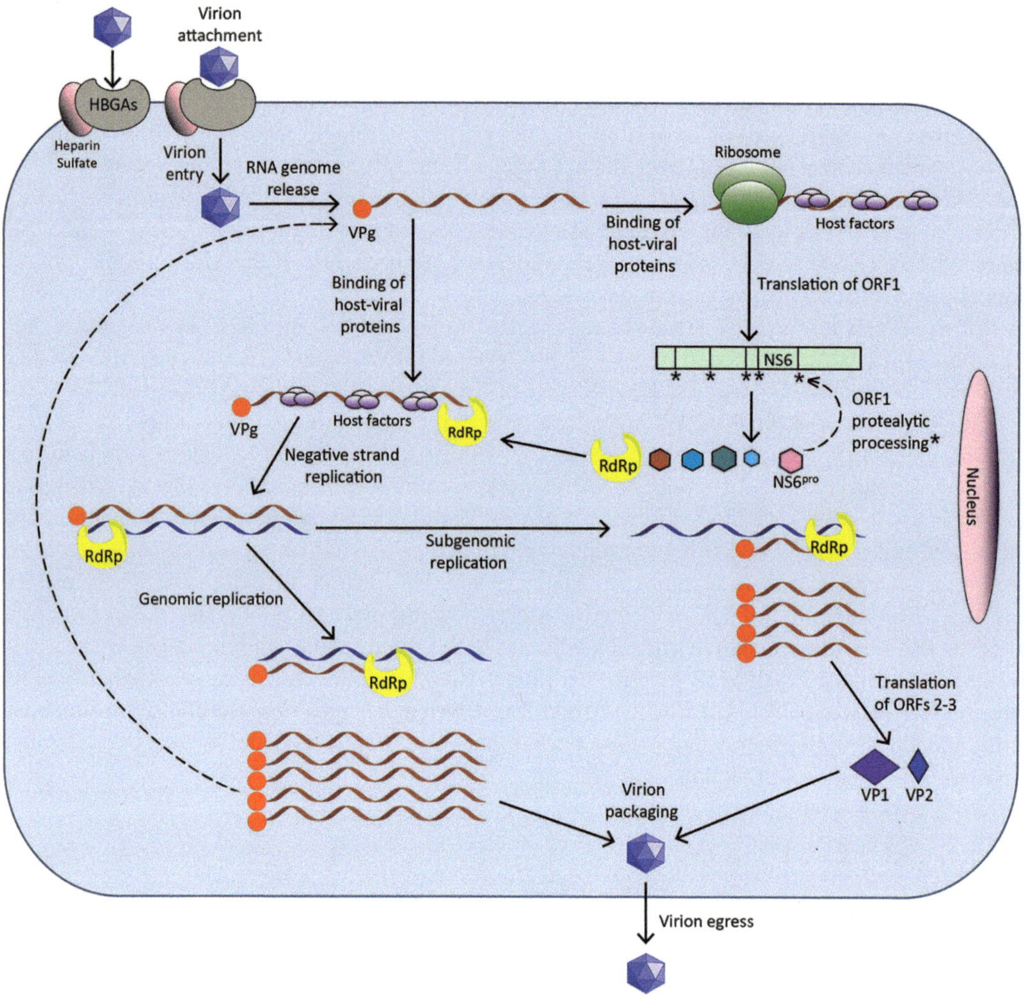

FIGURE 6.5 The NoV replication cycle. The NoV infectious particle attaches to histo-blood group antigens (HBGAs) and possibly to heparan sulfate (HS) expressed on the host cell surface. Attachment is followed by internalization through an unknown receptor to release the viral genome into the cell cytoplasm. VPg is covalently attached to the 5′ end of the viral genome and is used to recruit eIF3 and eIF4. This is required to initiate protein translation and expression of open reading frame 1 (ORF1). The ORF1-encoded polyprotein is cleaved by the viral protease, NS6 ($NS6^{pro}$) and its precursors, into mature nonstructural proteins. The viral RNA-dependent RNA polymerase NS7 ($NS7^{pol}$) synthesizes viral RNA to generate negative-sense RNA, followed by genomic replication. VPg acts as a primer that is constantly attached to genomic and subgenomic positive-sense RNA to facilitate the replication process. Subgenomic replication involves the generation of a shorter RNA genome that is later translated into major (VP1) and minor (VP2) capsid proteins. Newly synthesized genomes are encapsidated and assembled into mature particles, then released into the extracellular environment to infect other cells. (Image modified and adapted from Arias, A. et al. *Future Microbiol,* 8, 1475–1487, 2013.) [131].

patterns for different GII.4 strains. The evolution of GII.4 genotypes appears to be epochal, with a long stasis in population fitness followed by a sudden burst of evolution [69].

Of the six GII.4 pandemic strains identified over two decades ago, the four earliest variants, US95/96, Farmington Hills 2002, Hunter 2004, and Den Haag 2006b, likely evolved from their GII.4 predecessors through mutations within the hypervariable P2 domain. These mutations have been observed particularly within potential neutralizing epitope sites A, D, and E [50,76,134,135]. Interestingly, the more recent pandemic variants, New Orleans 2009 and Sydney 2012, have also demonstrated GII.4 intra-genotype recombination (antigenic shift) at the ORF1/ORF2 overlap as well as antigenic variation within the P2 domain [53], and have therefore evolved through two separate processes.

RNA Recombination

Recombination is an important process that leads to the emergence of novel NoV strains. This can be seen with the common identification of recombinant strains in molecular epidemiological studies [53,68,142–144]. Recombination provides evolutionary advantages for viral survival. It helps to generate and select advantageous traits and may also remove deleterious or detrimental genes from a viral population [145]. With NoV, it is proposed that homologous recombination involves parental RNAs that cross over at homologous sites [146]. This mechanism has also supported a copy-choice model of recombination, where the NoV RdRp complex switches from one RNA template to another during replication. It then continues to extend the new RNA strand precisely at the position where it left the previous RNA strand [145,147].

Researchers initially assumed that genotyping based on a single region (usually either the RdRp or capsid region) was sufficient to define the genotype of the entire genome [148–152]. This assumption was perceived on the basis of the nonsegmented nature of NoV genomes. However, more studies have recognized that recombination commonly occurs, particularly at the ORF1 and ORF2 overlapping region in NoV [53,143,153]. As a result of the recognition of recombinant NoVs, the current NoV nomenclature system consists of both RdRp and capsid genotypes (see 'Nomenclature for NoV Recombinants' section) [49].

NoV recombination most commonly occurs at the ORF1/ORF2 overlapping region, which is the most prevalent recombination breakpoint (Figure 6.6). As a result, the exchange of the ORF1

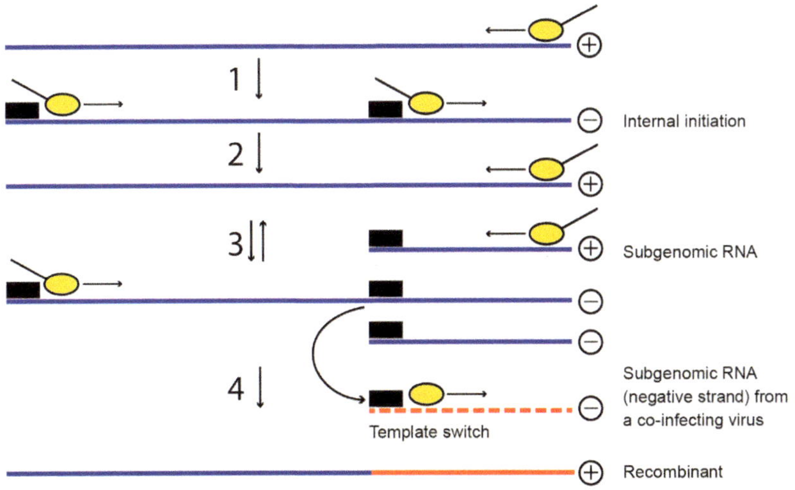

FIGURE 6.6 RNA recombination during the NoV transcription process. (1) A negative intermediate RNA strand (blue line with minus sign) is generated during the transcription process by the RdRp (yellow oval). (2) The RdRp binds to the nearly identical RNA sequences (putative RdRp promoter, with a short stem loop) at the 5′ end of the genome and at the ORF1/2 overlap (internal initiation) (filled black boxes) to generate positive stranded genomic RNA and shorter subgenomic RNA (straight blue lines). (3) These templates allow RNA synthesis from the 3′ end that leads to further generation of both negative-sense genomic RNA and negative-sense subgenomic RNA. (4) Recombination occurs when the RdRp initiates positive strand synthesis at the 3' end of the full-length negative strand, halts at the subgenomic RNA stem loop (black box), and then switches template to an available negative-sense subgenomic RNA species generated by a coinfecting virus (dashed red line). The net result is a recombinant virus that has acquired new ORF2 and ORF3 sequences (solid blue/red hybrid). (Image modified and adapted from Eden, J.S. et al., The evolutionary dynamics of norovirus, UNSW Australia, 2012.) [36]

fragment and the ORF2/ORF3 structural protein coding genes occurs between different NoV lineages that coinfect the same cell [143,153]. Apart from ORF1/ORF2 recombination, a secondary recombination breakpoint has been identified within the ORF2/ORF3 junction in GII.4 strains [53]. Although uncommon, some studies also found recombination breakpoints occurring within ORF2 [53,154,155]. The mechanism for recombination during NoV transcription is shown in Figure 6.6.

Recombination usually occurs between different genotypes (inter-genotype), particularly in the contemporary non-GII.4 NoVs. Some common recombinant NoVs include GII.Pb/GII.3, GII.P16/GII.13, GII.P6/GII.7, and GII.Pg/GII.12 [154,155]. The NoV GII.Pb/PII.3 strain has been one of the most prevalent recombinant strains since its identification in 2002 [78,156]. Unlike the NoV GII.4 strains, where a novel strain has emerged to displace its predecessor every 2–3 years, GII.Pb/GII.3 viruses continue to cause childhood infections across the globe [77,78]. The GII.b ORF1 lineage has also recombined naturally with at least three other capsid genotypes, including GII.1, GII.2, and GII.4 [157].

Nomenclature of NoV Recombinants

Nomenclature of recombinant NoV strains has recently been updated and now indicates the genotype of the RdRp followed by the capsid genotype [49]. For example, GII.P4/GII.10 indicates that the NoV strain contains GII.4-typed polymerase and GII.10-typed capsid. The Dutch National Institute for Public Health and the Environment (RIVM) has developed an automated NoV genotyping tool (http://www.rivm.nl/mpf/norovirus/typingtool) [46]. This web-based open-access tool is easy to use and provides a direct NoV typing system based on sequence analysis. It employs percentage pairwise identity in reference to standard strains and phylogenetic analysis to determine the genotype of NoV. The automated genotyping tool provides a platform that facilitates standardized genotyping and nomenclature. This is useful in supporting inter-laboratory comparisons that help to minimize any confusion or misinterpretation of NoV nomenclature in research publications.

NOROVIRUS CLINICAL DISEASE

NoV is the major cause of sporadic gastroenteritis cases and outbreaks worldwide. It is estimated that this virus causes more than one million hospitalizations and 220,000 deaths annually, mainly affecting young children in developing countries [2]. Ahmed et al. demonstrated that NoV is the most prevalent enteric pathogen globally, causing mild and severe gastroenteritis across all age groups [158].

Symptoms of NoV infection include projectile vomiting and loose or watery diarrhea, with some patients experiencing nausea, abdominal cramps, chills, low-grade fever, and/or malaise [11]. The viral incubation period is 24–48 h, and symptoms usually persist for 2–4 days in healthy adults [12]. Beyond the symptomatic phase, patients generally continue to shed the infectious viral particles for several weeks [159,160]. NoV also causes persistent infections in immunosuppressed patients, including immunodeficient and transplant patients [13–15]. Chronic infection can cause prolonged viral shedding for months to years [13,161,162]. These chronically shedding patients are potential reservoirs of antigenic variants as these viruses have been shown to evolve rapidly, resulting in changing antigenic profiles over time [13,162]. Young children and older adults are also very susceptible to NoV infections, and when severe dehydration occurs, this can sometimes result in fatalities [160,161,163]. Therefore, NoV infections can be severe in these vulnerable groups.

Pathogenesis

Human volunteer studies have played an important role in understanding NoV pathogenesis. Human volunteers who were challenged with GI.1 Norwalk virus developed gastroenteritis and histopathological lesions were seen in jejunum biopsies [164,165]. Broadening and blunting of the villi of the proximal small intestine was observed under light microscopy [164]. Also, pale and enlarged mitochondria, increased cytoplasmic vacuolization, and intercellular edema were seen in those acutely infected volunteers (reviewed in [166]).

Recent studies demonstrated that NoV is associated with debilitating illnesses other than gastroenteritis. These illnesses occur among vulnerable groups; for instance, NoV infections were implicated in cases of infantile seizures, a convulsive disorder [167], exacerbation of inflammatory bowel disease (IBD), including ulcerative colitis and Crohn's disease [168], as well as necrotizing enterocolitis outbreaks in children [169]. While reports of these clinical conditions associated with NoV infections are uncommon, NoV may manifest with varied clinical symptoms and should be considered as a potential etiological agent of disease other than gastroenteritis alone.

IMMUNITY

Little is known about immunity against human NoV. Volunteer challenge studies have demonstrated short-term antibody immunity against NoV [22,170,171]. These studies reported that most individuals were resistant to infection from the same virus strain if re-exposed in less than 6 months, while long-term immunity was not demonstrated. A long-term immunity study by Johnson et al. showed that individuals who were previously challenged were susceptible to infection 6–12 months post-NoV GI.1 exposure [172]. These studies suggested that NoV exposure confers short-term immunity but that the establishment of persistent, long-term immunity may be difficult to maintain, and that more studies are needed.

In contrast to these results, Parrino et al. demonstrated some protective immunity after NoV rechallenge [171]. The authors challenged and rechallenged 12 human volunteers with GI.1 Norwalk virus at an interval of 27–42 months [171]. Six out of the 12 subjects developed symptoms after the initial challenge and also became ill again upon rechallenge. Four of these volunteers underwent a third rechallenge after 4–8 weeks, and only one became ill. In contrast, the other six volunteers remained asymptomatic upon rechallenge. This study not only demonstrated the presence of a short-term immunity, but also suggested that some individuals were intrinsically resistant to Norwalk virus infection.

Virus–Host Interaction and Innate Immunity

Innate immunity is the first-line host immune defense against viral infections. Viral infection triggers the cellular innate response, which involves the activation of various signaling cascade pathways (reviewed in [173]). The innate response leads to the release of pro-inflammatory cytokines, including type I interferons (IFNs), and is followed by the activation of the adaptive immune response. Innate immunity is critical to control NoV replication. Studies using MNV in mice suggest the IFN and/or STAT-dependent immune responses are responsible for the control of NoV infection [44,174]. The critical role of the host innate immunity may be explained by the rapid onset of disease symptoms (~0.5–1 day) as well as the short course of illness (~3–5 days) that occurs before the induction of adaptive immunity (reviewed in [9]).

Viral infection is sensed by the host innate immune response in cells through the recognition of pathogen-associated molecular patterns (PAMPs), which are unique to each pathogen. For recognition of RNA viruses, two receptor families are commonly activated; these include RIG (retinoic acid–inducible gene)-I like receptors (RLRs) and Toll-like receptors (TLRs) (reviewed in [175]). Induction of RLRs and/or TLRs leads to the activation of a cascade of signaling pathways that result in the activation of various transcription factors, such as interferon regulatory factor 3 (IRF3), interferon regulatory factor 7 (IRF7), and NF-κB [176]. These transcription factors play an important role in mediating the expression of type I IFNs and pro-inflammatory cytokines [177,178]. IFNs and cytokines are important due to their antiviral properties. McCartney et al. demonstrated that

MNV infection of MDA5−/− dendritic cells (DCs) resulted in reduced cytokine production as well as higher titers of MNV *in vitro* and *in vivo* [179]. TLR3−/− mice also showed a slight increase in MNV titer upon infection. This suggests that MDA5 recognition of the virus is critical to control the replication of MNV.

Type I IFNs, including IFNα, IFNβ, play an important role in the innate immune response to viral invasion. They are also critical as key regulators that induce the adaptive immune response in order to clear viral infections [180]. However, viruses have evolved to develop various mechanisms and strategies in order to subvert the host innate immune response. Studies using MNV as a model for human NoV provide strong support for the role of the innate immune response, including both type I (IFN α/β) and type II (IFNγ) interferons in controlling the replication of MNV [44,174,181].

Apoptosis, another cellular response to viral infection, is programmed cell death following the activation of certain signaling pathways [182]. Apoptosis can be identified by several molecular characteristics, such as chromatin condensation from DNA fragmentation [183], cell shrinkage [182], membrane blebbing, and increased caspase activity [184,185]. A number of viruses induce apoptosis, such as human immunodeficiency virus (HIV), human papillomavirus, herpesvirus, adenovirus, and influenza virus. It is proposed that caliciviruses also induce apoptosis to facilitate the dissemination of viral progeny without activating the host innate immune response [186]. Studies have shown that MNV stimulates apoptosis in murine macrophages by down-regulation of survivin and activation of caspase [186,187]. Survivin [188] is a member of an apoptosis-inhibiting protein family, while caspase [189] is a protease that plays a role in apoptosis. Another study proposed that apoptosis is also required for NoV replication, as caspase-3 may be essential for the NoV proteolytic processing during polyprotein cleavage [190]. Therefore, apoptosis may be critical to facilitate the propagation and spread of NoV within-host cells.

HOST SUSCEPTIBILITY FACTORS

Studies have demonstrated the linkage between HBGA expression, antigen secretion, and NoV binding. Furthermore, HBGA genotype and the resulting HBGA phenotype are recognized as genetic factors that affect susceptibility or resistance of the host cell to NoV infection [92,93,191]. It is thought that NoV attaches to HBGAs in order to establish the first step of infection [192,193]. The three major HBGAs are ABO, secretor (H antigen), and Lewis antigens. ABO and Lewis antigens are complex carbohydrates that are commonly found on intestinal epithelial cells. The biosynthesis of these carbohydrates is regulated by the addition of a fucose residue to the precursor structure disaccharide of H type antigen by FUT2 and FUT3 enzymes [194]. The FUT2 enzyme is carried by secretor-positive individuals and the *FUT2* gene displays high levels of polymorphisms. As a result, 20% of the human population lack an active FUT2 enzyme and are therefore devoid of H antigen secretions [195]. These individuals are classified as nonsecretors, while the remaining 80% of the population are termed *secretors*. Host susceptibility to NoV has been associated with *FUT2* gene-encoded secretions of α1,2-linked fucose on HBGAs [196]. The H antigen, found in secretors, can be further modified by various enzymes to generate A and B antigens that form the ABO system [197] as well as the Lewis b (Le^b) and Lewis y (Le^y) antigens [198]. The *FUT3* gene also shows polymorphisms with null alleles and, as a result, 10%–30% of the population are Le^a and/or Le^b antigen deficient [194]. Combinations of polymorphisms between *ABO, FUT2*, and *FUT3* loci generate human subgroups that express distinct epithelial glycosylation patterns [197]. While secretor individuals have been shown to be significantly more susceptible to NoV infection with some strains, it has been demonstrated that other NoV strains will infect any person, regardless of secretor status, Lewis, or ABO phenotype [199].

Distinct NoV genogroups and genotypes have different HBGA binding affinities. Studies have shown that blood group B individuals have a relatively lower incidence of infection with NoV GI compared with other blood groups [200,201]. Another study has shown that individuals with blood group O appear to be more susceptible to NoV infection than the other blood groups [114]. There are also reports that NoV GI.1 Norwalk VLPs have a higher affinity to bind to the gastrointestinal epithelial cells of secretor individuals compared with nonsecretors [200,202–204]. Additional studies with other NoV genogroups indicate that their host receptor susceptibility to infection is more complicated than that of Norwalk GI.1 [104,115,205]. A study with NoV GII.2 (Snow Mountain virus) found that host susceptibility to infection did not correlate with the individual's blood group or secretor status [206]. Studies found that nonsecretor individuals had significantly lower antibody titers in response to NoV infection; hence it was proposed that they are likely to be more resistant to NoV infections compared with secretor individuals, regardless of blood group [200,207,208].

Overall, these observations suggest that NoV may recognize and infect subsets of the human population in different ways, based on these carbohydrate variations. As the susceptibility pattern is complex, the diverse HBGA binding profiles found within GI and GII viruses may affect nearly all individuals, particularly secretors [209], who are more susceptible to NoV infection [106]. This also suggests that HBGA binding affinities may be important in influencing the evolution of human NoV (reviewed in [75]). In summary, the evidence currently suggests that nonsecretor status is protective against NoV infection and reduces disease severity, with asymptomatic infections common in this group of individuals.

TRANSMISSION

NoV is classified as a category B biodefense agent in the United States because it is highly communicable with a small infectious dose (as low as 18 infectious viral particles) [210]. A more recent study, however, demonstrated that the 50% human infectious dose (HID_{50}) was approximately 1320 genomic equivalents, similar to other estimates for RNA viruses [211]. During acute infection in the symptomatic phase, approximately 10^8 viruses/gram of stool are shed [159,160,212]. NoV transmission mainly occurs via person-to-person, but can also occur through exposure to vomitus and fomites, ingestion of contaminated food, and contact with contaminated surfaces [38,213]. Of these transmission routes, person-to-person transmission has been linked to over 90% of health-care-related outbreaks [16,51,214].

In an outbreak, the illness generally starts from an individual infection, followed by rapid person-to-person transmission within the institution/setting. NoV is more resistant to common disinfectants compared with other viruses and is stable in the environment, remaining viable for over 2 weeks on environmental surfaces or for 2 months in water, thus enhancing its transmission properties (reviewed in [213]). Hence, isolation or containment of the disease is challenging and vulnerable groups, including older adults and the immunocompromised, are disproportionately affected with prolonged clinical symptoms [163]. The cycle of transmission, illustrated in Figure 6.7, outlines steps for reducing the risk of NoV infection.

Non-Foodborne Outbreak Settings

NoV has the ability to cause large-scale outbreaks of gastroenteritis. Outbreaks commonly occur in a diverse range of enclosed or semi-enclosed settings where people congregate, including nursing homes, hospitals, schools, restaurants, military camps, and cruise ships [215–217]. In Australia, approximately 50% of NoV-associated outbreaks occur in long-term care facilities [218].

The prevalence of different settings for NoV outbreaks may differ between countries. In England and Wales, between 1992 and 2000, 79% of reported outbreaks occurred in health-care institutions, including hospitals and residential care facilities [51]. A similar predisposition for NoV outbreaks in health-care facilities was documented between 1995 and 2000 in Spain (64%) and in the Netherlands (66%) [219]. In the United States, the majority of outbreaks (60%) between 2009 and 2013 were reported in aged-care facilities, followed by restaurants (10%) [16], with hospital outbreaks less common compared with other countries.

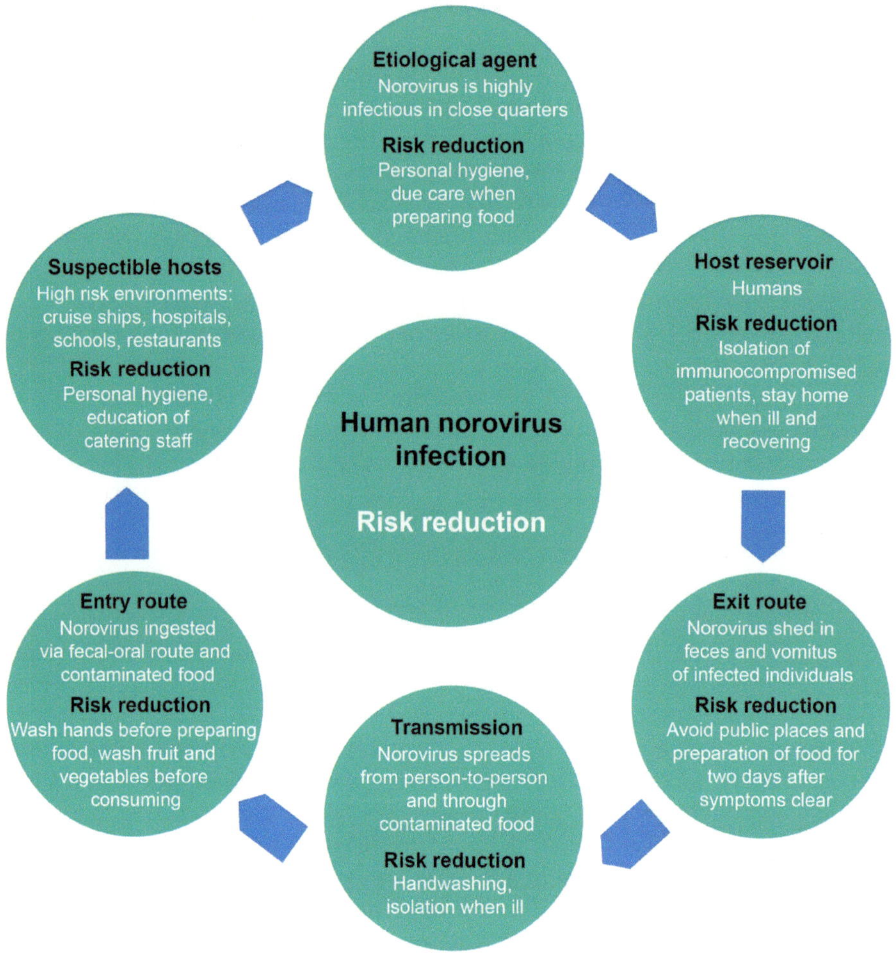

FIGURE 6.7 Risk reduction pathway for human NoV infection. The cycle of NoV transmission is illustrated, with recommended methods for reducing the risk of infection.

In summary, non-foodborne NoV outbreaks commonly occur in enclosed institutional settings, particularly health-care facilities such as nursing homes and hospitals.

Norovirus Transmission from Food and Water

NoV has been identified as one of the most common etiological agents attributed to foodborne outbreaks in the United States, accounting for up to 60% of AGE outbreaks reported [4,16,220–222]. One study in the United States evaluated data collected by the Centers for Disease Control and Prevention (CDC) from 520 suspected or confirmed NoV outbreaks between 2009 and 2012 [223]. The researchers discovered that the most common food-related outbreak settings were restaurants (64%), followed by catering and banquets (17%). They also found that 90% of foodborne NoV-associated outbreaks came from the food preparation setting [223]. In cases where the contamination source was identified, food workers were attributed with 70% of food contamination [223].

There are several potential steps for the transmission of NoV into the food chain before it touches the plate, from irrigation practices, preharvesting processes and washing steps through to food preparation (reviewed in [224,225]). Preharvest contamination can occur from farm effluent run-off onto fresh produce crops, or into riverbeds and marine environments where the virus can accumulate in shellfish farmed for consumption [224]. Contamination during harvesting and processing can be introduced through fruit pickers, contaminated equipment, or process water [224]. Potential post-harvest contamination occurs from washing fresh produce with contaminated water and insufficient hygiene practices in food handling and preparation. There are many potential stages for the introduction of NoV into the food chain (reviewed in [225]). While the main route of NoV transmission is from person-to-person [16], there are limited and often conflicting statistics available for foodborne NoV transmission (reviewed in [3]). This is partially due to the processes required to collect and track such data. These include the patient visiting a clinic, the collection of a stool sample, testing and identification of the pathogen, and reporting to health authorities [3,4]. Even when reported, it is difficult to attribute the illness to a specific event or meal, and outbreak data commonly link the food source to the cases [3]. A U.S.-based study by Scallan and colleagues estimated that 58% of all foodborne illness in the United States was caused by NoV [4]. Other studies have cited varying proportions of foodborne NoV transmission worldwide, ranging between 14% [226] and 46% [227]. While the true proportion of foodborne NoV infections has proved difficult to quantify, there have been many studies demonstrating the transmission of NoV into the food chain, in particular through fresh produce, catered meals, and seafood (reviewed in [3,225]).

A U.S.-based study analyzed over 2900 foodborne-related outbreaks of gastroenteritis reported to the CDC from 2001 to 2008, where NoV was either confirmed or suspected as the etiological agent responsible [227]. Over this period, an average of 365 reported NoV-related outbreaks occurred each year, affecting around 10,000 people [227]. In this study, 53% of outbreaks could be attributed to infected food handlers, although this is likely an underestimation [227]. In the United States, over 2000 foods have been implicated in foodborne illness and, in 2009, Painter et al. proposed a system to categorize food commodities into two classes: simple (e.g. a vegetable) and complex (e.g. a sandwich). The study evaluated cases reported in the United States between 1973 and 2006, with interesting findings that 57% of outbreaks were from simple foods and 43% from complex foods [228].

Another study by Painter et al. analyzed 13,352 reported outbreaks of foodborne-associated illnesses in the United States between 1998 and 2008 [229]. The study revealed that, out of the simple foods (usually single commodities), the majority of NoV cases were attributed to leafy vegetables (36%), followed by 15% to fruit and nuts, 12% to dairy products, 11% to meats (poultry, pork and beef), 7% to eggs, 6% to vines and stalks, 5% from root vegetables, 4% to grains and beans, 2% to mollusks, and 1% to fish, with the remaining proportion made up from oils, sugars, and crustaceans [229]. Interestingly, this study demonstrated that 70% (14/20) of NoV outbreaks that implicated dairy products identified cheese as the contaminated food item, highlighting the likelihood of food handlers introducing NoV into the cheese after pasteurization [229].

Transmission from Seafood

Studies have shown that NoV is environmentally resilient [230] with robust survival rates in water [231]. One study highlighted how NoV can remain detectable in groundwater for 3 years and infectious for at least 61 days [232]. Sewage contamination of water poses a significant health risk, especially when the water comes into contact with shellfish beds through damaged sewerage systems, for example. In this case, the consumption of pathogen-contaminated bivalve mollusks such as oysters and mussels can lead to large outbreaks, as shellfish filter-feed and concentrate microorganisms within their flesh [233].

NoV is recognized as the most common etiological agent responsible for shellfish-related cases of gastroenteritis [234]. Oysters, in particular, are a high-risk food item linked to NoV infection. It has been suggested that NoV can selectively interact with carbohydrate ligands in oysters, further concentrating the virus within the flesh [235].

The first report of a food-related NoV gastroenteritis outbreak in Australia was linked to oyster consumption back in 1978 [37]. More recently, in 2013, an NoV outbreak affecting 306 people in Tasmania was also linked to the consumption of oysters [234]. The subsequent investigation widened to investigate a further 219 cases from Victoria and New South Wales. Out of 10 fecal samples collected from the patients in Tasmania, seven tested positive for NoV GII.1 [234]. These two examples are mirrored in all countries around the world, demonstrating the increasing health risks resulting from human waste reaching waters where shellfish are farmed.

While oysters are often eaten raw, studies have shown that cooking shellfish is not always sufficient to deactivate the virus. NoV is relatively heat resistant and remains infectious even after cooking for up to 30 min at 60°C [236]. Cooking oysters for a long enough period and at a high enough temperature to deactivate the virus decreases the quality of the food and is therefore usually avoided [237]. As shellfish are often lightly steamed and considered cooked once the shell is open [238], the virus can remain infectious when consumed. Somewhat counterintuitively, one study showed that the likelihood of NoV infection was similar whether the oysters were consumed cooked or raw [239].

One study in China detected NoV RNA in 112 of 840 (13%) shellfish samples collected from seven coastal cities between 2009 and 2011 [240]. The shellfish species with the highest proportion of NoV detected were oysters (19%), followed by mussels (17%), clams (16%), ark shells (14%), razor clams (9%), and scallops (6%) [240]. This demonstrates how NoV can accumulate in the flesh of many species of filter-feeding mollusks, not only oysters, and can also enter the food chain if harvested. In another example from the Netherlands, the presence of NoV RNA in 42 local and imported samples of shellfish was determined using RT-PCR [241]. The results revealed that, from within the local Dutch-farmed shellfish, 5% (1/21) of oysters had detectable NoV [241]. Among the samples imported into the Netherlands, 38% (5/13) of mussels and 13% (1/8) of oysters had detectable levels of NoV RNA [241].

In summary, shellfish are at risk of sewage contamination from effluent run-off into marine environments, damaged sewerage infrastructure, and other reasons, which can result in the accumulation of pathogens in their flesh. Shellfish, in particular, have been widely implicated in NoV-related outbreaks.

Norovirus Transmission from Food Handlers

In this section, we use a number of examples of foodborne NoV outbreaks described in the literature to highlight the main recognized transmission modalities that facilitate viral spread. These usually involve contamination of the food at source, during processing, or through an infected food handler.

The transmission of NoV is facilitated by a lack of hygiene when handling food and the consequences can be widespread, particularly when large-scale catering is involved. There have been many documented cases demonstrating how lapses in hygiene by kitchen/catering staff can introduce the virus into several food types, including salads, rolls, sandwiches, and burgers [242]. One American bakery caused several outbreaks, infecting an estimated 2700 out of 7169 individuals over a single weekend. Initially, 332 wedding guests from 46 separate weddings reported gastroenteritis symptoms following the consumption of the bakery's wedding cakes [243]. The source was finally traced to two bakery employees who had been infected with NoV the week before the weddings [243].

In Finland, several outbreaks of gastroenteritis occurred in 2006 across 10 cafeterias that shared a common raw vegetable processing plant [244]. A subsequent patient survey revealed that the attack rate was 55% (150/273) of respondents. Following the outbreak, environmental, water, and food samples were collected in addition to patient fecal samples from three of the workplace canteens. The samples were tested for the presence of enteric pathogens, including NoV. Although raw vegetables were significantly associated with the outbreak, no single contaminated vegetable commodity was identified in the study as a source. NoV GII.1 was detected in all patient samples tested, although none of the water, food, or environmental samples had detectable NoV [244]. This example highlights an important transmission route for enteric pathogens, namely, where contaminated water is used to wash food.

In 2007, three unrelated groups of up to 13 people reported gastroenteritis symptoms to the Dutch Food and Consumer Product Safety Authority around 30 hours after dining at the same restaurant [245]. Seven of the restaurant's staff had recently suffered from gastroenteritis. A total of 31 diners and the aforementioned 7 restaurant staff became ill. Fecal samples from 13 patients and 6 food handling staff were collected and environmental swabs were taken from several surfaces and the hands of 3 food-handling employees. Real-time RT-PCR was used to detect NoV GI.6 RNA in 17 of the 19 fecal samples. NoV was also detected in 4 of the 9 environmental swabs, including from one employee's hands. Sequencing linked the NoV detected in the environmental samples to the NoV sequence in the patient samples, with a 100% nucleotide match [245]. This study highlights the need for education of kitchen staff and stringent hygiene practices in food preparation.

Transmission from Contaminated Water

In addition to the risks posed by contaminated water affecting shellfish for consumption, NoV outbreaks have been directly attributed to the contamination of municipal drinking water supplies, even when the water was chlorinated. In 2008, there was an outbreak of 1699 reported cases in the Montenegrin city of Podgorica, the population of which was around 136,000 at that time [246]. A matched case-control study was carried out to assess the water supply system, municipal water samples, and patient fecal samples. The attack rate was estimated to be close to 10% of the population, with up to 15,000 people affected. Analysis of patient samples revealed that all those affected had most likely consumed unboiled, chlorinated municipal water [246]. Several defects in the water distribution system were identified following the outbreak. Six NoV genotypes were detected in 55% of the fecal samples (21/38), together with other viruses [246], highlighting the risk of viral contamination of city water supplies.

Transmission from Fresh Produce

NoV can remain viable and infectious for extended periods in groundwater [232]. Often fresh produce has undergone little processing before being consumed raw, and fertilization of food crops with manure and contaminated irrigation water allow opportunities for pathogens, such as NoV, to enter the food chain (reviewed in [247]).

Campaigns to encourage the population to eat more fresh fruit and vegetables as part of a healthy diet are popular in developed nations. These campaigns may have contributed to an increase in cases of produce-related gastroenteritis reported to the CDC, rising from 0.7% in the 1970s to 6% in the 1990s [248]. A study using CDC data collected between 1973 and 1997 analyzed outbreak data from 190 produce-associated outbreaks, 16,058 illnesses, 8 deaths, and 598 hospitalizations [248]. Fifty-four percent of cases could be attributed to a known pathogen. The study identified salad, lettuce, and mixed fruit as the foods most frequently implicated in NoV infections [248]. An increasingly complex, global supply system for fresh produce has developed in recent decades, meaning that the control of NoV outbreaks, particularly those that transverse the globe, will remain a challenge.

Lettuce has been implicated in foodborne NoV outbreaks in several studies. From a total of 260 outbreaks of gastroenteritis in Denmark in 2010, 11 were linked to contaminated lettuce sourced from France [249]. Patients tested positive for NoV in 7 out of the 11 gastroenteritis outbreaks, and NoV was also directly detected in the imported lettuce, most likely caused by contact with sewage-contaminated water [249]. Viruses are often detected in fresh produce, including lettuce and berries, where there is limited processing before consumption (reviewed in [224]). Raspberries, in particular, have caused a number of NoV outbreaks in Europe, with contamination introduced during preharvest irrigation, fruit pickers, or freezing processes that lack stringent hygiene practices (reviewed in [224]).

In Finland, a series of outbreaks affecting approximately 200 people was recorded in 2009 [250]. The outbreaks were linked to a single batch of imported, frozen raspberries supplied to catering businesses and subsequently served as garnishes for cakes and mixed into other foods. Samples taken from the berries and fecal samples of two patients were analyzed, and NoV GI.4 RNA was detected. Partial NoV capsid sequences found in the berries and patient stool samples were identical, confirming the source of the infection [250].

A widespread outbreak of up to 11,000 people in Germany in 2012 was linked to contaminated frozen strawberries. Real-time RT-PCR was used to detect NoV in 64% (7/11) of the strawberries analyzed [251]. Three different NoV genotypes were detected in the berries, including a recombinant strain (NoV GII.P16/GII.13). As multiple genotypes were detected in the berries, it was suggested that sewage contamination of the strawberries was a more likely scenario than introduction via one infected food handler [251].

The studies described in this section highlight infection risks associated with fresh food preparation, particularly with fresh berries [250]. While the majority of NoV infections are spread person-to-person, another major NoV transmission route is through contaminated food, in particular from fresh produce, seafood, and foods prepared by infected food handlers. There are many opportunities for NoV to enter the human food chain from preharvest through to food serving, and the introduction of NoV into food commodities by food handlers lacking sufficient hygiene practices appears to be one of the predominant causes.

NOV PREVENTION AND CONTROL

NoV has been increasingly recognized as a major public health burden due to the debilitating diarrhea and chronic gastroenteritis it causes in vulnerable groups. In the absence of an effective vaccine or antiviral agent to control NoV-associated AGE, a number of strategies have been described to help health-care professionals manage NoV infections [1,2,252]. Infection control remains the first-line defense in the effort to mitigate the spread of NoV, which is important to prevent vulnerable and immunocompromised populations from being infected in health-care institutions [253]. Generally, isolation of infected patients and limiting the movement of health-care workers between isolation wards are measures deployed in the effort to control the spread of NoV [254]. Hygienic practices such as frequent handwashing with soap and disinfection or disposal of contaminated surfaces or materials also help to reduce the spread of NoV infection [215,255]. In the absence of specific NoV drugs to tackle the virus, oral fluid replacement and electrolyte treatment are commonly used to prevent dehydration. Severe cases of NoV-associated illness can be treated with nutritional supplementation to alleviate symptoms [256,257].

Vaccine Development

There is no vaccine available to prevent NoV infection at present, though current development and trials of several vaccine candidates have shown promising results. These vaccine candidates are described in this section.

There are two NoV vaccine contenders in the early phases of human clinical trials. The first to undergo clinical trials was a nasal spray containing monovalent GI.1 virus-like particles (VLPs), developed by LigoCyte Pharmaceuticals (which was later taken over by Takeda, in 2012). VLPs are noninfectious and do not contain viral genetic material. The vaccine induced a moderate level of protection; 69% of volunteers who received the placebo developed gastroenteritis compared with 37% of vaccine recipients [202]. Intranasal delivery has the advantage of ease of use, and induces robust systemic and mucosal protection with significant serum IgG and fecal IgA responses, respectively [258]. However, as the nasal vaccine delivery device (Bespak) experienced several malfunctions [202], Takeda also pursued a second vaccine with an injectable formulation. This version is injected intramuscularly and is currently in phase II clinical trials.

A second vaccine candidate, developed by Bernstein et al. [259], is a bivalent vaccine injected intramuscularly for protection against both GI.1 and GII.4 NoVs. The vaccine contains both GI.1 and GII.4 VLPs and is adjuvanted with 3-O-desacyl-4′-monophosphoryl lipid A (MPL) and alum [184,259]. This bivalent vaccine was shown to be less effective than the aforementioned nasal vaccine, with 54% of vaccine recipients developing gastroenteritis upon live NoV challenge compared with 62.5% of placebo recipients [259]. However, fewer vaccine recipients (20%) than placebo recipients (41.7%) reported the gastroenteritis symptoms of vomiting and/or diarrhea after challenge with a GII.4 virus.

Vaccines for NoV are still in early stage development, and more research and clinical trials are needed before they become realized as a useful preventive measure against NoV infection.

NOROVIRUS: CONCLUSIONS

NoV is recognized as one of the leading causes of AGE worldwide, as demonstrated through active NoV epidemiological research and surveillance. Over the last two decades, at least six NoV-associated gastroenteritis pandemics have occurred, with GII.4 viruses implicated and accounting for more than 70% of NoV infections overall. Given the fact that a new NoV GII.4 variant emerges approximately every 3 years, it is important that the predominant strain of NoV is continually monitored for replacement by another emerging GII.4 variant. As NoV continues to cause outbreaks and AGE cases worldwide, management and control of NoV infections are crucial.

NoV is highly infectious, often causing outbreaks in enclosed settings such as nursing homes, hospitals, restaurants, cruise ships, childcare centers, and schools. This has highlighted the need for advanced approaches to control NoV infections, either through vaccines or antivirals. NoV is widely spread through contaminated food and water, and public health professionals are now faced with the new challenge of an increasingly complex, global food distribution system from many seasonally varying sources. Careful control and constant monitoring of food sources at different points of harvest and preparation are important in highlighting the risks and preventing NoV and other enteric pathogens from entering the food chain. Education across relevant industries, including catering, farming, processing, transportation and storage, will help to raise awareness to prevent and manage NoV outbreaks.

SAPOVIRUS

Introduction

Sapovirus (SaV) is a significant cause of gastroenteritis in humans and animals. Initially, human SaV was thought to predominantly infect infants, but recent molecular epidemiological studies have shown that SaV also infects adult populations, particularly those over 60 years of age [260]. Similar to human norovirus, SaV has been detected in potable water, river samples, and shellfish prepared for human consumption. Taken together, it can be concluded that SaV is an important pathogen and its implication in food chain contamination is now gaining increased recognition. Sapovirus is a genus within the *Caliciviridae* family and, as such, human SaV shares many similarities with human NoV. This section of the chapter reviews SaV and outlines its modes of transmission, including the relevant foodborne route.

CLASSIFICATION AND HISTORY

Along with *Norovirus*, *Sapovirus* is one of the five currently accepted genera within the *Caliciviridae* family. The prototype strain of human SaV, the Sapporo virus, was originally discovered from an outbreak of gastroenteritis at an orphanage in Sapporo, Japan, in 1977. In that study, Chiba

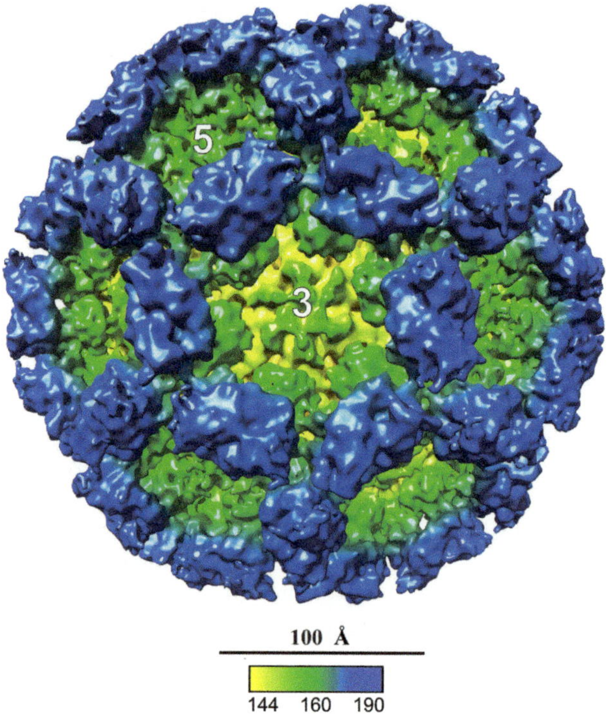

FIGURE 6.8 Cryo-electron microscopy map of a chimeric SaV VLP determined at 8.5 Å resolution. The graded color code represents the radius: 144 Å (yellow), 160 Å (green), and 190 Å (marine). The graded color code corresponds to the partial N-terminus (hidden from view), shell domain (yellow), and protruding domains (green and marine). The 3- and 5-fold axes are labeled. Scale bar = 100 Å. (Modified and adapted from Miyazaki, N. et al. *J Virol,* 90, 2664–2675, 2015.)

et al. [261] identified viruses with the typical calicivirus morphology using EM. SaV virions are 41–46 nm in diameter and have cup-shaped depressions on the virion surface (also known as Star-of-David morphology) and/or 10 protruding capsid spikes. The SaV virion is formed by 180 copies of the single capsid protein VP1 and organized into 90 dimers that form an icosahedron. The capsid protein can be divided into the shell (S) and protruding (P) domains [262] (Figure 6.8). The S domain forms protection around the RNA genome, while the P domain is the most exposed region and likely interacts with host receptors for cell entry. SaV capsid morphology is illustrated by a cryo-EM map of a chimeric SaV VLP in Figure 6.8.

Currently, SaVs are divided into five accepted genogroups [264], although several new genogroups have recently been proposed (GVI–GXIV) based on capsid nucleotide sequences [265]. These genogroups can be further subdivided into numerous genotypes [264]. SaVs belonging to GI, GII, GIV, and GV infect humans, GIII infects swine, while genetically distinct SaVs have been detected in other species including mink, bats, and canines, and may represent additional genotypes (reviewed in [264]). One study showed that the genetic and antigenic relationship between human SaV genogroups corresponded with capsid genogrouping and genotyping [266].

GENOME ORGANIZATION

The SaV genome, like NoV, is a single-stranded, positive-sense RNA molecule of approximately 7.5 kb, flanked by VPg at the 5′ end and polyadenylated at the 3′ end. The SaV genome contains two or three ORFs (Figure 6.9). While SaV GII and GIII genomes contain only two ORFs, SaVs from GI, GIV, and GV contain a third putative ORF [267,268]. ORF1 encodes both the nonstructural

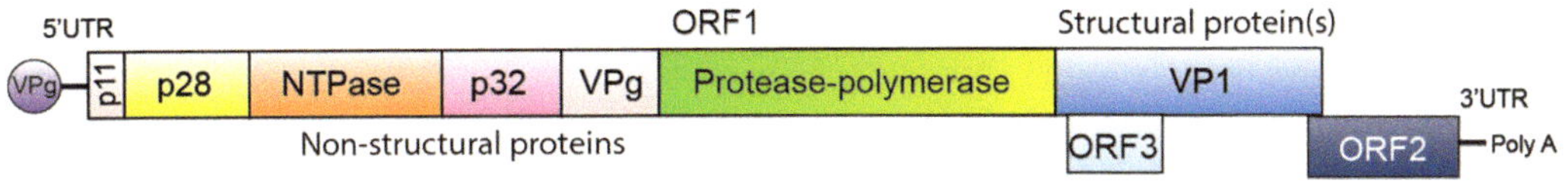

FIGURE 6.9 Genome organization of SaV GI. SaV GI viruses have three open reading frames as depicted, ORFs 1–3, flanked by 5′- and 3′-UTRs. Encoded protein name and function is shown for each region. A polyadenylated tail follows the 3′-UTR end and the 5′-UTR is attached to VPg, encoded by the viral genome. (Adapted and modified from Tse, H. et al. *PLoS One,* 7, e34987, 2012; ViralZone. Sapovirus, on Swiss Institute of Bioinformatics. http://viralzone.expasy.org/all_by_species/196.html, 2015. Accessed November 21, 2015.)

proteins and the major structural protein, while ORF2 (and ORF3 in GI, GIV and GV) encodes a small protein(s) whose function(s) has yet to be elucidated [268]. The nonstructural protein domains at the 5′ end of ORF1 include a putative NTPase, VPg, and the protease-polymerase domains. Three regions encode proteins of unknown function, with molecular masses of 11, 28, and 32 kDa, respectively (Figure 6.9) [269,270]. The 3′ end of ORF1 encodes the major capsid protein VP1. ORF2 and ORF3 partially and fully overlap ORF1, respectively (Figure 6.9).

Similar to human NoVs, SaVs frequently undergo genetic recombination, often at the recombination hotspot located at the RdRp capsid junction [39,271–273]. As is often the case for NoV, recombinant strains of SaV can emerge and cause outbreaks as they are able to evade the current host population immunity [271].

DETECTION METHODS

SaV was first detected using EM in the late 1970s. However, this technique is time consuming, as virus particles are difficult to identify and the sensitivity of the method is generally low. Enzyme-linked immunosorbent assays (ELISAs) have been used to screen for both SaV antibodies and antigens, and are amenable to screening large numbers of samples [261,274,275]. However, generating information on the genogroup or genotype using ELISAs is difficult, unless the ELISA is genotype specific. The most common method of SaV detection currently in use is RT-PCR. A number of groups have designed primers that can detect a broad range of SaV strains [276–280] and most primers are directed against the 5′ end of the capsid gene. One advantage of RT-PCR is that the products can be used for further genetic analysis, including genotyping. Real-time or quantitative RT-PCR (qRT-PCR) methods to detect human SaV have further increased the utility of this method [281–287]. The advantage of qRT-PCR over traditional RT-PCR is that it gives rapid results and can be used to determine RNA levels in patient stool samples.

PATHOGENESIS AND CLINICAL MANIFESTATION

Increasing evidence has emerged in recent years that human NoVs bind to HBGAs [205,288]. These carbohydrate epitopes are present in mucosal secretions and throughout many tissues of the human body, including the small intestine. So far, there is no evidence that SaV binds to HBGAs, suggesting a different mechanism of cell attachment compared with human NoVs. A recent study investigating cell-receptor binding of porcine SaV implicated the involvement of sialic acids linked to host cell glycoproteins as the porcine SaV cellular receptors [289]. However, little is known about host cell binding of human SaV, which to date cannot be cultured in cells; more work is therefore needed to establish cell entry requirements for the virus.

Once inside the cell, the incubation period of human SaV infections usually ranges from 1 to 4 days (reviewed in [264]). Most commonly, symptoms include vomiting, diarrhea, nausea, and cramps. These major symptoms are often accompanied by malaise, chills, headaches, and myalgia [264]. Fever is unusual and, in some rare cases, blood or mucus is detected in stool samples [201]. Infections generally last between 2 and 3 days, but there are some reports describing symptoms lasting more than 1 week [290]. While SaV infections are not usually associated with mortality, outbreaks within U.S. older adult care facilities have occasionally resulted in hospitalization and death [291]. Medical treatment usually involves orally administered fluids and electrolyte replacement therapy. No vaccines or specific antivirals currently exist for human SaV [264].

ENVIRONMENTAL PERSISTENCE

Like many enteric viral pathogens, SaV is nonenveloped and therefore inherently resistant in the environment [292]. SaVs have been detected in environmental waters such as rivers and wastewater treatment plants [293–295], and in coastal waters around Spain, Italy, the United States, South Africa, Central America and Japan [294–302]. SaV infectivity has been tested using the culturable porcine SaV to determine its resistance to various treatments. The effects of varying pH, ethanol, chlorine, and temperature were tested on SaV survival [292]. SaV was found to be stable between pH 3.0 and 8.0, and demonstrated similar resistance to heat at 56°C and chlorine treatment compared with NoV. Additionally, SaV was more resistant to ethanol treatment (60%–70% v/v) than NoV [292]. This highlights the resistance of SaVs, allowing them to persist in the environment and likely facilitating transmission to new hosts.

EPIDEMIOLOGY

Like NoV infections, SaV outbreaks occur all year round but are most common in the winter months in temperate climates [260,303]. SaVs cause AGE across all age groups worldwide, but particularly in infants, young children [304], and adults over 60 years of age [260,305]. The incidence, detection, and overall prevalence of SaV infections vary in each country and setting, as statistics are likely affected by the diagnostic techniques used in each region [219]. Although much less prevalent than NoV, the incidence of SaV detection in cases of AGE from a number of settings has reportedly increased in recent years [306].

SaV spreads easily in closed settings, and outbreaks are common in long-term care facilities [291], kindergartens [293], day care centers [307,308], schools, hospitals, restaurants [309], canteens, and cruise ships (reviewed in [264]). Studies on the number of gastroenteritis outbreaks attributable to SaV vary worldwide, but report between 1.3% (9 of 676) [310] and 8.0% (7 of 88) of outbreaks [311].

Genogroup I, genotype 2 (GI.2) SaV strains with a high level of sequence homology have been detected in outbreaks across different countries including Sweden, Slovenia, Taiwan, Japan, and Russia, which could indicate a shared origin followed by rapid spread [260], likely facilitated by international travel.

SaV GI and GII have been reported as the predominant genogroups in circulation in Japan [312]. However, in 2006, a GI.1 intragenotype recombinant SaV rapidly became the predominant strain detected in Japanese children [313]. This highlights how viruses use mechanisms such as recombination to evade host immunity and increase viral fitness [53].

TRANSMISSION OF INFECTION

Like NoV, transmission of human SaV is predominately via the fecal–oral route and spread by person-to-person contact (reviewed in [264]). Transmission also occurs through ingestion of contaminated food or water or contact with fomites and vomitus, and is easily spread in closed settings such as schools, prisons, and hospitals.

FOODBORNE TRANSMISSION

Studies have reported the detection of SaV in several food types destined for human consumption, and SaV outbreaks have been linked to contaminated shellfish [293,309,314].

As with many other enteric pathogens, the poor hygiene practices of food handlers present a significant risk of introducing SaV into the food chain. Extensive SaV outbreaks have been documented in Japan, mainly linked to catering. A large-scale outbreak in 2010 was eventually linked to a catering company that delivered boxed lunches throughout several Japanese prefectures [306]. SaV was detected in 13.5% (7 of 52) of food handlers at the catering company, which may itself have had a small SaV outbreak at the time. As a result, 17.1% (655 of 3827) of individuals who ate the contaminated lunches developed AGE symptoms. RT-PCR and sequencing were used to identify SaV GI.2 as the strain responsible [306].

Boxed lunches were implicated in another outbreak at a wedding hall in 2007 in Ehime, Japan [315]. One hundred and nine people developed AGE symptoms after eating the lunches. RT-PCR and sequencing were used to detect SaV in stool samples from two asymptomatic food handlers. Stool samples were also collected from 56 of the 109 cases, and 35.7% (20 of 56) tested positive for SaV GIV RNA. Sequence identity from positive RT-PCT products revealed a single source of the outbreak, further highlighting the importance of hygienic food preparation and handling [315].

Seafood is often implicated in gastroenteritis cases, as shellfish can bioaccumulate pathogenic microbes in their flesh during filter feeding [233]. This is compounded by the fact that many consumers prefer to eat shellfish raw or lightly steamed, so any enteric pathogens present are likely to remain infectious upon consumption. Raw or undercooked clams are a common transmission source for SaV [314,316]. Indeed, SaV has been detected in clam flesh from markets in Japan using RT-PCR [316]. Nested RT-PCR has been used to detect multiple SaV genotypes in frozen clams intended for human consumption [314]. In another study, qRT-PCR was used to detect SaV in 11.7% (9 of 77) of shellfish samples collected from two coastal areas in Morocco between 2006 and 2010 [317]. This study detected SaV in 14.7% (5 of 34) of oysters, 21.4% (3 of 14) of cockles, and 3.4% (1 of 29) of clam samples tested [317]. These examples illustrate the presence of SaV but arc limited in determining the infectious nature of the detected viruses, so caution must be taken. However, there is a clear risk of SaV infection from eating seafood, particularly if shellfish beds sourced for human consumption are contaminated with human sewage or waste.

In an outbreak linked to a Japanese restaurant in 2008, mixed infections with both NoV and SaV were detected in stool samples from 17 diners and one asymptomatic, food-handling kitchen employee [309]. RT-PCR and qRT-PCR were used to detect SaV and NoV in leftover shellfish samples and shellfish packaging. Sequence analysis revealed that they contained almost identical partial capsid sequences (99.3%–100%) to SaV sequences detected in the patient stool samples, indicating that the shellfish were the likely source of the outbreak [309].

As SaVs are relatively resistant to environmental forces [292], fecal contamination of water used to irrigate crops for human consumption presents another significant health risk [247]. Particularly problematic are foods that are eaten raw with little processing, such as berries and leafy vegetables. In one study that assessed the effect of pH on the attachment of SaV to lettuce, it was found that viral particles remained infectious on lettuce leaves for a week when stored at 4°C [292].

These examples demonstrate the risk of foodborne SaV transmission. Poor hygiene in food handling and preparation practices or contamination of food before harvest can increase the risk of infection from several enteric pathogens, including SaV.

SAPOVIRUS: CONCLUSIONS

Like NoV, SaV causes AGE worldwide, mostly affecting infants, young children, and older adults. SaV is easily spread from person-to-person in semiclosed settings such as hospitals, schools, and cruise ships. Although fecal–oral spread is the most widespread mode of transmission, foodborne transmission is also common. Raising awareness of the importance of stringent hygiene when preparing and serving food is paramount to prevent SaV from entering the food chain and causing outbreaks of gastroenteritis.

REFERENCES

1. Glass RI, Parashar UD, Estes MK. 2009. Norovirus gastroenteritis. *N Engl J Med* 361:1776–1785.
2. Patel MM, Widdowson MA, Glass RI, Akazawa K, Vinje J, Parashar UD. 2008. Systematic literature review of role of noroviruses in sporadic gastroenteritis. *Emerg Infect Dis* 14:1224–1231.
3. Moore MD, Goulter RM, Jaykus LA. 2015. Human norovirus as a foodborne pathogen: Challenges and developments. *Annu Rev Food Sci Technol* 6:411–433.
4. Scallan E, Hoekstra RM, Angulo FJ, Tauxe RV, Widdowson MA, Roy SL, Jones JL, Griffin PM. 2011. Foodborne illness acquired in the United States-major pathogens. *Emerg Infect Dis* 17:7–15.
5. Hall G, Kirk MD, Becker N, Gregory JE, Unicomb L, Millard G, Stafford R, Lalor K, Group OW. 2005. Estimating foodborne gastroenteritis, Australia. *Emerg Infect Dis* 11:1257–1264.
6. Greening GE, Hewitt J, Rivera-Aban M, Croucher D. 2012. Molecular epidemiology of norovirus gastroenteritis outbreaks in New Zealand from 2002–2009. *J Med Virol* 84:1449–1458.
7. Thomas MK, Murray R, Flockhart L, Pintar K, Pollari F, Fazil A, Nesbitt A, Marshall B. 2013. Estimates of the burden of foodborne illness in Canada for 30 specified pathogens and unspecified agents, circa 2006. *Foodborne Pathog Dis* 10:639–648.
8. Duizer E, Schwab KJ, Neill FH, Atmar RL, Koopmans MP, Estes MK. 2004. Laboratory efforts to cultivate noroviruses. *J Gen Virol* 85:79–87.
9. Wobus CE, Thackray LB, Virgin HW IV. 2006. Murine norovirus: A model system to study norovirus biology and pathogenesis. *J Virol* 80:5104–5112.
10. Taube S, Kolawole AO, Hohne M, Wilkinson JE, Handley SA, Perry JW, Thackray LB, Akkina R, Wobus CE. 2013. A mouse model for human norovirus. *MBio* 4:e00450-13.
11. Graham DY, Jiang X, Tanaka T, Opekun AR, Madore HP, Estes MK. 1994. Norwalk virus infection of volunteers: New insights based on improved assays. *J Infect Dis* 170:34–43.
12. Green KY, Chanock RM, Kapikian AZ. 2001. *Human Calicivirus*. Lippincott Williams & Wilkins, Philadelphia, PA.
13. Bull RA, Eden JS, Luciani F, McElroy K, Rawlinson WD, White PA. 2012. Contribution of intra- and interhost dynamics to norovirus evolution. *J Virol* 86:3219–3229.
14. Siebenga JJ, Beersma MF, Vennema H, van Biezen P, Hartwig NJ, Koopmans M. 2008. High prevalence of prolonged norovirus shedding and illness among hospitalized patients: A model for *in vivo* molecular evolution. *J Infect Dis* 198:994–1001.
15. Gallimore CI, Lewis D, Taylor C, Cant A, Gennery A, Gray JJ. 2004. Chronic excretion of a norovirus in a child with cartilage hair hypoplasia (CHH). *J Clin Virol* 30:196–204.
16. Vega E, Barclay L, Gregoricus N, Shirley SH, Lee D, Vinje J. 2014. Genotypic and epidemiologic trends of norovirus outbreaks in the United States, 2009–2013. *J Clin Microbiol* 52:147–155.
17. Lucas R. 1930. Epidemic vomiting or intestinal influenza. *New Orleans M & SJ* 83:213–214.
18. Zahorsky J. 1929. Hyperemesis hiemis or the winter vomiting disease. *Arch of Paed* 46:391–395.
19. Wildman HA. 1933. Polytropous enteronitis (acute infectious gastro-enteritis, Spencer's disease): Is it a form of influenza? *Arch Int Med* 52:959–977.
20. Dolin R, Blacklow NR, DuPont H, Formal S, Buscho RF, Kasel JA, Chames RP, Hornick R, Chanock RM. 1971. Transmission of acute infectious nonbacterial gastroenteritis to volunteers by oral administration of stool filtrates. *J Infect Dis* 123:307–312.
21. Gordon I, Ingraham HS, Korns RF. 1947. Transmission of epidemic gastroenteritis to human volunteers by oral administration of fecal filtrates. *J Exp Med* 86:409–422.
22. Wyatt RG, Dolin R, Blacklow NR, DuPont HL, Buscho RF, Thornhill TS, Kapikian AZ, Chanock RM. 1974. Comparison of three agents of acute infectious nonbacterial gastroenteritis by cross-challenge in volunteers. *J Infect Dis* 129:709–714.
23. Adler JL, Zickl R. 1969. Winter vomiting disease. *J Infect Dis* 119:668–673.
24. Thornhill TS, Wyatt RG, Kalica AR, Dolin R, Chanock RM, Kapikian AZ. 1977. Detection by immune electron microscopy of 26- to 27-nm virus like particles associated with two family outbreaks of gastroenteritis. *J Infect Dis* 135:20–27.
25. Kapikian AZ. 2000. The discovery of the 27-nm Norwalk virus: An historic perspective. *J Infect Dis* 181 Suppl 2:S295–S302.
26. Cubitt WD, Jiang XJ, Wang J, Estes MK. 1994. Sequence similarity of human caliciviruses and small round structured viruses. *J Med Virol* 43:252–258.
27. Glass RI, Noel J, Ando T, Fankhauser R, Belliot G, Mounts A, Parashar UD, Bresee JS, Monroe SS. 2000. The epidemiology of enteric caliciviruses from humans: A reassessment using new diagnostics. *J Infect Dis* 181 Suppl 2:S254–S261.

28. Monroe SS, Ando T, Glass RI. 2000. Introduction: Human enteric caliciviruses—an emerging pathogen whose time has come. *J Infect Dis* 181 Suppl 2:S249–S251.
29. Jiang X, Wang M, Wang K, Estes MK. 1993. Sequence and genomic organization of Norwalk virus. *Virology* 195:51–61.
30. Mayo MA. 2002. A summary of taxonomic changes recently approved by ICTV. *Arch Virol* 147:1655–1656.
31. Parashar UD, Hummelman EG, Bresee JS, Miller MA, Glass RI. 2003. Global illness and deaths caused by rotavirus disease in children. *Emerg Infect Dis* 9:565–572.
32. Patel M, Pedreira C, De Oliveira LH, Tate J, Orozco M, Mercado J, Gonzalez A, et al. 2009. Association between pentavalent rotavirus vaccine and severe rotavirus diarrhea among children in Nicaragua. *JAMA* 301:2243–2251.
33. Patel MM, Steele D, Gentsch JR, Wecker J, Glass RI, Parashar UD. 2011. Real-world impact of rotavirus vaccination. *Pediatr Infect Dis J* 30:S1–S5.
34. Tate JE, Mutuc JD, Panozzo CA, Payne DC, Cortese MM, Cortes JE, Yen C, et al. 2011. Sustained decline in rotavirus detections in the United States following the introduction of rotavirus vaccine in 2006. *Pediatr Infect Dis J* 30:S30–S34.
35. Tate JE, Panozzo CA, Payne DC, Patel MM, Cortese MM, Fowlkes AL, Parashar UD. 2009. Decline and change in seasonality of US rotavirus activity after the introduction of rotavirus vaccine. *Pediatrics* 124:465–471.
36. Eden JS. 2012. The evolutionary dynamics of norovirus. PhD thesis. UNSW Australia.
37. Murphy AM, Grohmann GS, Christopher PJ, Lopez WA, Davey GR, Millsom RH. 1979. An Australia-wide outbreak of gastroenteritis from oysters caused by Norwalk virus. *Med J Aust* 2:329–333.
38. Estes MK, Prasad BV, Atmar RL. 2006. Noroviruses everywhere: Has something changed? *Curr Opin Infect Dis* 19:467–474.
39. Hansman GS, Oka T, Katayama K, Takeda N. 2007. Human sapoviruses: Genetic diversity, recombination, and classification. *Rev Med Virol* 17:133–141.
40. Zheng DP, Ando T, Fankhauser RL, Beard RS, Glass RI, Monroe SS. 2006. Norovirus classification and proposed strain nomenclature. *Virology* 346:312–323.
41. Mesquita JR, Barclay L, Nascimento MS, Vinje J. 2010. Novel norovirus in dogs with diarrhea. *Emerg Infect Dis* 16:980–982.
42. Tse H, Lau SK, Chan WM, Choi GK, Woo PC, Yuen KY. 2012. Complete genome sequences of novel canine noroviruses in Hong Kong. *J Virol* 86:9531–9532.
43. Liu BL, Lambden PR, Gunther H, Otto P, Elschner M, Clarke IN. 1999. Molecular characterization of a bovine enteric calicivirus: Relationship to the Norwalk-like viruses. *J Virol* 73:819–825.
44. Karst SM, Wobus CE, Lay M, Davidson J, Virgin HW IV. 2003. STAT1-dependent innate immunity to a Norwalk-like virus. *Science* 299:1575–1578.
45. Wang QH, Han MG, Cheetham S, Souza M, Funk JA, Saif LJ. 2005. Porcine noroviruses related to human noroviruses. *Emerg Infect Dis* 11:1874–1881.
46. Kroneman A, Vennema H, Deforche K, vd Avoort H, Penaranda S, Oberste MS, Vinje J, Koopmans M. 2011. An automated genotyping tool for enteroviruses and noroviruses. *J Clin Virol* 51:121–125.
47. Siebenga JJ, Vennema H, Zheng DP, Vinje J, Lee BE, Pang XL, Ho EC, et al. 2009. Norovirus illness is a global problem: Emergence and spread of norovirus GII.4 variants, 2001–2007. *J Infect Dis* 200:802–812.
48. Zheng DP, Widdowson MA, Glass RI, Vinje J. 2010. Molecular epidemiology of genogroup II-genotype 4 noroviruses in the United States between 1994 and 2006. *J Clin Microbiol* 48:168–177.
49. Kroneman A, Vega E, Vennema H, Vinje J, White PA, Hansman G, Green K, Martella V, Katayama K, Koopmans M. 2013. Proposal for a unified norovirus nomenclature and genotyping. *Arch Virol* 158:2059–2068.
50. Bull RA, White PA. 2011. Mechanisms of GII.4 norovirus evolution. *Trends Microbiol* 19:233–240.
51. Lopman BA, Adak GK, Reacher MH, Brown DW. 2003. Two epidemiologic patterns of norovirus outbreaks: Surveillance in England and Wales, 1992–2000. *Emerg Infect Dis* 9:71–77.
52. Atmar RL, Estes MK. 2006. The epidemiologic and clinical importance of norovirus infection. *Gastroenterol Clin North Am* 35:275–290.
53. Eden JS, Tanaka MM, Boni MF, Rawlinson WD, White PA. 2013. Recombination within the pandemic norovirus GII.4 lineage. *J Virol* 87:6270–6282.
54. Eden JS, Hewitt J, Lim KL, Boni MF, Merif J, Greening G, et al. 2014. The emergence and evolution of the novel epidemic norovirus GII.4 variant Sydney 2012. *Virology* 450–451:106–113.
55. van Beek J, Ambert-Balay K, Botteldoorn N, Eden JS, Fonager J, Hewitt J, Iritani N, et al. 2013. Indications for worldwide increased norovirus activity associated with emergence of a new variant of genotype II.4, late 2012. *Euro Surveill* 18:8–9.

56. Noel JS, Fankhauser RL, Ando T, Monroe SS, Glass RI. 1999. Identification of a distinct common strain of "Norwalk-like viruses" having a global distribution. *J Infect Dis* 179:1334–1344.
57. White PA, Hansman GS, Li A, Dable J, Isaacs M, Ferson M, McIver CJ, Rawlinson WD. 2002. Norwalk-like virus 95/96-US strain is a major cause of gastroenteritis outbreaks in Australia. *J Med Virol* 68:113–118.
58. Fankhauser RL, Noel JS, Monroe SS, Ando T, Glass RI. 1998. Molecular epidemiology of "Norwalk-like viruses" in outbreaks of gastroenteritis in the United States. *J Infect Dis* 178:1571–1578.
59. Lopman B, Vennema H, Kohli E, Pothier P, Sanchez A, Negredo A, Buesa J, et al. 2004. Increase in viral gastroenteritis outbreaks in Europe and epidemic spread of new norovirus variant. *Lancet* 363:682–688.
60. Widdowson MA, Cramer EH, Hadley L, Bresee JS, Beard RS, Bulens SN, Charles M, et al. 2004. Outbreaks of acute gastroenteritis on cruise ships and on land: Identification of a predominant circulating strain of norovirus—United States, 2002. *J Infect Dis* 190:27–36.
61. Bull RA, Tu ET, McIver CJ, Rawlinson WD, White PA. 2006. Emergence of a new norovirus genotype II.4 variant associated with global outbreaks of gastroenteritis. *J Clin Microbiol* 44:327–333.
62. Kroneman A, Vennema H, Harris J, Reuter G, von Bonsdorff CH, Hedlund KO, Vainio K, et al. 2006. Increase in norovirus activity reported in Europe. *Euro Surveill* 11:E061214 061211.
63. Ho EC, Cheng PK, Wong DA, Lau AW, Lim WW. 2006. Correlation of norovirus variants with epidemics of acute viral gastroenteritis in Hong Kong. *J Med Virol* 78:1473–1479.
64. Johansen K, Mannerqvist K, Allard A, Andersson Y, Burman LG, Dillner L, Hedlund KO, et al. 2008. Norovirus strains belonging to the GII.4 genotype dominate as a cause of nosocomial outbreaks of viral gastroenteritis in Sweden 1997–2005: Arrival of new variants is associated with large nation-wide epidemics. *J Clin Virol* 42:129–134.
65. Ramirez S, De Grazia S, Giammanco GM, Milici M, Colomba C, Ruggeri FM, Martella V, Arista S. 2006. Detection of the norovirus variants GGII.4 hunter and GGIIb/hilversum in Italian children with gastroenteritis. *J Med Virol* 78:1656–1662.
66. Wu FT, Oka T, Katayama K, Wu HS, Donald Jiang DS, Miyamura T, Takeda N, Hansman GS. 2006. Genetic diversity of noroviruses in Taiwan between November 2004 and March 2005. *Arch Virol* 151:1319–1327.
67. Tu ET, Bull RA, Greening GE, Hewitt J, Lyon MJ, Marshall JA, McIver CJ, Rawlinson WD, White PA. 2008. Epidemics of gastroenteritis during 2006 were associated with the spread of norovirus GII.4 variants 2006a and 2006b. *Clin Infect Dis* 46:413–420.
68. Eden JS, Bull RA, Tu E, McIver CJ, Lyon MJ, Marshall JA, Smith DW, Musto J, Rawlinson WD, White PA. 2010. Norovirus GII.4 variant 2006b caused epidemics of acute gastroenteritis in Australia during 2007 and 2008. *J Clin Virol* 49:265–271.
69. Siebenga JJ, Vennema H, Renckens B, de Bruin E, van der Veer B, Siezen RJ, Koopmans M. 2007. Epochal evolution of GGII.4 norovirus capsid proteins from 1995 to 2006. *J Virol* 81:9932–9941.
70. Puustinen L, Blazevic V, Salminen M, Hämäläinen M, Räsänen S, Vesikari T. 2011. Noroviruses as a major cause of acute gastroenteritis in children in Finland, 2009–2010. *Scand J Infect Dis* 43:804–808.
71. Vega E, Barclay L, Gregoricus N, Williams K, Lee D, Vinjé J. 2011. Novel surveillance network for norovirus gastroenteritis outbreaks, United States. *Emerg Infect Dis* 17:1389.
72. Yen C, Wikswo ME, Lopman BA, Vinje J, Parashar UD, Hall AJ. 2011. Impact of an emergent norovirus variant in 2009 on norovirus outbreak activity in the United States. *Clin Infect Dis* 53:568–571.
73. Bull RA, Eden JS, Rawlinson WD, White PA. 2010. Rapid evolution of pandemic noroviruses of the GII.4 lineage. *PLoS Pathog* 6:e1000831.
74. Debbink K, Lindesmith LC, Donaldson EF, Baric RS. 2012. Norovirus immunity and the great escape. *PLoS Pathog* 8:e1002921.
75. Donaldson EF, Lindesmith LC, Lobue AD, Baric RS. 2010. Viral shape-shifting: Norovirus evasion of the human immune system. *Nat Rev Microbiol* 8:231–241.
76. Lindesmith LC, Donaldson EF, Baric RS. 2011. Norovirus GII.4 strain antigenic variation. *J Virol* 85:231–242.
77. Boon D, Mahar JE, Abente EJ, Kirkwood CD, Purcell RH, Kapikian AZ, Green KY, Bok K. 2011. Comparative evolution of GII.3 and GII.4 norovirus over a 31-year period. *J Virol* 85:8656–8666.
78. Mahar JE, Kirkwood CD. 2011. Characterization of norovirus strains in Australian children from 2006 to 2008: Prevalence of recombinant strains. *J Med Virol* 83:2213–2219.
79. Xi JN, Graham DY, Wang KN, Estes MK. 1990. Norwalk virus genome cloning and characterization. *Science* 250:1580–1583.

80. Belliot G, Sosnovtsev SV, Mitra T, Hammer C, Garfield M, Green KY. 2003. In vitro proteolytic processing of the MD145 norovirus ORF1 nonstructural polyprotein yields stable precursors and products similar to those detected in calicivirus-infected cells. *J Virol* 77:10957–10974.
81. Herbert TP, Brierley I, Brown TD. 1997. Identification of a protein linked to the genomic and subgenomic mRNAs of feline calicivirus and its role in translation. *J Gen Virol* 78 (Pt 5):1033–1040.
82. Morales M, Barcena J, Ramirez MA, Boga JA, Parra F, Torres JM. 2004. Synthesis *in vitro* of rabbit hemorrhagic disease virus subgenomic RNA by internal initiation on (-)sense genomic RNA: Mapping of a subgenomic promoter. *J Biol Chem* 279:17013–17018.
83. Lin X, Thorne L, Jin Z, Hammad LA, Li S, Deval J, Goodfellow IG, Kao CC. 2015. Subgenomic promoter recognition by the norovirus RNA-dependent RNA polymerases. *Nucleic Acids Res* 43:446–460.
84. Hyde JL, Mackenzie JM. 2010. Subcellular localization of the MNV-1 ORF1 proteins and their potential roles in the formation of the MNV-1 replication complex. *Virology* 406:138–148.
85. Choi JM, Hutson AM, Estes MK, Prasad BV. 2008. Atomic resolution structural characterization of recognition of histo-blood group antigens by Norwalk virus. *Proc Natl Acad Sci USA* 105:9175–9180.
86. Baker ES, Luckner SR, Krause KL, Lambden PR, Clarke IN, Ward VK. 2012. Inherent structural disorder and dimerisation of murine norovirus NS1-2 protein. *PLoS One* 7:e30534.
87. Pfister T, Wimmer E. 2001. Polypeptide p41 of a Norwalk-like virus is a nucleic acid-independent nucleoside triphosphatase. *J Virol* 75:1611–1619.
88. Rawlings ND, Barrett AJ. 1994. Families of cysteine peptidases. *Methods Enzymol* 244:461–486.
89. Choe SS, Dodd DA, Kirkegaard K. 2005. Inhibition of cellular protein secretion by picornaviral 3A proteins. *Virology* 337:18–29.
90. Sharp TM, Guix S, Katayama K, Crawford SE, Estes MK. 2010. Inhibition of cellular protein secretion by Norwalk virus nonstructural protein p22 requires a mimic of an endoplasmic reticulum export signal. *PLoS One* 5:e13130.
91. Goodfellow I. 2011. The genome-linked protein VPg of vertebrate viruses: A multifaceted protein. *Curr Opin Virol* 1:355–362.
92. Daughenbaugh KF, Fraser CS, Hershey JW, Hardy ME. 2003. The genome-linked protein VPg of the Norwalk virus binds eIF3, suggesting its role in translation initiation complex recruitment. *EMBO J* 22:2852–2859.
93. Goodfellow I, Chaudhry Y, Gioldasi I, Gerondopoulos A, Natoni A, Labrie L, Laliberte JF, Roberts L. 2005. Calicivirus translation initiation requires an interaction between VPg and eIF 4 E. *EMBO Rep* 6:968–972.
94. Someya Y, Takeda N, Miyamura T. 2000. Complete nucleotide sequence of the chiba virus genome and functional expression of the 3C-Like protease in Escherichia coli. *Virology* 278:490–500.
95. Blakeney SJ, Cahill A, Reilly PA. 2003. Processing of Norwalk virus nonstructural proteins by a 3C-like cysteine proteinase. *Virology* 308:216–224.
96. Porter AG. 1993. Picornavirus nonstructural proteins: Emerging roles in virus replication and inhibition of host cell functions. *J Virol* 67:6917–6921.
97. Kuyumcu-Martinez M, Belliot G, Sosnovtsev SV, Chang K-O, Green KY, Lloyd RE. 2004. Calicivirus 3C-like proteinase inhibits cellular translation by cleavage of poly (A)-binding protein. *J Virol* 78:8172–8182.
98. Hogbom M, Jager K, Robel I, Unge T, Rohayem J. 2009. The active form of the norovirus RNA-dependent RNA polymerase is a homodimer with cooperative activity. *J Gen Virol* 90:281–291.
99. Fukushi S, Kojima S, Takai R, Hoshino FB, Oka T, Takeda N, Katayama K, Kageyama T. 2004. Poly(A)- and primer-independent RNA polymerase of Norovirus. *J Virol* 78:3889–3896.
100. Barr JN, Fearns R. 2010. How RNA viruses maintain their genome integrity. *J Gen Virol* 91:1373–1387.
101. Steinhauer DA, Domingo E, Holland JJ. 1992. Lack of evidence for proofreading mechanisms associated with an RNA virus polymerase. *Gene* 122:281–288.
102. Hardy ME. 2005. Norovirus protein structure and function. *FEMS Microbiol Lett* 253:1–8.
103. Prasad BV, Hardy ME, Dokland T, Bella J, Rossmann MG, Estes MK. 1999. X-ray crystallographic structure of the Norwalk virus capsid. *Science* 286:287–290.
104. Tan M, Jiang X. 2005. Norovirus and its histo-blood group antigen receptors: An answer to a historical puzzle. *Trends Microbiol* 13:285–293.
105. Cao S, Lou Z, Tan M, Chen Y, Liu Y, Zhang Z, Zhang XC, Jiang X, Li X, Rao Z. 2007. Structural basis for the recognition of blood group trisaccharides by norovirus. *J Virol* 81:5949–5957.
106. Donaldson EF, Lindesmith LC, Lobue AD, Baric RS. 2008. Norovirus pathogenesis: Mechanisms of persistence and immune evasion in human populations. *Immunol Rev* 225:190–211.

107. Tan M, Hegde RS, Jiang X. 2004. The P domain of norovirus capsid protein forms dimer and binds to histo-blood group antigen receptors. *J Virol* 78:6233–6242.
108. Tan M, Huang P, Meller J, Zhong W, Farkas T, Jiang X. 2003. Mutations within the P2 domain of norovirus capsid affect binding to human histo-blood group antigens: Evidence for a binding pocket. *J Virol* 77:12562–12571.
109. Bertolotti-Ciarlet A, Crawford SE, Hutson AM, Estes MK. 2003. The 3′ end of Norwalk virus mRNA contains determinants that regulate the expression and stability of the viral capsid protein VP1: A novel function for the VP2 protein. *J Virol* 77:11603–11615.
110. Wirblich C, Thiel H-J, Meyers G. 1996. Genetic map of the calicivirus rabbit hemorrhagic disease virus as deduced from *in vitro* translation studies. *J Virol* 70:7974–7983.
111. Glass PJ, White LJ, Ball JM, Leparc-Goffart I, Hardy ME, Estes MK. 2000. Norwalk virus open reading frame 3 encodes a minor structural protein. *J Virol* 74:6581–6591.
112. Sosnovtsev SV, Green KY. 2000. Identification and genomic mapping of the ORF3 and VPg proteins in feline calicivirus virions. *Virology* 277:193–203.
113. McFadden N, Bailey D, Carrara G, Benson A, Chaudhry Y, Shortland A, Heeney J, et al. 2011. Norovirus regulation of the innate immune response and apoptosis occurs via the product of the alternative open reading frame 4. *PLoS Pathog* 7:e1002413.
114. Hutson AM, Atmar RL, Graham DY, Estes MK. 2002. Norwalk virus infection and disease is associated with ABO histo-blood group type. *J Infect Dis* 185:1335–1337.
115. Tan M, Jiang X. 2011. Norovirus–host interaction: Multi-selections by human histo-blood group antigens. *Trends Microbiol* 19:382–388.
116. Murakami K, Kurihara C, Oka T, Shimoike T, Fujii Y, Takai-Todaka R, Park Y, et al. 2013. Norovirus binding to intestinal epithelial cells is independent of histo-blood group antigens. *PLoS One* 8:e66534.
117. Pendu JL. 2004. Section IV Innate and acquired immune systems of human milk: Histo-blood group antigen and human milk oligosaccharides—Genetic polymorphism and risk of infectious diseases. *Adv Exp Med Biol* 554:135–144.
118. Dietrich CP, Nader HB, Straus AH. 1983. Structural differences of heparan sulfates according to the tissue and species of origin. *Biochem Biophys Res Commun* 111:865–871.
119. Hedman K, Kurkinen M, Alitalo K, Vaheri A, Johansson S, Höök M. 1979. Isolation of the pericellular matrix of human fibroblast cultures. *J Cell Biol* 81:83–91.
120. Tamura M, Natori K, Kobayashi M, Miyamura T, Takeda N. 2004. Genogroup II noroviruses efficiently bind to heparan sulfate proteoglycan associated with the cellular membrane. *J Virol* 78:3817–3826.
121. Chang KO. 2009. Role of cholesterol pathways in norovirus replication. *J Virol* 83:8587–8595.
122. Gerondopoulos A, Jackson T, Monaghan P, Doyle N, Roberts LO. 2010. Murine norovirus-1 cell entry is mediated through a non-clathrin-, non-caveolae-, dynamin- and cholesterol-dependent pathway. *J Gen Virol* 91:1428–1438.
123. Perry JW, Wobus CE. 2010. Endocytosis of murine norovirus 1 into murine macrophages is dependent on dynamin II and cholesterol. *J Virol* 84:6163–6176.
124. Tan M, Jiang X. 2010. Virus–host interaction and celluar receptors of caliciviruses. *In* Hansman GS, Jiang XJ, Green KY (eds), *Caliciviruses Molecular and Cellular Virology*, 1st ed., pp. 111–129. Caister Academic Press, Norwich, United Kingdom.
125. Gamarnik AV, Andino R. 1998. Switch from translation to RNA replication in a positive-stranded RNA virus. *Genes Dev* 12:2293–2304.
126. Lai MM. 1998. Cellular factors in the transcription and replication of viral RNA genomes: A parallel to DNA-dependent RNA transcription. *Virology* 244:1–12.
127. Thorne LG, Goodfellow IG. 2014. Norovirus gene expression and replication. *J Gen Virol* 95:278–291.
128. Ettayebi K, Hardy ME. 2003. Norwalk virus nonstructural protein p48 forms a complex with the SNARE regulator VAP-A and prevents cell surface expression of vesicular stomatitis virus G protein. *J Virol* 77:11790–11797.
129. Hyde JL, Sosnovtsev SV, Green KY, Wobus C, Virgin HW, Mackenzie JM. 2009. Mouse norovirus replication is associated with virus-induced vesicle clusters originating from membranes derived from the secretory pathway. *J Virol* 83:9709–9719.
130. Hyde JL, Gillespie LK, Mackenzie JM. 2012. Mouse norovirus 1 utilizes the cytoskeleton network to establish localization of the replication complex proximal to the microtubule organizing center. *J Virol* 86:4110–4122.
131. Arias A, Emmott E, Vashist S, Goodfellow I. 2013. Progress towards the prevention and treatment of norovirus infections. *Future Microbiol* 8:1475–1487.

132. Eden J-S, Hewitt J, Lim KL, Boni MF, Merif J, Greening G, Ratcliff RM, Holmes EC, Tanaka MM, Rawlinson WD. 2014. The emergence and evolution of the novel epidemic norovirus GII. 4 variant Sydney 2012. *Virology* 450:106–113.
133. Koelle K, Cobey S, Grenfell B, Pascual M. 2006. Epochal evolution shapes the phylodynamics of interpandemic influenza A (H3N2) in humans. *Science* 314:1898–1903.
134. Debbink K, Donaldson EF, Lindesmith LC, Baric RS. 2012. Genetic mapping of a highly variable norovirus GII.4 blockade epitope: Potential role in escape from human herd immunity. *J Virol* 86:1214–1226.
135. Lindesmith LC, Costantini V, Swanstrom J, Debbink K, Donaldson EF, Vinje J, Baric RS. 2013. Emergence of a norovirus GII.4 strain correlates with changes in evolving blockade epitopes. *J Virol* 87:2803–2813.
136. Chavez-Bueno S, Bozdogan B, Katz K, Bowlware KL, Cushion N, Cavuoti D, Ahmad N, McCracken GH, Jr., Appelbaum PC. 2005. Inducible clindamycin resistance and molecular epidemiologic trends of pediatric community-acquired methicillin-resistant *Staphylococcus aureus* in Dallas, Texas. *Antimicrob Agents Chemother* 49:2283–2288.
137. Fankhauser RL, Monroe SS, Noel JS, Humphrey CD, Bresee JS, Parashar UD, Ando T, Glass RI. 2002. Epidemiologic and molecular trends of "Norwalk-like viruses" associated with outbreaks of gastroenteritis in the United States. *J Infect Dis* 186:1–7.
138. Allen DJ, Noad R, Samuel D, Gray JJ, Roy P, Iturriza-Gomara M. 2009. Characterisation of a GII-4 norovirus variant-specific surface-exposed site involved in antibody binding. *Virol J* 6:150.
139. Lindesmith LC, Beltramello M, Donaldson EF, Corti D, Swanstrom J, Debbink K, Lanzavecchia A, Baric RS. 2012. Immunogenetic mechanisms driving norovirus GII. 4 antigenic variation. *PLoS Pathog* 8:e1002705.
140. Lindesmith LC, Donaldson EF, Lobue AD, Cannon JL, Zheng DP, Vinje J, Baric RS. 2008. Mechanisms of GII.4 norovirus persistence in human populations. *PLoS Med* 5:e31.
141. Allen DJ, Gray JJ, Gallimore CI, Xerry J, Iturriza-Gomara M. 2008. Analysis of amino acid variation in the P2 domain of the GII-4 norovirus VP1 protein reveals putative variant-specific epitopes. *PLoS One* 3:e1485.
142. Arana A, Cilla G, Montes M, Gomariz M, Perez-Trallero E. 2014. Genotypes, recombinant forms, and variants of norovirus GII.4 in Gipuzkoa (Basque Country, Spain), 2009–2012. *PLoS One* 9:e98875.
143. Bull RA, Tanaka MM, White PA. 2007. Norovirus recombination. *J Gen Virol* 88:3347–3359.
144. Chhabra P, Walimbe AM, Chitambar SD. 2010. Molecular characterization of three novel intergenotype norovirus GII recombinant strains from western India. *Virus Res* 147:242–246.
145. Worobey M, Holmes EC. 1999. Evolutionary aspects of recombination in RNA viruses. *J Gen Virol* 80 (Pt 10):2535–2543.
146. Lai MM. 1992. RNA recombination in animal and plant viruses. *Microbiol Rev* 56:61–79.
147. Cooper P, Steiner-Pryor A, Scotti P, Delong D. 1974. On the nature of poliovirus genetic recombinants. *J Gen Virol* 23:41–49.
148. Ando T, Mulders MN, Lewis DC, Estes MK, Monroe SS, Glass RI. 1994. Comparison of the polymerase region of small round structured virus strains previously classified in three antigenic types by solid-phase immune electron microscopy. *Arch Virol* 135:217–226.
149. Green J, Vinje J, Gallimore CI, Koopmans M, Hale A, Brown DW, Clegg JC, Chamberlain J. 2000. Capsid protein diversity among Norwalk-like viruses. *Virus Genes* 20:227–236.
150. Hardy ME, Kramer SF, Treanor JJ, Estes MK. 1997. Human calicivirus genogroup II capsid sequence diversity revealed by analyses of the prototype Snow Mountain agent. *Arch Virol* 142:1469–1479.
151. Kageyama T, Shinohara M, Uchida K, Fukushi S, Hoshino FB, Kojima S, Takai R, Oka T, Takeda N, Katayama K. 2004. Coexistence of multiple genotypes, including newly identified genotypes, in outbreaks of gastroenteritis due to Norovirus in Japan. *J Clin Microbiol* 42:2988–2995.
152. Vinje J, Hamidjaja RA, Sobsey MD. 2004. Development and application of a capsid VP1 (region D) based reverse transcription PCR assay for genotyping of genogroup I and II noroviruses. *J Virol Methods* 116:109–117.
153. Bull RA, Hansman GS, Clancy LE, Tanaka MM, Rawlinson WD, White PA. 2005. Norovirus recombination in ORF1/ORF2 overlap. *Emerg Infect Dis* 11:1079–1085.
154. Rohayem J, Munch J, Rethwilm A. 2005. Evidence of recombination in the norovirus capsid gene. *J Virol* 79:4977–4990.
155. Waters A, Coughlan S, Hall WW. 2007. Characterisation of a novel recombination event in the norovirus polymerase gene. *Virology* 363:11–14.

156. Hoa Tran TN, Trainor E, Nakagomi T, Cunliffe NA, Nakagomi O. 2013. Molecular epidemiology of noroviruses associated with acute sporadic gastroenteritis in children: Global distribution of genogroups, genotypes and GII.4 variants. *J Clin Virol* 56:269–277.
157. Reuter G, Vennema H, Koopmans M, Szucs G. 2006. Epidemic spread of recombinant noroviruses with four capsid types in Hungary. *J Clin Virol* 35:84–88.
158. Ahmed SM, Hall AJ, Robinson AE, Verhoef L, Premkumar P, Parashar UD, Koopmans M, Lopman BA. 2014. Global prevalence of norovirus in cases of gastroenteritis: A systematic review and meta-analysis. *Lancet Infect Dis* 14:725–730.
159. Atmar RL, Opekun AR, Gilger MA, Estes MK, Crawford SE, Neill FH, Graham DY. 2008. Norwalk virus shedding after experimental human infection. *Emerg Infect Dis* 14:1553–1557.
160. Tu ET, Bull RA, Kim MJ, McIver CJ, Heron L, Rawlinson WD, White PA. 2008. Norovirus excretion in an aged-care setting. *J Clin Microbiol* 46:2119–2121.
161. Murata T, Katsushima N, Mizuta K, Muraki Y, Hongo S, Matsuzaki Y. 2007. Prolonged norovirus shedding in infants < or = 6 months of age with gastroenteritis. *Pediatr Infect Dis J* 26:46–49.
162. Nilsson M, Hedlund KO, Thorhagen M, Larson G, Johansen K, Ekspong A, Svensson L. 2003. Evolution of human calicivirus RNA in vivo: Accumulation of mutations in the protruding P2 domain of the capsid leads to structural changes and possibly a new phenotype. *J Virol* 77:13117–13124.
163. Mattner F, Sohr D, Heim A, Gastmeier P, Vennema H, Koopmans M. 2006. Risk groups for clinical complications of norovirus infections: An outbreak investigation. *Clin Microbiol Infect* 12:69–74.
164. Agus SG, Dolin R, Wyatt RG, Tousimis A, Northrup RS. 1973. Acute infectious nonbacterial gastroenteritis: Intestinal histopathology histologic and enzymatic alterations during illness produced by the Norwalk agent in man. *Ann Intern Med* 79:18–25.
165. Schreiber DS, Blacklow NR, Trier JS. 1973. The mucosal lesion of the proximal small intestine in acute infectious nonbacterial gastroenteritis. *New Engl J Med* 288:1318–1323.
166. Karst SM. 2010. Pathogenesis of noroviruses, emerging RNA viruses. *Viruses* 2:748–781.
167. Chen SY, Tsai CN, Lai MW, Chen CY, Lin KL, Lin TY, Chiu CH. 2009. Norovirus infection as a cause of diarrhea-associated benign infantile seizures. *Clin Infect Dis* 48:849–855.
168. Khan RR, Lawson AD, Minnich LL, Martin K, Nasir A, Emmett MK, Welch CA, Udall JN, Jr. 2009. Gastrointestinal norovirus infection associated with exacerbation of inflammatory bowel disease. *J Pediatr Gastroenterol Nutr* 48:328–333.
169. Turcios-Ruiz RM, Axelrod P, St John K, Bullitt E, Donahue J, Robinson N, Friss HE. 2008. Outbreak of necrotizing enterocolitis caused by norovirus in a neonatal intensive care unit. *J Pediatr* 153:339–344.
170. Gray JJ, Cunliffe C, Ball J, Graham DY, Desselberger U, Estes MK. 1994. Detection of immunoglobulin M (IgM), IgA, and IgG Norwalk virus-specific antibodies by indirect enzyme-linked immunosorbent assay with baculovirus-expressed Norwalk virus capsid antigen in adult volunteers challenged with Norwalk virus. *J Clin Microbiol* 32:3059–3063.
171. Parrino TA, Schreiber DS, Trier JS, Kapikian AZ, Blacklow NR. 1977. Clinical immunity in acute gastroenteritis caused by Norwalk agent. *N Engl J Med* 297:86–89.
172. Johnson PC, Mathewson JJ, DuPont HL, Greenberg HB. 1990. Multiple-challenge study of host susceptibility to Norwalk gastroenteritis in US adults. *J Infect Dis* 161:18–21.
173. Medzhitov R. 2007. Recognition of microorganisms and activation of the immune response. *Nature* 449:819–826.
174. Changotra H, Jia Y, Moore TN, Liu G, Kahan SM, Sosnovtsev SV, Karst SM. 2009. Type I and type II interferons inhibit the translation of murine norovirus proteins. *J Virol* 83:5683–5692.
175. Kim Y, Lovell S, Tiew KC, Mandadapu SR, Alliston KR, Battaile KP, Groutas WC, Chang KO. 2012. Broad-spectrum antivirals against 3C or 3C-like proteases of picornaviruses, noroviruses, and coronaviruses. *J Virol* 86:11754–11762.
176. Honda K, Takaoka A, Taniguchi T. 2006. Type I inteferon gene induction by the interferon regulatory factor family of transcription factors. *Immunity* 25:349–360.
177. Collins T, Read M, Neish A, Whitley M, Thanos D, Maniatis T. 1995. Transcriptional regulation of endothelial cell adhesion molecules: NF-kappa B and cytokine-inducible enhancers. *FASEB J* 9:899–909.
178. Tamura T, Yanai H, Savitsky D, Taniguchi T. 2008. The IRF family transcription factors in immunity and oncogenesis. *Annu Rev Immunol* 26:535–584.
179. McCartney SA, Thackray LB, Gitlin L, Gilfillan S, Virgin HW, Colonna M. 2008. MDA-5 recognition of a murine norovirus. *PLoS Pathog* 4:e1000108.
180. Basler CF, Garcia-Sastre A. 2002. Viruses and the type I interferon antiviral system: Induction and evasion. *Int Rev Immunol* 21:305–337.

181. Wobus CE, Karst SM, Thackray LB, Chang KO, Sosnovtsev SV, Belliot G, Krug A, Mackenzie JM, Green KY, Virgin HW. 2004. Replication of Norovirus in cell culture reveals a tropism for dendritic cells and macrophages. *PLoS Biol* 2:e432.
182. Chen Z, Sosnovtsev SV, Bok K, Parra GI, Makiya M, Agulto L, Green KY, Purcell RH. 2013. Development of Norwalk virus-specific monoclonal antibodies with therapeutic potential for the treatment of Norwalk virus gastroenteritis. *J Virol* 87:9547–9557.
183. Bartsch SM, Lopman BA, Hall AJ, Parashar UD, Lee BY. 2012. The potential economic value of a human norovirus vaccine for the United States. *Vaccine* 30:7097–7104.
184. Bernstein DI, Atmar J, Lyon M, Treanor J, Chen WH, Frenck R, Jiang X, et al. 2013. An intramuscular (IM) bivalent Norovirus GI.1/GII.4 virus like particle (VLP) vaccine protects against vomiting and diarrhea in an experimental human GII.4 oral challenge study., abstr IDWeek 2013, San Francisco, CA., October 2–6.
185. Tamminen K, Lappalainen S, Huhti L, Vesikari T, Blazevic V. 2013. Trivalent combination vaccine induces broad heterologous immune responses to norovirus and rotavirus in mice. *PloS One* 8:e70409.
186. Bok K, Prikhodko VG, Green KY, Sosnovtsev SV. 2009. Apoptosis in murine norovirus-infected RAW264.7 cells is associated with downregulation of survivin. *J Virol* 83:3647–3656.
187. Furman LM, Maaty WS, Petersen LK, Ettayebi K, Hardy ME, Bothner B. 2009. Cysteine protease activation and apoptosis in murine norovirus infection. *Virol J* 6:139.
188. Ambrosini G, Adida C, Altieri DC. 1997. A novel anti-apoptosis gene, survivin, expressed in cancer and lymphoma. *Nat Med* 3:917–921.
189. Cohen G. 1997. Caspases: The executioners of apoptosis. *Biochem J* 326:1–16.
190. Sosnovtsev SV, Belliot G, Chang KO, Prikhodko VG, Thackray LB, Wobus CE, Karst SM, Virgin HW, Green KY. 2006. Cleavage map and proteolytic processing of the murine norovirus nonstructural polyprotein in infected cells. *J Virol* 80:7816–7831.
191. Harrington PR, Lindesmith L, Yount B, Moe CL, Baric RS. 2002. Binding of Norwalk virus-like particles to ABH histo-blood group antigens is blocked by antisera from infected human volunteers or experimentally vaccinated mice. *J Virol* 76:12335–12343.
192. Hutson AM, Atmar RL, Marcus DM, Estes MK. 2003. Norwalk virus-like particle hemagglutination by binding to h histo-blood group antigens. *J Virol* 77:405–415.
193. Bu W, Mamedova A, Tan M, Xia M, Jiang X, Hegde RS. 2008. Structural basis for the receptor binding specificity of Norwalk virus. *J Virol* 82:5340–5347.
194. Marionneau S, Cailleau-Thomas A, Rocher J, Le Moullac-Vaidye B, Ruvoen N, Clement M, Le Pendu J. 2001. ABH and Lewis histo-blood group antigens, a model for the meaning of oligosaccharide diversity in the face of a changing world. *Biochimie* 83:565–573.
195. Kelly RJ, Rouquier S, Giorgi D, Lennon GG, Lowe JB. 1995. Sequence and expression of a candidate for the human secretor blood group alpha(1,2)fucosyltransferase gene (FUT2). Homozygosity for an enzyme-inactivating nonsense mutation commonly correlates with the non-secretor phenotype. *J Biol Chem* 270:4640–4649.
196. Thorven M, Grahn A, Hedlund KO, Johansson H, Wahlfrid C, Larson G, Svensson L. 2005. A homozygous nonsense mutation (428G–>A) in the human secretor (FUT2) gene provides resistance to symptomatic norovirus (GGII) infections. *J Virol* 79:15351–15355.
197. Yamamoto F-i, Clausen H, White T, Marken J, Hakomori S-i. 1990. Molecular genetic basis of the histo-blood group ABO system. *Nature* 345:229–233.
198. Orntoft T, Holmes E, Johnson P, Hakomori S, Clausen H. 1991. Differential tissue expression of the Lewis blood group antigens: Enzymatic, immunohistologic, and immunochemical evidence for Lewis a and b antigen expression in Le (ab-) individuals. *Blood* 77:1389–1396.
199. Nordgren J, Kindberg E, Lindgren PE, Matussek A, Svensson L. 2010. Norovirus gastroenteritis outbreak with a secretor-independent susceptibility pattern, Sweden. *Emerg Infect Dis* 16:81–87.
200. Lindesmith L, Moe C, Marionneau S, Ruvoen N, Jiang X, Lindblad L, Stewart P, LePendu J, Baric R. 2003. Human susceptibility and resistance to Norwalk virus infection. *Nat Med* 9:548–553.
201. Rockx BH, Vennema H, Hoebe CJ, Duizer E, Koopmans MP. 2005. Association of histo-blood group antigens and susceptibility to norovirus infections. *J Infect Dis* 191:749–754.
202. Atmar RL, Bernstein DI, Harro CD, Al-Ibrahim MS, Chen WH, Ferreira J, Estes MK, Graham DY, Opekun AR, Richardson C, Mendelman PM. 2011. Norovirus vaccine against experimental human Norwalk virus illness. *N Engl J Med* 365:2178–2187.
203. Tan M, Huang P, Xia M, Fang PA, Zhong W, McNeal M, Wei C, Jiang W, Jiang X. 2011. Norovirus P particle, a novel platform for vaccine development and antibody production. *J Virol* 85:753–764.
204. Fang H, Tan M, Xia M, Wang L, Jiang X. 2013. Norovirus P particle efficiently elicits innate, humoral and cellular immunity. *PLoS One* 8:e63269.

205. Huang P, Farkas T, Marionneau S, Zhong W, Ruvoën-Clouet N, Morrow AL, Altaye M, et al. 2003. Noroviruses bind to human ABO, Lewis, and secretor histo-blood group antigens: Identification of 4 distinct strain-specific patterns. *J Infect Dis* 188:19–31.
206. Lindesmith L, Moe C, Lependu J, Frelinger JA, Treanor J, Baric RS. 2005. Cellular and humoral immunity following Snow Mountain virus challenge. *J Virol* 79:2900–2909.
207. Larsson MM, Rydell GE, Grahn A, Rodriguez-Diaz J, Akerlind B, Hutson AM, Estes MK, Larson G, Svensson L. 2006. Antibody prevalence and titer to norovirus (genogroup II) correlate with secretor (FUT2) but not with ABO phenotype or Lewis (FUT3) genotype. *J Infect Dis* 194:1422–1427.
208. Hutson AM, Airaud F, LePendu J, Estes MK, Atmar RL. 2005. Norwalk virus infection associates with secretor status genotyped from sera. *J Med Virol* 77:116–120.
209. Le Pendu J, Ruvoën-Clouet N, Kindberg E, Svensson L. 2006. Mendelian resistance to human norovirus infections. *Semin Immunol* 18:375–386.
210. Teunis PF, Moe CL, Liu P, Miller SE, Lindesmith L, Baric RS, Le Pendu J, Calderon RL. 2008. Norwalk virus: How infectious is it? *J Med Virol* 80:1468–1476.
211. Atmar RL, Opekun AR, Gilger MA, Estes MK, Crawford SE, Neill FH, Ramani S, Hill H, Ferreira J, Graham DY. 2014. Determination of the 50% human infectious dose for Norwalk virus. *J Infect Dis* 209:1016–1022.
212. Chan MC, Sung JJ, Lam RK, Chan PK, Lee NL, Lai RW, Leung WK. 2006. Fecal viral load and norovirus-associated gastroenteritis. *Emerg Infect Dis* 12:1278–1280.
231. Lopman B, Gastañaduy P, Park GW, Hall AJ, Parashar UD, Vinjé J. 2012. Environmental transmission of norovirus gastroenteritis. *Curr Opin Virol* 2:96–102.
214. Kroneman A, Verhoef L, Harris J, Vennema H, Duizer E, van Duynhoven Y, Gray J, et al. 2008. Analysis of integrated virological and epidemiological reports of norovirus outbreaks collected within the Foodborne Viruses in Europe network from 1 July 2001 to 30 June 2006. *J Clin Microbiol* 46:2959–2965.
215. Hall AJ, Vinje J, Lopman B, Park GW, Yen C, Gregoricus N, Parashar U. 2011. *Updated Norovirus Outbreak Management and Disease Prevention Guidelines.* National Center for Immunization and Respiratory Diseases, Centers for Disease Control and Prevention.
216. Isakbaeva ET, Widdowson M-A, Beard RS, Bulens SN, Mullins J, Monroe SS, Bresee J, Sassano P, Cramer EH, Glass RI. 2005. Norovirus transmission on cruise ship. *Emerg Infect Dis* 11:154.
217. Mayet A, Andréo V, Bédubourg G, Victorion S, Plantec J, Soullié B, Meynard J, Dedieu J, Polvèche P, Migliani R. 2011. Food-borne outbreak of norovirus infection in a French military parachuting unit, April 2011. *Euro Surveill* 16:1–3.
218. Kirk MD, Fullerton KE, Hall GV, Gregory J, Stafford R, Veitch MG, Becker N. 2010. Surveillance for outbreaks of gastroenteritis in long-term care facilities, Australia, 2002–2008. *Clin Infect Dis* 51:907–914.
219. Lopman BA, Reacher MH, Van Duijnhoven Y, Hanon FX, Brown D, Koopmans M. 2003. Viral gastroenteritis outbreaks in Europe, 1995–2000. *Emerg Infect Dis* 9:90–96.
220. Hall AJ, Wikswo ME, Manikonda K, Roberts VA, Yoder JS, Gould LH. 2013. Acute gastroenteritis surveillance through the National Outbreak Reporting System, United States. *Emerg Infect Dis* 19:1305–1309.
221. Lopman B, van Duynhoven Y, Hanon FX, Reacher M, Koopmans M, Brown D, Consortium on Foodborne Viruses in Europe. 2002. Laboratory capability in Europe for foodborne viruses. *Euro Surveill* 7:61–65.
222. Verhoef L, Vennema H, van Pelt W, Lees D, Boshuizen H, Henshilwood K, Koopmans M, Food-Borne Viruses in Europe Network. 2010. Use of norovirus genotype profiles to differentiate origins of foodborne outbreaks. *Emerg Infect Dis* 16:617–624.
223. Hall AJ, Wikswo ME, Pringle K, Gould LH, Parashar UD, Division of Viral Diseases NCfI, Respiratory Diseases CDC. 2014. Vital signs: Foodborne norovirus outbreaks—United States, 2009–2012. *MMWR Morb Mortal Wkly Rep* 63:491–495.
224. Richards GP. 2001. Enteric virus contamination of foods through industrial practices: A primer on intervention strategies. *J Ind Microbiol Biotechnol* 27:117–125.
225. Mathijs E, Stals A, Baert L, Botteldoorn N, Denayer S, Mauroy A, Scipioni A, et al. 2012. A review of known and hypothetical transmission routes for noroviruses. *Food Environ Virol* 4:131–152.
226. Verhoef L, Hewitt J, Barclay L, Ahmed SM, Lake R, Hall AJ, Lopman B, et al. 2015. Norovirus genotype profiles associated with foodborne transmission, 1999–2012. *Emerg Infect Dis* 21:592–599.
227. Hall AJ, Eisenbart VG, Etingue AL, Gould LH, Lopman BA, Parashar UD. 2012. Epidemiology of foodborne norovirus outbreaks, United States, 2001–2008. *Emerg Infect Dis* 18:1566–1573.

228. Painter JA, Ayers T, Woodruff R, Blanton E, Perez N, Hoekstra RM, Griffin PM, Braden C. 2009. Recipes for foodborne outbreaks: A scheme for categorizing and grouping implicated foods. *Foodborne Pathog Dis* 6:1259–1264.
229. Painter JA, Hoekstra RM, Ayers T, Tauxe RV, Braden CR, Angulo FJ, Griffin PM. 2013. Attribution of foodborne illnesses, hospitalizations, and deaths to food commodities by using outbreak data, United States, 1998–2008. *Emerg Infect Dis* 19:407–415.
230. Cheesbrough JS, Green J, Gallimore CI, Wright PA, Brown DW. 2000. Widespread environmental contamination with Norwalk-like viruses (NLV) detected in a prolonged hotel outbreak of gastroenteritis. *Epidemiol Infect* 125:93–98.
231. Rusinol M, Fernandez-Cassi X, Timoneda N, Carratala A, Abril JF, Silvera C, Figueras MJ, et al. 2015. Evidence of viral dissemination and seasonality in a Mediterranean river catchment: Implications for water pollution management. *J Environ Manage* 159:58–67.
232. Seitz SR, Leon JS, Schwab KJ, Lyon GM, Dowd M, McDaniels M, Abdulhafid G, et al. 2011. Norovirus infectivity in humans and persistence in water. *Appl Environ Microbiol* 77:6884–6888.
233. Robertson LJ. 2007. The potential for marine bivalve shellfish to act as transmission vehicles for outbreaks of protozoan infections in humans: A review. *Int J Food Microbiol* 120:201–216.
234. Lodo KL, Veitch MG, Green ML. 2014. An outbreak of norovirus linked to oysters in Tasmania. *Commun Dis Intell Q Rep* 38:E16–E19.
235. Le Guyader FS, Atmar RL, Le Pendu J. 2012. Transmission of viruses through shellfish: When specific ligands come into play. *Curr Opin Virol* 2:103–110.
236. Dolin R, Blacklow NR, DuPont H, Buscho RF, Wyatt RG, Kasel JA, Hornick R, Chanock RM. 1972. Biological properties of Norwalk agent of acute infectious nonbacterial gastroenteritis. *Proc Soc Exp Biol Med* 140:578–583.
237. Webby RJ, Carville KS, Kirk MD, Greening G, Ratcliff RM, Crerar SK, Dempsey K, et al. 2007. Internationally distributed frozen oyster meat causing multiple outbreaks of norovirus infection in Australia. *Clin Infect Dis* 44:1026–1031.
238. Koff RS, Sear HS. 1967. Internal temperature of steamed clams. *N Engl J Med* 276:737–739.
239. McDonnell S, Kirkland KB, Hlady WG, Aristeguieta C, Hopkins RS, Monroe SS, Glass RI. 1997. Failure of cooking to prevent shellfish-associated viral gastroenteritis. *Arch Intern Med* 157:111–116.
240. Ma LP, Zhao F, Yao L, Li XG, Zhou DQ, Zhang RL. 2013. The presence of genogroup II norovirus in retail shellfish from seven coastal cities in China. *Food Environ Virol* 5:81–86.
241. Boxman IL, Tilburg JJ, Te Loeke NA, Vennema H, Jonker K, de Boer E, Koopmans M. 2006. Detection of noroviruses in shellfish in the Netherlands. *Int J Food Microbiol* 108:391–396.
242. Mattison K. 2011. Chapter 1—Norovirus as a foodborne disease hazard. *In* Taylor S (ed), *Advances in Food and Nutrition Research*, vol 62, pp. 1–39. Academic Press, Burlington, MA.
243. Friedman DS, Heisey-Grove D, Argyros F, Berl E, Nsubuga J, Stiles T, Fontana J, et al. 2005. An outbreak of norovirus gastroenteritis associated with wedding cakes. *Epidemiol Infect* 133:1057–1063.
244. Makary P, Maunula L, Niskanen T, Kuusi M, Virtanen M, Pajunen S, Ollgren J, Tran Minh NN. 2009. Multiple norovirus outbreaks among workplace canteen users in Finland, July 2006. *Epidemiol Infect* 137:402–407.
245. Boxman I, Dijkman R, Verhoef L, Maat A, van Dijk G, Vennema H, Koopmans M. 2009. Norovirus on swabs taken from hands illustrate route of transmission: A case study. *J Food Prot* 72:1753–1755.
246. Werber D, Lausevic D, Mugosa B, Vratnica Z, Ivanovic-Nikolic L, Zizic L, Alexandre-Bird A, et al. 2009. Massive outbreak of viral gastroenteritis associated with consumption of municipal drinking water in a European capital city. *Epidemiol Infect* 137:1713–1720.
247. Tyrrel SF, Quinton JN. 2003. Overland flow transport of pathogens from agricultural land receiving faecal wastes. *J Appl Microbiol* 94 Suppl:87S–93S.
248. Sivapalasingam S, Friedman CR, Cohen L, Tauxe RV. 2004. Fresh produce: A growing cause of outbreaks of foodborne illness in the United States, 1973 through 1997. *J Food Prot* 67:2342–2353.
249. Ethelberg S, Lisby M, Böttiger B, Schultz AC, Villif A, Jensen T, Olsen KE, Scheutz F, Kjelsø C, Müller L. 2010. Outbreaks of gastroenteritis linked to lettuce, Denmark, January 2010. *Euro Surveill* 15(6):pii–19484.
250. Maunula L, Roivainen M, Keranen M, Makela S, Soderberg K, Summa M, von Bonsdorff CH, et al. 2009. Detection of human norovirus from frozen raspberries in a cluster of gastroenteritis outbreaks. *Euro Surveill* 14(49):pii–19435.
251. Made D, Trubner K, Neubert E, Hohne M, Johne R. 2013. Detection and typing of norovirus from frozen strawberries involved in a large-scale gastroenteritis outbreak in Germany. *Food Environ Virol* 5(3): 162–168.

252. Patel MM, Hall AJ, Vinje J, Parashar UD. 2009. Noroviruses: A comprehensive review. *J Clin Virol* 44:1–8.
253. MacCannell T, Umscheid CA, Agarwal RK, Lee I, Kuntz G, Stevenson KB, Healthcare Infection Control Practices Advisory Committee H. 2011. Guideline for the prevention and control of norovirus gastroenteritis outbreaks in healthcare settings. *Infect Control Hosp Epidemiol* 32:939–969.
254. Weinstein RA, Said MA, Perl TM, Sears CL. 2008. Gastrointestinal flu: Norovirus in health care and long-term care facilities. *Clin Infect Dis* 47:1202–1208.
255. Pittet D, Allegranzi B, Sax H, Dharan S, Pessoa-Silva CL, Donaldson L, Boyce JM. 2006. Evidence-based model for hand transmission during patient care and the role of improved practices. *Lancet Infect Dis* 6:641–652.
256. Dennehy PH. 2011. Viral gastroenteritis in children. *Pediatr Infect Dis J* 30:63–64.
257. Bok K, Green KY. 2012. Norovirus gastroenteritis in immunocompromised patients. *N Engl J Med* 367:2126–2132.
258. Guerrero RA, Ball JM, Krater SS, Pacheco SE, Clements JD, Estes MK. 2001. Recombinant Norwalk virus-like particles administered intranasally to mice induce systemic and mucosal (fecal and vaginal) immune responses. *J Virol* 75:9713–9722.
259. Bernstein DI, Atmar RL, Lyon GM, Treanor JJ, Chen WH, Jiang X, Vinjé J, et al. 2015. Norovirus vaccine against experimental human GII.4 virus illness: A challenge study in healthy adults. *J Infect Dis* 211:870–878.
260. Svraka S, Vennema H, van der Veer B, Hedlund KO, Thorhagen M, Siebenga J, Duizer E, Koopmans M. 2010. Epidemiology and genotype analysis of emerging sapovirus-associated infections across Europe. *J Clin Microbiol* 48:2191–2198.
261. Chiba S, Nakata S, Numata-Kinoshita K, Honma S. 2000. Sapporo virus: History and recent findings. *J Infect Dis* 181 Suppl 2:S303–S308.
262. Chen R, Neill JD, Noel JS, Hutson AM, Glass RI, Estes MK, Prasad BV. 2004. Inter- and intragenus structural variations in caliciviruses and their functional implications. *J Virol* 78:6469–6479.
263. Miyazaki N, Taylor DW, Hansman GS, Murata K. 2015. Antigenic and cryo-electron microscopy structure analysis of a chimeric sapovirus capsid. *J Virol* 90:2664–2675.
264. Oka T, Wang Q, Katayama K, Saif LJ. 2015. Comprehensive review of human sapoviruses. *Clin Microbiol Rev* 28:32–53.
265. Scheuer KA, Oka T, Hoet AE, Gebreyes WA, Molla BZ, Saif LJ, Wang Q. 2013. Prevalence of porcine noroviruses, molecular characterization of emerging porcine sapoviruses from finisher swine in the United States, and unified classification scheme for sapoviruses. *J Clin Microbiol* 51:2344–2353.
266. Hansman GS, Oka T, Sakon N, Takeda N. 2007. Antigenic diversity of human sapoviruses. *Emerg Infect Dis* 13:1519–1525.
267. Farkas T, Zhong WM, Jing Y, Huang PW, Espinosa SM, Martinez N, Morrow AL, Ruiz-Palacios GM, Pickering LK, Jiang X. 2004. Genetic diversity among sapoviruses. *Arch Virol* 149:1309–1323.
268. Clarke IN, Lambden PR. 2000. Organization and expression of calicivirus genes. *J Infect Dis* 181 Suppl 2:S309–S316.
269. Tse H, Chan WM, Li KS, Lau SK, Woo PC, Yuen KY. 2012. Discovery and genomic characterization of a novel bat sapovirus with unusual genomic features and phylogenetic position. *PLoS One* 7:e34987.
270. ViralZone. 2015. Sapovirus, on Swiss Institute of Bioinformatics. http://viralzone.expasy.org/all_by_species/196.html. Accessed November 21, 2015.
271. Katayama K, Miyoshi T, Uchino K, Oka T, Tanaka T, Takeda N, Hansman GS. 2004. Novel recombinant sapovirus. *Emerg Infect Dis* 10:1874–1876.
272. Hansman GS, Takeda N, Oka T, Oseto M, Hedlund KO, Katayama K. 2005. Intergenogroup recombination in sapoviruses. *Emerg Infect Dis* 11:1916–1920.
293. Hansman GS, Ishida S, Yoshizumi S, Miyoshi M, Ikeda T, Oka T, Takeda N. 2007. Recombinant sapovirus gastroenteritis, Japan. *Emerg Infect Dis* 13:786–788.
274. Nakata S, Chiba S, Terashima H, Sakuma Y, Kogasaka R, Nakao T. 1983. Microtiter solid-phase radioimmunoassay for detection of human calicivirus in stools. *J Clin Microbiol* 17:198–201.
275. Nakata S, Estes MK, Chiba S. 1988. Detection of human calicivirus antigen and antibody by enzyme-linked immunosorbent assays. *J Clin Microbiol* 26:2001–2005.
276. Hansman GS, Katayama K, Maneekarn N, Peerakome S, Khamrin P, Tonusin S, Okitsu S, Nishio O, Takeda N, Ushijima H. 2004. Genetic diversity of norovirus and sapovirus in hospitalized infants with sporadic cases of acute gastroenteritis in Chiang Mai, Thailand. *J Clin Microbiol* 42:1305–1307.
277. Okada M, Shinozaki K, Ogawa T, Kaiho I. 2002. Molecular epidemiology and phylogenetic analysis of Sapporo-like viruses. *Arch Virol* 147:1445–1451.

278. Vinje J, Deijl H, van der Heide R, Lewis D, Hedlund KO, Svensson L, Koopmans MP. 2000. Molecular detection and epidemiology of Sapporo-like viruses. *J Clin Microbiol* 38:530–536.
279. Jiang X, Huang PW, Zhong WM, Farkas T, Cubitt DW, Matson DO. 1999. Design and evaluation of a primer pair that detects both Norwalk- and Sapporo-like caliciviruses by RT-PCR. *J Virol Methods* 83:145–154.
280. Okada M, Yamashita Y, Oseto M, Shinozaki K. 2006. The detection of human sapoviruses with universal and genogroup-specific primers. *Arch Virol* 151:2503–2509.
281. Chan MC, Sung JJ, Lam RK, Chan PK, Lai RW, Leung WK. 2006. Sapovirus detection by quantitative real-time RT-PCR in clinical stool specimens. *J Virol Methods* 134:146–153.
282. Gunson RN, Bennett S, Maclean A, Carman WF. 2008. Using multiplex real time PCR in order to streamline a routine diagnostic service. *J Clin Virol* 43:372–375.
283. Gunson RN, Collins TC, Carman WF. 2006. The real-time detection of sapovirus. *J Clin Virol* 35:321–322.
284. Logan C, O'Leary JJ, O'Sullivan N. 2007. Real-time reverse transcription PCR detection of norovirus, sapovirus and astrovirus as causative agents of acute viral gastroenteritis. *J Virol Methods* 146:36–44.
285. Mauroy A, Van der Poel WH, der Honing RH, Thys C, Thiry E. 2012. Development and application of a SYBR green RT-PCR for first line screening and quantification of porcine sapovirus infection. *BMC Vet Res* 8:193.
286. Oka T, Katayama K, Hansman GS, Kageyama T, Ogawa S, Wu FT, White PA, Takeda N. 2006. Detection of human sapovirus by real-time reverse transcription-polymerase chain reaction. *J Med Virol* 78:1347–1353.
287. Yan H, Yagyu F, Okitsu S, Nishio O, Ushijima H. 2003. Detection of norovirus (GI, GII), Sapovirus and astrovirus in fecal samples using reverse transcription single-round multiplex PCR. *J Virol Methods* 114:37–44.
288. Harrington PR, Vinje J, Moe CL, Baric RS. 2004. Norovirus capture with histo-blood group antigens reveals novel virus–ligand interactions. *J Virol* 78:3035–3045.
289. Kim DS, Hosmillo M, Alfajaro MM, Kim JY, Park JG, Son KY, Ryu EH, et al. 2014. Both alpha2,3- and alpha2,6-linked sialic acids on O-linked glycoproteins act as functional receptors for porcine Sapovirus. *PLoS Pathog* 10:e1004172.
290. Lee RM, Lessler J, Lee RA, Rudolph KE, Reich NG, Perl TM, Cummings DA. 2013. Incubation periods of viral gastroenteritis: A systematic review. *BMC Infect Dis* 13:446.
291. Lee LE, Cebelinski EA, Fuller C, Keene WE, Smith K, Vinje J, Besser JM. 2012. Sapovirus outbreaks in long-term care facilities, Oregon and Minnesota, USA, 2002–2009. *Emerg Infect Dis* 18:873–876.
292. Wang Q, Zhang Z, Saif LJ. 2012. Stability of and attachment to lettuce by a culturable porcine sapovirus surrogate for human caliciviruses. *Appl Environ Microbiol* 78:3932–3940.
293. Hansman GS, Saito H, Shibata C, Ishizuka S, Oseto M, Oka T, Takeda N. 2007. Outbreak of gastroenteritis due to sapovirus. *J Clin Microbiol* 45:1347–1349.
294. Kitajima M, Oka T, Haramoto E, Katayama H, Takeda N, Katayama K, Ohgaki S. 2010. Detection and genetic analysis of human sapoviruses in river water in Japan. *Appl Environ Microbiol* 76:2461–2467.
295. Murray TY, Mans J, Taylor MB. 2013. First detection of human sapoviruses in river water in South Africa. *Water Sci Technol* 67:2776–2783.
296. Haramoto E, Katayama H, Phanuwan C, Ohgaki S. 2008. Quantitative detection of sapoviruses in wastewater and river water in Japan. *Lett Appl Microbiol* 46:408–413.
297. Haramoto E, Kitajima M, Kishida N, Katayama H, Asami M, Akiba M. 2012. Occurrence of viruses and protozoa in drinking water sources of Japan and their relationship to indicator microorganisms. *Food Environ Virol* 4:93–101.
298. Hassan-Rios E, Torres P, Munoz E, Matos C, Hall AJ, Gregoricus N, Vinje J. 2013. Sapovirus gastroenteritis in preschool center, Puerto Rico, 2011. *Emerg Infect Dis* 19:174–175.
299. Kitajima M, Haramoto E, Phanuwan C, Katayama H. 2011. Genotype distribution of human sapoviruses in wastewater in Japan. *Appl Environ Microbiol* 77:4226–4229.
300. Kiulia NM, Netshikweta R, Page NA, Van Zyl WB, Kiraithe MM, Nyachieo A, Mwenda JM, Taylor MB. 2010. The detection of enteric viruses in selected urban and rural river water and sewage in Kenya, with special reference to rotaviruses. *J Appl Microbiol* 109:818–828.
301. Mans J, van Zyl WB, Taylor MB, Page NA, Sobsey MD, Barnard TG, Potgieter N. 2014. Applicability of bio-wipes for the collection of human faecal specimens for detection and characterisation of enteric viruses. *Trop Med Int Health* 19:293–300.
302. Sano D, Perez-Sautu U, Guix S, Pinto RM, Miura T, Okabe S, Bosch A. 2011. Quantification and genotyping of human sapoviruses in the Llobregat river catchment, Spain. *Appl Environ Microbiol* 77:1111–1114.

303. Dey SK, Phan GT, Nishimura S, Mizuguchi M, Okitsu S, Ushijima H. 2010. Molecular and epidemiological trend of sapovirus, and astrovirus infection in Japan. *J Trop Pediatr* 56:205–207.
304. Pang XL, Honma S, Nakata S, Vesikari T. 2000. Human caliciviruses in acute gastroenteritis of young children in the community. *J Infect Dis* 181 Suppl 2:S288–S294.
305. Pang XL, Lee BE, Tyrrell GJ, Preiksaitis JK. 2009. Epidemiology and genotype analysis of sapovirus associated with gastroenteritis outbreaks in Alberta, Canada: 2004–2007. *J Infect Dis* 199:547–551.
306. Kobayashi S, Fujiwara N, Yasui Y, Yamashita T, Hiramatsu R, Minagawa H. 2012. A foodborne outbreak of sapovirus linked to catered box lunches in Japan. *Arch Virol* 157:1995–1997.
307. Matson DO, Estes MK, Tanaka T, Bartlett AV, Pickering LK. 1990. Asymptomatic human calicivirus infection in a day care center. *Pediatr Infect Dis J* 9:190–196.
308. Matson DO, Estes MK, Glass RI, Bartlett AV, Penaranda M, Calomeni E, Tanaka T, Nakata S, Chiba S. 1989. Human calicivirus-associated diarrhea in children attending day care centers. *J Infect Dis* 159:71–78.
309. Iizuka S, Oka T, Tabara K, Omura T, Katayama K, Takeda N, Noda M. 2010. Detection of sapoviruses and noroviruses in an outbreak of gastroenteritis linked genetically to shellfish. *J Med Virol* 82:1247–1254.
310. Hedlund KO, Rubilar-Abreu E, Svensson L. 2000. Epidemiology of calicivirus infections in Sweden, 1994–1998. *J Infect Dis* 181 Suppl 2:S275–S280.
311. Iritani N, Kaida A, Abe N, Kubo H, Sekiguchi J, Yamamoto SP, Goto K, Tanaka T, Noda M. 2014. Detection and genetic characterization of human enteric viruses in oyster-associated gastroenteritis outbreaks between 2001 and 2012 in Osaka City, Japan. *J Med Virol* 86:2019–2025.
312. Phan TG, Trinh QD, Yagyu F, Okitsu S, Ushijima H. 2007. Emergence of rare sapovirus genotype among infants and children with acute gastroenteritis in Japan. *Eur J Clin Microbiol Infect Dis* 26:21–27.
313. Phan TG, Khamrin P, Quang TD, Dey SK, Takanashi S, Okitsu S, Maneekarn N, Ushijima H. 2007. Emergence of intragenotype recombinant sapovirus in Japan. *Infect Genet Evol* 7:542–546.
314. Iizuka S, Takai-Todaka R, Ohshiro H, Kitajima M, Wang Q, Saif LJ, Wakita T, Noda M, Katayama K, Oka T. 2013. Detection of multiple human sapoviruses from imported frozen individual clams. *Food Environ Virol* 5:119–125.
315. Yamashita Y, Ootsuka Y, Kondo R, Oseto M, Doi M, Miyamoto T, Ueda T, et al. 2010. Molecular characterization of Sapovirus detected in a gastroenteritis outbreak at a wedding hall. *J Med Virol* 82:720–726.
316. Hansman GS, Oka T, Okamoto R, Nishida T, Toda S, Noda M, Sano D, et al. 2007. Human sapovirus in clams, Japan. *Emerg Infect Dis* 13:620–622.
317. Benabbes L, Ollivier J, Schaeffer J, Parnaudeau S, Rhaissi H, Nourlil J, Le Guyader FS. 2013. Norovirus and other human enteric viruses in moroccan shellfish. *Food Environ Virol* 5:35–40.

7 Hepatitis A Virus

Michael J. Casteel

CONTENTS

INTRODUCTION

Hepatitis in humans has been described throughout history, but it was not until the twentieth century that distinct forms of the disease were characterized and ascribed to specific infectious agents. Studies involving humans and nonhuman primates (NHPs) and retrospective analyses of outbreaks and cases from the 1940s to the 1960s revealed distinct forms of infectious hepatitis. One form of the disease was transmitted by the fecal–oral route with a relatively short incubation period, and a second form of the disease was transmitted parenterally. These diseases were later defined as hepatitis A and B, respectively. In the early 1970s, virus-like particles in the stools from human patients with hepatitis A were observed by immune electron microscopy (IEM) and presumptively called hepatitis A virus (HAV). In the late 1970s, a major development occurred with the demonstration that HAV could be propagated in cultured cells following serial passage in marmosets. Molecular cloning and complete sequencing of the HAV genome in the 1980s were followed by the licensure of hepatitis A vaccines in the United States in 1995. Due to childhood vaccinations, the number of cases of hepatitis A in the United States has declined dramatically in recent years.

CLASSIFICATION AND BIOLOGY

Taxonomy and Morphology

HAV was first visualized by IEM in 1973 as icosahedral-shaped virions approximately 27 nm in diameter [1]. HAV is included within the *Picornaviridae* family of nonenveloped, positive-sense, single-stranded RNA viruses, although there are enough unique properties of HAV to merit placement in its own genus, *Hepatovirus* [2]. HAV is unusually resistant to heat [3,4], and compared with other picornaviruses displays limited nucleotide homology [5] and different growth properties in cultured cells. In addition, the composition and assembly of the HAV capsid may be different than that of other members of *Picornaviridae* [6], and some evidence suggests that the HAV capsid lacks the so-called canyon feature (the site of cellular receptor binding on other picornaviruses) [7].

Genetic Composition

Molecular cloning [8] and complete sequencing [9] showed that the genome of HAV is approximately 7.5 kb in length. The genome consists of a 5′ untranslated region of (UTR) about 735 nucleotides containing an internal ribosomal entry site, followed by structural (VP1–VP4) and nonstructural protein–encoding regions (e.g. RNA-dependent RNA polymerase), and a 3′ UTR with a terminal poly(A) tract [5,10] (Figure 7.1). Like other members of the *Picornaviridae* family, the 5′ end of the HAV genome does not have a cap structure but instead has a small, covalently-bound, virus-coded protein, designated VPg. The genome organization of HAV is shown in Figure 7.1.

Various strains of HAV share up to 90% similarity at the nucleotide sequence level and 98% similarity at the amino acid level [11]. While only one serotype has been identified [12], genetic analysis of 152 HAV strains from humans and NHPs has identified seven genotypes (four human—I, II, III, VII and three simian—IV, V, and VI), based on nucleotide differences of 15%–20% in the VP1/2A encoding region [13–15]. These studies have shown that circulating human strains of HAV are genetically closely related. Most human HAV strains belong to either genotype I (80% of human strains) or III. Each genotype has two subgenotypes designated IA, IB, IIIA, or IIIB. More recent phylogenetic studies of HAV, using full-length VP1, VP2, and VP3 nucleotide and amino

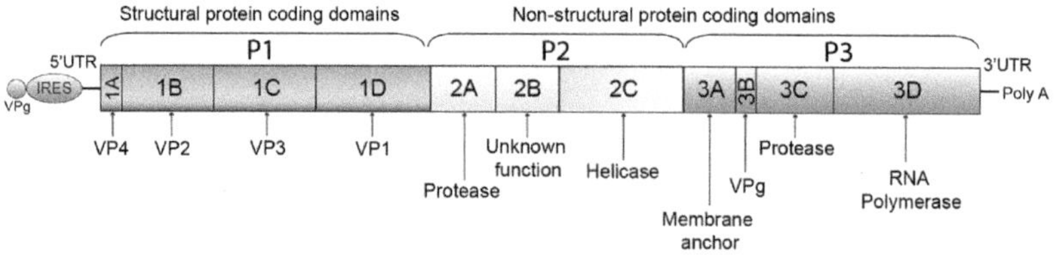

FIGURE 7.1 Hepatitis A virus genome organization. The HAV genome is a single-stranded, positive-sense RNA molecule, approximately 7.5 kb in size. The genome is flanked at the 5′ end by VPg, an internal ribosomal binding site (IRES), and the 5′-UTR, and at the 3′ end by the 3′-UTR and a polyA tail. HAV comprises one open reading frame (ORF) encoding a single polyprotein made up of three functionally distinct segments (P1–P3, segments indicated above the genome). The P1 segment functional domains include the capsid encoding regions (1A–1D), which are translated into the four capsid proteins (VP1–VP4), while P2 and P3 segments comprise the nonstructural protein–encoding regions essential for replication. These segments encode a putative protease (2A), a domain of unknown function (2B), a putative helicase (2C), a membrane anchor (3A), VPg (3B), a protease (3C), and an RNA-dependent RNA polymerase (3D).

acid sequences from 81 HAV isolates from Europe and Central and South America, suggest that a minor reclassification of this system is needed [16–18].

Laboratory-Adapted Strains

There are at least 20 well-characterized strains of HAV, distinguishable from each other in terms of growth properties, nucleotide sequence, or geographical origin [10]. Laboratory-adapted variants of these strains, some of which are currently used for vaccine production and laboratory studies, originate from experimentally and naturally infected humans and NHPs. Experimental infectivity studies of HAV in humans were performed in the middle of the last century, and in NHPs from about that time to the present. The human studies include oral and parenteral studies in volunteers [19], the mentally disabled, institutionalized children [20], and prisoners [21]. It was not until the 1970s that HAV was successfully transmitted to NHPs, including marmosets (*Saguinus mystax*), chimpanzees (*Pan troglodytes*), and owl monkeys (*Aotus trivirgatus*).

Today, isolation of wild-type HAV is possible in primary African green monkey kidney (AGMK) cells, but growth is slow with little or no evidence of cytopathic effects (CPE) and low virus yields [22,23]. For several years following the identification of HAV by IEM, it was thought incorrectly that the virus could not be propagated in cultured cells [24]. Hence, access to adequate amounts of HAV for laboratory studies and vaccine development remained problematic compared with the success of poliovirus culture. An important breakthrough occurred when Provost and Hilleman [22] achieved propagation of HAV *in vitro*, using a marmoset-adapted strain to infect a cloned cell line of fetal rhesus kidney-derived cells (FRhK-6). This was followed by reports of propagation of HAV in other cell lines.

With each subsequent passage, HAV eventually adapts to cell culture, with faster accumulation of intracellular antigen and higher virus yields. Several attenuated, cytopathic, and other variants have been selected after numerous passages in NHPs and/or cells. The most widely used of these variants (HM175 and CR326) originate from infected humans, and strain HM175 has the widest range of the cell-adapted variants [25]. In contrast to wild-type HAV, the cytopathic variants of HM175 typically reach titers of about 10^7 tissue culture infectious doses ($TCID_{50}$) per milliliter (mL) in extracts from infected cells. A classical plaque technique was developed [26,27] and is now widely used in laboratory studies. Attenuated variants of HM175 and CR326 are currently used as the material for inactivated vaccines used in the United States (see the "Cell Infectivity Methods" section).

Environmental Stability

HAV's stability against chemical and physical disinfection plays a role in its persistence and spread in the environment. HAV is stable at pH 1.0 for 2 h at room temperature and retains its infectivity for up to 5 h, which explains how HAV, as an enteric virus, can pass through the human or primate stomach. Other picornaviruses do not fare so well, and lose infectivity after 2 h at pH 1.0. HAV is resistant to some level and duration of heat. The virus is resistant to 60°C for 1 h, partially inactivated at 60°C for 10–12 h [3,28], and inactivated within minutes at 98°C–100°C [29]. HAV retains infectivity when dried and stored at 25°C and 42% humidity for 1 month [30], and may remain infective indefinitely when stored at −20°C. Environments where HAV may survive for months or longer are artificially contaminated fresh water, seawater, wastewater, soils, marine sediments and oysters [31]. HAV is inactivated by a variety of chemicals or mechanisms including ultraviolet radiation, autoclaving, formalin, iodine, or chlorine, with specific thresholds for duration and intensity [10]. Inactivation by chlorine is of particular interest because water treatment plants often use chlorine to reduce the pathogen load in drinking water. For water disinfection, HAV requires 10–15 parts per million (ppm) of residual chlorine over 30 min, or free residual chlorine at 2.0–2.5 ppm for 15 min [32].

INFECTIOUS DOSE, CLINICAL FEATURES, DIAGNOSIS, AND PATHOGENESIS

Infectious Dose

The infectious dose is the concentration of virus required to illicit an immune response and/or cause disease in a percentage of exposed hosts, which is important information for vaccine development and quantitative microbial risk assessment. The concentrations of HAV in stool and serum inoculums from the early human and NHP studies described in the "Laboratory-Adapted Strains" section are unknown; however, some useful information was obtained. The incubation period was found to decrease with a larger amount of inoculum [19], and Krugman et al. [20] reported that 0.1 g of infected stool constituted the minimal human infectious dose. More recent information on the infectivity of HAV has been generated from infectivity studies in NHPs, vaccine studies in humans, and from epidemiological data.

In NHPs, the onset and duration of viremia and the antibody response were found to be dependent on the infectious dose [33]. Routes of exposure and infectivity of HAV were investigated in *S. mystax* and *P. troglodytes* by Purcell et al. [34], who showed that wild-type HAV (in acute phase stool from an infected human) was 32,000 times less infectious by the oral route compared with parenteral administration; however, *S. mystax* and *P. troglodytes* fed the equivalent of 0.0001 or 0.00001 g, respectively, seroconverted and showed elevated liver enzymes. Seronegative adult humans were inoculated with $10^{4.1}$, $10^{5.2}$, $10^{6.1}$, or $10^{7.3}$ $TCID_{50}$ of a variant of HAV CR326; 6 months after immunization, antibody to HAV was detected in 20%, 40%, 60%, and 100% of the vaccine recipients, respectively [24]. Using circumstantial outbreak information, an inverse relationship was found between the numbers of contaminated sandwiches and clams consumed and the incubation period for ill individuals [35]. Taken together, these data demonstrate a dose–response relationship for HAV infection in humans and NHPs, although the minimal human oral infectious dose of wild-type HAV remains undefined [24].

Pathogenesis

HAV is a hepatotropic virus; most of the virus appears to be produced in the liver, and the liver is the site of pathology. While the exact mechanism for liver damage during HAV infection remains poorly characterized, the disease is thought to arise as a result of immunologically mediated responses (e.g., stimulation of nonspecific inflammatory cells to virus-infected hepatocytes) rather than a direct CPE of the virus [7]. Following ingestion of HAV-contaminated water, food, or feces, HAV must survive stomach acidity, and it is assumed that virions transit to the liver in the portal blood [36]. Although some experimental studies suggest that initial replication of HAV may occur in some extrahepatic sites such as the oral pharynx, intestinal epithelial cells [37], and crypt cells of NHPs [38], such data are equivocal.

During the incubation period of the disease, HAV levels increase in blood, followed by secretion of the virus from the liver into the bile. HAV is present in the feces 1–2 weeks before onset of symptoms or 2–7 weeks postexposure, and fecal shedding may last for months after clinical symptoms have ended [39,40]. Peak fecal shedding of HAV occurs just before the onset of injury to hepatocytes, and titers may reach 9 $\log_{10}$ infectious virions per gram of feces [41]. HAV is occasionally found in urine, oropharyngeal (including saliva) secretions, and semen [42]. However, there is a lack of convincing evidence to suggest that these substances play a major role in the transmission of HAV, and the concentration of virus they contain is appreciably lower compared with feces and serum.

Clinical Features

HAV causes acute hepatitis and may be categorized as having four distinct clinical phases, consisting of incubation, preicteric and icteric stages, and a convalescent period [43]. HAV has a median incubation of approximately 1 month, but it may range from 15 to 50 days. The disease typically lasts 2 weeks in mild cases, but severe cases may debilitate patients for months. Signs and symptoms of HAV infections include jaundice, dark urine, clay-colored stools, anorexia, nausea, malaise, fever, abdominal discomfort, and headaches. Subclinical and anicteric infections are common, particularly in children. Two-thirds of clinically-defined cases occur in children and young adults in the United States, while approximately 70% of deaths from the disease occur in individuals >50 years of age.

Diagnosis

A clinical case of hepatitis is defined as an acutely ill individual with discrete onset of symptoms and jaundice or elevated serum aminotransferase levels [44]. Serologic testing (see "Immunoassays and Molecular Methods" section) is required to distinguish HAV from other forms of viral hepatitis. The laboratory criterion for diagnosis is the detection of immunoglobulin M (IgM) antibody to HAV. False negative IgM results can occur within the first days of the appearance of symptoms because IgM titers may be low. However, IgM titers rise quickly and remain high for 4–6 months after infection [45]. Immunoglobulin G (IgG) antibody appears soon after anti-HAV IgM. Testing for IgM anti-HAV in the serum of acute or subclinical recent infection (<6 months) is used to confirm a clinical diagnosis. A confirmed case of hepatitis A is one that meets the clinical case definition and is laboratory confirmed, or is a case that meets the clinical case definition and occurs in a person who has an epidemiologic link with a person who has laboratory-confirmed hepatitis A during the 15–50 days before the onset of symptoms [44]. Testing for total antibodies to HAV (total anti-HAV) can be performed as prevaccination screening for individuals who may have been previously exposed to HAV, or for studying prior exposure to HAV in a population.

EPIDEMIOLOGY

Transmission

Enteric Transmission

HAV is transmitted primarily through the enteric (fecal–oral) route, either from person-to-person contact or from environmental vehicles such as fomites, food, and water. A number of NHP species can be infected with HAV, but humans are the only significant reservoir. The transmission of HAV from infected to susceptible hosts within a household is the predominant way of spreading the disease, where sequential infections occur about one incubation period apart [43]. The transmission of HAV by fecally contaminated food or water is also well described, and there are many reports linking HAV to such vehicles using retrospective epidemiology. While some common-source outbreaks have been well studied, such occurrences accounted for a small percentage of all reported HAV cases in the United States.

Foodborne Transmission

Any food or water type can serve as a vehicle for HAV; contamination of drinking water sources by sewage is well documented [46], and fecal contamination of food can occur at any level from growing, harvesting, processing, preparation, or value added-production. However, many instances of foodborne HAV have been traced to infected food handlers [47]. Individuals may contaminate food during handling if basic handwashing practices are not followed. Infectious HAV survives on hands for up to 4 h, and HAV can be transferred from fingers to inert surfaces [48], including lettuce [49]. Transmission usually occurs in the late incubation period when a food handler is asymptomatic, but when fecal shedding of the virus is at its peak. In two separate occurrences, a single infected food handler was thought to transmit HAV to 133 and 230 individuals who had become ill after consuming salads and sandwiches, respectively [50,51]. Transmission risks of HAV may increase when foods are consumed raw or partially cooked; shellfish, fruits, and vegetables are representative of such foods and have been implicated in numerous multifocal outbreaks of HAV. Likewise, transmission risks of HAV increase when drinking water is consumed untreated. Most documented waterborne outbreaks in the United States have been associated with the consumption of water that was not filtered or chlorinated.

Parenteral Transmission

HAV has occasionally been transmitted by the parenteral route, through receipt of contaminated blood products from pooled donor plasma (e.g., coagulation factor concentrates, interleukin-2, and lymphokine activated killer cells) [52] or through intravenous drug use(rs) (IVDU). In the former category, viremic blood donors were the probable source of virus, which was present in pooled donor plasma and survived plasma processing procedures to contaminate final products. In the latter case, needle sharing has not been clearly demonstrated as the mode of transmission in outbreaks among IVDU, a group frequently associated with poor hygienic conditions [53]. However, studies performed using humans and NHPs have clearly demonstrated parenteral transmission of HAV [34] and, at present, inactivated vaccines are administered by injection (see "Control and Prevention" section).

Global Distribution

HAV is distributed in human populations worldwide [54], but varying epidemiologic patterns are observed. In Africa, Asia, and Latin America, where crowding and poor hygiene are prevalent, asymptomatic seroconversion in children is widespread and most adults are immune to HAV. An increase in the mean age of disease onset has been observed in people living in developing regions of the world with steadily improving sanitation and public health education programs, such as some regions of southern and eastern Europe and the Middle East. In areas such as northern and western Europe, Japan, Australia and the United States, children generally remain unexposed to HAV and therefore do not develop anti-HAV antibodies. Hence, a large proportion of unvaccinated adolescents and adults in these and other low-endemicity regions are susceptible to infection with HAV, particularly if they travel to endemic regions. It is estimated that 31% of people in the United States have been infected with HAV [55], and HAV accounts for nearly two-thirds of all viral hepatitis cases in the United States [56]. The majority of human infections are attributed to genotype I or III, and genotype I strains are the most widespread globally [14,15].

Epidemiologically Significant Outbreaks

Numerous outbreaks of HAV have been epidemiologically linked to the consumption of fecally contaminated water, food, and various blood products from pooled donor plasma. Some of these outbreaks have been large and multifocal, affecting dozens to hundreds of thousands of individuals. However, in most of these reported outbreaks, HAV was detected in either the implicated vehicle or in clinical samples from case-patients, but not both. Gravelle and colleagues [57] were the first to use IEM and serology to confirm HAV-like particles in stools of infected individuals involved in a foodborne outbreak. Hutin and colleagues [58] used reverse transcriptase polymerase chain reaction (RT-PCR) and genomic sequencing to identify identical HAV sequences in over 100 case-patients in the United States who had consumed frozen strawberries, but there were no reported attempts at recovery and detection of HAV in the implicated fruit. In contrast, several sources of HAV have been directly linked to case-patients through the genetic relatedness of HAV isolated from clinical specimens and environmental vehicles. These include contaminated ground (well) water [59,60], blueberries [61], shellfish [62], sandwiches prepared by an infected food handler [63], and coagulation Factor VIII [64,65].

At-Risk Groups

In 2005 and 2006, the most reported risk factor for HAV in the United States was international travel (15% of cases), and about four out of five infected travelers had visited Mexico or Central/South America [66,67]. Other risk factors that are typically reported in 10% or fewer cases are men who have sex with men (MSM), children/employees at day care (or contact with one), common source outbreaks, and IVDU, as shown in Table 7.1. Occupational exposure to raw sewage in a wastewater treatment plant may be another risk factor, although given less attention [68].

Following surveillance of reported viral hepatitis across 50 American states in 2013, the Centers for Disease Control (CDC) released data on the behaviors and risks associated with HAV infection [69]. The report found that the most common reported risk factor associated with HAV infection was foodborne transmission 12.8%, followed by travel outside North America (6.2%) and contact with a child or employee from day care (6.2%) [69]. Table 7.1 presents the reported behaviors and risk exposures during the incubation period (2–6 weeks prior to symptom onset) gathered by the CDC in 2013 [69].

Control and Prevention

HAV is controlled by simple hygienic measures, effective drinking water treatment, and proper disposal of excreta. Handwashing and disinfection practices in the food preparation, health-care, and service settings are important barriers to the transmission of HAV and other human enteric viruses [70]. Vigorous handwashing procedures (using hospital handwashing agents) reduce levels

TABLE 7.1
Reported Behaviors and Risk Exposures Associated with HAV Infection

Behavior/Risk Exposure	Number of Cases Including Risk Information from 1781 Surveyed	Reported Exposure to HAV		Missing Data (from 1781 total surveyed[a])	Percentage of Reported HAV with Associated Risk
		Yes	No		
Food/waterborne outbreak	734	94	640	1047	12.8
Travel outside the United States/Canada	771	48	723	1010	6.2
Contact with day-care child or employee	779	48	731	1002	6.2
Sexual/household contact with HAV-infected person	744	42	702	1037	5.6
Men who have sex with men	73	4	69	791[b]	5.5
Intravenous drug use	628	25	603	1153	4.0
Child/employee in a day-care center	881	30	851	900	3.4
Other contact with a HAV-infected person	744	12	732	1037	1.6

Source: Centers for Disease Control and Prevention (CDC). Viral hepatitis surveillance—United States, 2013. 2013. http://www.cdc.gov/hepatitis/statistics/2013surveillance/pdfs/2013hepsurveillancerpt.pdf. Accessed February 2016.

Note: More than one risk exposure/behavior may be included in each case report.

[a] No risk data reported.

[b] A total of 864 of 1781 hepatitis A cases were reported among males in 2013.

of HAV and other enteric viruses on hands by 1–2 $\log_{10}$ [71], indicating that surface disinfection does not completely remove or inactivate HAV. Methods for control of HAV in water, food, and other material include the application of the various chemical and physical processes discussed in the "Environmental Stability" section. Such procedures also include chlorination of water used for washing minimally processed fruits and vegetables [72] and specific time/temperature conditions for some foods [28].

While there is no specific antiviral for patients with HAV, immune globulin derived from plasma may be administered during and after suspected outbreaks [73] as a means of passive immunization. If administered within 2 weeks of exposure to HAV, immunoglobulin is 80%–90% effective in preventing disease for postexposure prevention of HAV [74]. Prevention of HAV is best provided by several inactivated vaccines. Vaccines have been licensed in the United States since 1995 and include HAVRIX (GlaxoSmithKline, Philadelphia, PA) and VAQTA (Merck & Co. Inc., West Point, PA); AVAXIM (Sanofi Pasteur, Lyon, France) and EPAXAL (Berna Biotech Ltd, Bern, Switzerland), which are used in Europe, Canada, and elsewhere [75]. These vaccines originate from some of the strains described in the "Laboratory-Adapted Strains" section; for example, VAQTA and HAVRIX are derived from the CR326F and HM175 strains, respectively.

In 1996, the CDC's Advisory Committee on Immunization Practices (ACIP) recommended routine vaccination for persons at increased risk of contracting HAV. Routine vaccination for children living in regions of the United States with a HAV prevalence from 10 to ≥20 cases per 100,000 population was recommended by ACIP in 1999, and in 2006 was expanded to include routine vaccination of children in all 50 states. Others recommended vaccination of the at-risk individuals discussed in the "At-Risk Groups" section, in addition to blood transfusion recipients, military personnel, health-care workers, sewage workers, food handlers, day-care assistants, institutionalized subjects, drug addicts, and liver transplantees [76].

Clinical Specimen and Environmental Sample Processing

HAV and other associated analytes of interest (e.g., HAV antigens, HAV RNA, and anti-HAV) must be recovered, concentrated, purified, and/or extracted from a clinical specimen or environmental sample prior to detection. This is because the levels of HAV and its associated analytes may vary widely in specimens and samples containing substances known to interfere with or inhibit various detection procedures. For example, IgG in oral fluid is believed to be 800- to 1000-fold lower than in serum levels [77]. Furthermore, the titer of HAV in blood is 3–5 $\log_{10}$ infectious units per mL, significantly lower than HAV in feces with 9 $\log_{10}$ infectious units per gram [41]. Clinical specimens typically range in amount or volume from a few grams to a few mLs, or less. In contrast, levels of HAV in most types of environmental samples are relatively low and more variable, necessitating large sample volumes (e.g., tens to thousands of liters of water; dozens to hundreds of grams of food). Blood, serum, feces, and environmental materials are all known inhibitors of immunological and molecular assays, and many background substances in these materials are cytotoxic. Regardless of the sample or specimen type, methods for HAV recovery from clinical specimens, water, and food are intended to reduce sample volumes and to separate HAV in a small, purified volume of liquid compatible with a detection assay. There is extensive literature describing such procedures for HAV and other enteric viruses, and therefore they will not be discussed here.

Clinical Specimens

Serum is the most common type of clinical specimen used by state public health laboratories and the CDC for laboratory confirmation of clinical cases. Specimens besides serum (e.g., feces, liver tissue, or saliva) are not typically collected or tested by such organizations. Nevertheless, such specimens are widely used in research and epidemiological studies, and published guidelines are available for their collection, transport, and storage [78]. Typically, whole blood is collected by venipuncture in appropriate vacuum containers or is captured from finger or heel punctures on filter paper. Saliva for anti-HAV testing or recovery of RNA may be collected using sterile swabs or pads followed by immersion in a transport medium supplemented with antimicrobial and anti-proteolytic substances. Further steps include the separation of serum from whole blood by centrifugation or the elution of saliva and dried blood from swabs or filter paper. Dilution of serum or eluates is usually performed prior to the detection of immunoglobulin or virus. The preparation of feces and liver tissue for recovery of HAV generally involves suspension or homogenization of a small amount (e.g., ≤1 g) in saline or another buffer, followed by centrifugation and recovery of the virus-containing supernatant. The original stool and tissue specimens, and subsequent suspensions thereof, may be processed several times using such procedures. Fluorocarbon (solvent) extraction steps from crude stool or tissue suspensions may be used for further purification of virions.

Environmental Samples

Primary recovery methods for waterborne HAV typically utilize some type of filtration procedure including ultrafiltration or capture of viruses to positively or negatively charged filters. Water samples analyzed for HAV in this manner include drinking water, environmental water sources (fresh or surface water and ground or well water; marine and estuarine water), and municipal wastewater or sewage. Where levels of HAV are relatively concentrated, such as in sewage, direct concentration may be achieved by ultracentrifugation or by direct adsorption to glass beads. The U.S. Environmental Protection Agency (EPA) protocol [79], based on methods developed by Sobsey and colleagues [80], is perhaps the most widely used method for the recovery of HAV from water; the method specifies the use of a 1MDS positively charged cartridge filter followed by elution and flocculation to recover and concentrate HAV from various water types.

Foods contaminated with HAV may be collected using procedures described by the U.S. Food and Drug Administration [81] or by the International Commission on Microbiological Specifications for Foods [82]. Recovery and purification regimes for HAV depend on the type of food; foods may be superficially contaminated (e.g., whole fruits, vegetables and deli meats) or contaminated internally, such as shellfish, food composites, and sauces. Samples obtained during an investigation involving an infected food handler may be of either or both types. Once the food type is determined, various techniques may be employed, including homogenization and/or elution, followed by secondary concentration and purification techniques ("Extraction and Purification of HAV and Viral RNA" section).

Extraction and Purification of HAV and Viral RNA

Following the primary concentration procedures described in the "Clinical Specimens" and "Environmental Samples" sections, additional steps are usually required to further concentrate and purify HAV virions. A variety of techniques are available for such procedures, including additional filtration and adsorption/elution methods, centrifugation, chemical precipitations, and antibody-capture methods. Extraction of HAV RNA is also required before detection using molecular methods and, in most cases, further concentration and purification of the RNA is required in order to maximize sensitivity and reduce PCR inhibitors. The simplest method for extracting viral RNA is the heat release procedure, whereby HAV is heated to 99°C for several minutes to degrade the virion capsid and expose naked RNA. Commonly used chemical extraction techniques involve combinations of chemical and physical procedures for lysing virions and binding naked RNA to silica, followed by washing steps to remove impurities and elution of RNA from silica [83]; a variety of commercially available kits utilize these principles. Another RNA extraction method is the classic phenol-chloroform procedure [84], which can be performed with common laboratory reagents; however, this procedure is more time consuming than spin glass column-based methods. Following extraction, further steps may be employed to concentrate and purify RNA using ethanol precipitation. Other simple methods for the reduction of inhibition in RNA extracts include dilution in molecular-grade water, although this diminishes the amount of RNA template available.

DETECTION AND CHARACTERIZATION

Following recovery and purification procedures for HAV and its associated analytes, some type of detection method is performed. The choice of detection method is mainly driven by the speed of the results and the depth of information required, and can include immunoassays, nucleic acid assays, and cell culture infectivity assays, either alone or in combination. Immunoassays for detecting antibodies (or antigens in feces) are simple, fast, and less expensive to perform than other techniques and are the predominant method for clinical diagnosis. Molecular methods such as RT-PCR and sequencing are becoming more widespread, and can characterize the genetic relatedness of strains circulating in a population for molecular epidemiology or detect low levels of viral RNA in contaminated samples. Although molecular methods are more expensive, they are beginning to be used for clinical diagnostic purposes. They can also provide detailed and nuanced information that can enhance immunoassay results. Cell culture methods are useful for directed laboratory studies (e.g., disinfection) involving laboratory-adapted strains, or for primary isolation of wild-type HAV. Most studies on the occurrence and levels of HAV in the environment report a percentage of samples positive (or negative) for the presence of HAV. When reported, levels of wild-type HAV (either infectious levels or numbers of virions) are usually only semiquantitative or are estimated from various assay data, such as the number of genomic copies per unit volume.

Immunoassays and Molecular Methods

Antibody-based detection is a specific way to detect anti-HAV in clinical specimens and HAV in environmental samples, provided there are adequate numbers of these analytes present. Several types of anti-HAV IgM and IgG assays have been developed, and the most widely used are capture enzyme immunoassays (EIAs) and radioimmunoassays (RIAs). HAV antigen may be detected in serum or in the cytoplasm of infected cells, feces, or tissues using such methods. Although useful for clinical diagnostic purposes, antigen tests are relatively insensitive, with a lower limit of detection of about 6 $\log_{10}$ HAV particles per mL. Another limitation of antigenic detection is that genetically distinct strains of HAV will display indistinguishable serotypes (because there is only one serotype), and therefore appear identical.

The advent of molecular methods, notably gene-probe hybridization and PCR, led to the development of sensitive techniques for detection of HAV. Methods for conventional RT-PCR and real-time RT-PCR detection of HAV have been previously reported [58,85,86]. Such techniques have been successfully applied in the detection and characterization of wild-type HAV in both clinical specimens [45] and environmental samples, and for the detection of laboratory-adapted HAV in experimentally contaminated environmental samples with various reported levels of detection [87,88]. Molecular techniques also provide the basic tools for molecular epidemiology studies. Tools such as RT-PCR and sequencing can assist in linking seemingly unrelated cases of viral hepatitis across time and space, because elevated viral levels in serum and feces leave a long-lasting record of transmission that, in connection with traditional epidemiologic investigations, can overcome patient recall bias during the long incubation period of the disease [45]. The application of RT-PCR and sequencing has been used effectively in outbreaks to definitively link various sources, such as green onions [89], groundwater [59,60], and contaminated blood products [64] administered to case-patients.

Generating nucleotide sequences of HAV outbreak strains for molecular epidemiology is straightforward. A region (or multiple regions) of the HAV genome is amplified by RT-PCR, and the DNA amplicons are purified and sent to a commercial or other sequencing facility. The most common sequencing method is the Sanger chain termination method [90], although newer methods are promising tools for large-scale DNA sequencing and metagenomics. Sanger sequence results are returned to researchers in two forms: a file containing the nucleotide bases and the dye-terminator read, so that researchers can manually modify the sequence results. Sequences are aligned using commercial software (e.g., Geneious, www.geneious.com) and are easily compared with a large database of known sequences such as GenBank, the database operated by the National Center for Biotechnology Information. Phylogenetic trees with bootstrapped values are typically generated to explain genetic relatedness of known and unknown strains, where related strains cluster closely.

Cell Infectivity Methods

Routine cell infectivity determination for wild-type HAV is impractical for clinical diagnosis, though *in vitro* infectivity assays for wild-type and cytopathic variants of HAV are often used in research laboratories (see "Laboratory-Adapted Strains" section). *In vitro* infectivity and propagation methods of cytopathic HAV are well described by Cromeans and colleagues [26,27]. In summary, these procedures consist of inoculating HAV onto confluent FRhK-4 cell monolayers in plates, flasks, or bottles followed by a short period (60 min) of incubation at 37°C in a humidified chamber containing CO_2. For enumerative (plaque) infectivity assays, cell monolayers are overlaid with molten, electrophoretic-grade agarose containing cell maintenance media and other components (e.g., nonessential amino acids, newborn calf serum, antifungal and antibacterial agents, and buffers). For virus propagation, fresh maintenance media is added (instead of agarose) to infected monolayers in flasks or roller bottles. Dishes, flasks, or bottles are then incubated for about 7 days. For plaque assays, a second agarose overlay identical to the first (but containing neutral red solution) is added and plates are returned to the incubator for another 3–4 days. Clear areas of lysis against the neutral red background are counted and recorded as plaque forming units; alternatively, infected monolayers may be stained using crystal violet on the eighth or ninth day postinfection. Propagated HAV may be recovered from the cell culture fluid and from infected cells when the CPE is >95%. Primary isolation of wild-type HAV in AGMK cells uses procedures similar to the ones described here [91].

Other Detection Methods

Other approaches for the detection of HAV include combinations or modifications of the aforementioned methods to improve detection sensitivity and specificity. For example, cell culture and nucleic acid detection methods have been combined (termed *cell culture RT-PCR*) and used to detect HAV and other fastidious enteric viruses in water [92].

CONCLUSIONS AND FUTURE PERSPECTIVES

An appreciable amount of knowledge on the biology and epidemiology of HAV has been obtained during the past six decades, culminating in the availability of highly efficacious, inactivated vaccines. Because of vaccination campaigns, particularly for children, the rate of reported HAV cases in the United States has decreased by 88% from 1995 to 2005 and is now less than two cases per 100,000 [93]. The application of various detection methods has elucidated the distribution and determinants (risk factors) of HAV in human populations and has greatly contributed to our understanding of the occurrence and control of HAV in the environment. These activities represent major public health and medical achievements.

Despite these successes, a number of challenges remain. Although HAV is a disease in decline in the United States, approximately 50% of HAV cases remain unattributable to any known source or risk factor [94]. Less than 50% of children aged 24–35 months in the United States are vaccinated [95], and self-reported HAV vaccination coverage among adults aged 18–49 years was only 12.1% in 2007 [96]. The disease is still highly endemic in many regions of the world. Improvements in the economic and living conditions of communities with moderate endemicity shifts the age of acquiring HAV infection from early childhood to adolescence and young adulthood. Hence, a leading risk factor for citizens residing in areas with low endemicity is travel to endemic regions, and areas of moderate endemicity are actually at greater risk for experiencing large outbreaks of HAV.

Even if people do not travel abroad, the foods they eat do; food is increasingly grown abroad and shipped to the United States all year-round, and so continued transmission of HAV from foods grown or produced in countries where the disease is endemic should be expected. Person-to-person transmission of HAV also remains an important risk factor, either within households or in day-care or foodservice settings. However, relatively little data are available that definitive link sources with infected individuals during cases and outbreaks of HAV. Such data are critical, because inaccurate identification of potential sources does nothing to control or prevent the disease, reduces availability of immune serum globulin administered unnecessarily, and increases the already appreciable economic burden [97] associated with HAV, which is now estimated to range from $443 to $773 million (USD) annually in the United States.

Resolving the issue of source-based association of viral contamination is of critical importance for the remediation of foodborne and person-to-person outbreaks. Defining critical control points for HAV in foods will require a better understanding of how and when contamination occurs [47], in addition to an improved understanding and application of disinfection and sanitization procedures used in agriculture. One reason the scientific community has difficulty addressing this question is that published peer-reviewed articles blur the line between causal modes of pathogen contamination and transmission. To resolve this issue, viral outbreak investigations should attempt to include common methods. To these ends, the CDC has issued a web-based *Outbreak Investigation Toolkit* providing protocols for sample collection, case identification, a patient questionnaire, and a form for state or local health departments to report foodborne outbreaks. Unfortunately, the CDC model lacks advice on how to collect and process foods implicated in viral outbreaks, as do many state and local health departments [98], and while considerable progress has been made in method development, there is generally a lack of consensus or standardization of procedures among various research laboratories.

In conclusion, future control and prevention efforts should continue to include vaccination, continued improvements in sanitation and hygiene, and access to safe drinking water. Future activities should also include the continued study of the pathogenesis and infectivity of HAV, the development of robust, sensitive methods to detect HAV in the environment, active disease surveillance, and sharing of research findings. If global immunization against HAV is the goal, an inexpensive, attenuated, and live vaccine that is administered orally and that replicates in the gastrointestinal tract (stimulating a secretory antibody response) may be required [99]. Future efforts should continue to include the application of molecular methods for detection and characterization of HAV in specimens and samples collected during outbreaks and cases. In a limited number of exemplary studies, such methods were used to definitively link contaminated sources to infected individuals; in some instances, this information was used to prevent continued cases of disease from occurring. As various laboratories continue to develop the capacity to detect human viruses such as HAV using molecular methods, they should incorporate viral sequencing and sequence sharing as a priority. Lastly, prospective nucleic acid testing and laboratory assessment of inactivation procedures for the detection and control of HAV in targeted, high-risk environmental and other vehicles may contribute to the reduction in transmission risks. Such procedures should continue to be developed and evaluated using laboratory adapted strains.

REFERENCES

1. Feinstone SM, Kapikian AZ, Purcell RH. 1973. Hepatitis A: Detection by immune electron microscopy of a viruslike antigen associated with acute illness. *Science* 182:1026–1028.
2. Minor PD. 1995. Family picornaviridae. *In Virus Taxonomy, Sixth Report of the International Committee for the Taxonomy of Viruses*, pp. 329–336. Springer-Verlag, New York.
3. Siegl G, Weitz M, Kronauer G. 1984. Stability of hepatitis A virus. *Intervirology* 22:218–226.
4. Murphy P, Nowak T, Lemon SM, Hilfenhaus J. 1993. Inactivation of hepatitis A virus by heat treatment in aqueous solution. *J Med Virol* 41:61–64.
5. Cohen JI, Ticehurst JR, Purcell RH, Buckler-White A, Baroudy B. 1987. Complete nucleotide sequence of wild-type hepatitis A virus: Comparison with different strains of hepatitis A virus and other picornaviruses. *J Virol* 61:50–59.
6. Lemon S.M. aMA. 2005. Structure and molecular virology. *In* Thomas H, Lemon S, and Zuckerman A (eds), *Viral Hepatitis*, 3rd edn, vol 79, pp. 79–91. Blackwell Publishing, Malden, MA.
7. Martin A, Lemon SM. 2006. Hepatitis A virus: From discovery to vaccines. *Hepatology* 43:S164–S172.
8. Ticehurst JR, Racaniello VR, Baroudy BM, Baltimore D, Purcell RH, Feinstone SM. 1983. Molecular cloning and characterization of hepatitis A virus cDNA. *Proc Natl Acad Sci USA* 80:5885–5889.
9. Najarian R, Caput D, Gee W, Potter SJ, Renard A, Merryweather J, Van Nest G, Dina D. 1985. Primary structure and gene organization of human hepatitis A virus. *Proc Natl Acad Sci USA* 82:2627–2631.
10. Hollinger FB, Ticehurst JR. 1996. Hepatitis A virus. *In* Fields BN, Knipe DM, and Howley PM (eds), *Fields Virology*, pp. 735–782. Lippincott-Raven, Philadelphia, PA.
11. Lemon SM, Chao S-F, Jansen RW, Binn L, LeDuc J. 1987. Genomic heterogeneity among human and nonhuman strains of hepatitis A virus. *J Virol* 61:735–742.
12. Lemon SM, Jansen RW, Brown EA. 1992. Genetic, antigenic and biological differences between strains of hepatitis A virus. *Vaccine* 10:S40–S44.
13. Jansen RW, Siegl G, Lemon SM. 1990. Molecular epidemiology of human hepatitis A virus defined by an antigen-capture polymerase chain reaction method. *Proc Natl Acad Sci USA* 87:2867–2871.
14. Robertson BH, Khanna B, Nainan OV, Margolis HS. 1991. Epidemiologic patterns of wild-type hepatitis A virus determined by genetic variation. *J Infect Dis* 163:286–292.
15. Robertson BH, Jansen RW, Khanna B, Totsuka A, Nainan OV, Siegl G, Widell A, Margolis HS, Isomura S, Ito K. 1992. Genetic relatedness of hepatitis A virus strains recovered from different geographical regions. *J Gen Virol* 73:1365–1377.
16. Costa-Mattioli M, Cristina J, Romero H, Perez-Bercof R, Casane D, Colina R, Garcia L, Vega I, Glikman G, Romanowsky V. 2002. Molecular evolution of hepatitis A virus: A new classification based on the complete VP1 protein. *J Virol* 76:9516–9525.
17. Costa-Mattioli M, Di Napoli A, Ferré V, Billaudel S, Perez-Bercoff R, Cristina J. 2003. Genetic variability of hepatitis A virus. *J Gen Virol* 84:3191–3201.
18. Lu L, Ching KZ, de Paula VS, Nakano T, Siegl G, Weitz M, Robertson BH. 2004. Characterization of the complete genomic sequence of genotype II hepatitis A virus (CF53/Berne isolate). *J Gen Virol* 85:2943–2952.
19. Paul JR, Havens W, Sabin A, Philip C. 1945. Transmission experiments in serum jaundice and infectious hepatitis. *JAMA* 128:911–915.
20. Krugman S, Ward R, Giles JP. 1962. The natural history of infectious hepatitis. *Am J Med* 32:717–728.
21. Boggs JD, Melnick JL, Conrad ME, Felsher BF. 1970. Viral hepatitis: Clinical and tissue culture studies. *JAMA* 214:1041–1046.
22. Provost PJ, Hilleman MR. 1979. Propagation of human hepatitis A virus in cell culture *in vitro. Exp Biol Med* 160:213–221.
23. Gust ID, Feinstone SM. 1988. *Hepatitis A*. CRC Press, Boca Raton, FL.
24. Lemon SM, Schultz DE, Shaffer DR. 1997. The molecular basis of attentuation of hepatitis A virus. *In* Harrison TJ, and Zuckerman AJ (eds), *The Molecular Medicine of Viral Hepatitis*, vol 3, pp. 3–31. Wiley, Chichester, England.
25. Gust I, Lehmann N, Crowe S, McCrorie M, Locarnini S, Lucas C. 1985. The origin of the HM175 strain of hepatitis A virus. *J Infect Dis* 151(2):365–367.
26. Cromeans T, Sobsey MD, Fields HA. 1987. Development of a plaque assay for a cytopathic, rapidly replicating isolate of hepatitis A virus. *J Med Virol* 22:45–56.
27. Cromeans T, Fields HA, Sobsey MD. 1989. Replication kinetics and cytopathic effect of hepatitis A virus. *J Gen Virol* 70:2051–2062.

28. Parry J, Mortimer P. 1984. The heat sensitivity of hepatitis A virus determined by a simple tissue culture method. *J Med Virol* 14:277–283.
29. Provost PJ, Wolanski BS, Miller WJ, Ittensohn OL, McAleer WJ, Hilleman MR. 1975. Physical, chemical and morphologic dimensions of human hepatitis A virus strain CR326. *Exp Biol Med* 148:532–539.
30. McCaustland K, Bond W, Bradley D, Ebert J, Maynard J. 1982. Survival of hepatitis A virus in feces after drying and storage for 1 month. *J Clin Microbiol* 16:957–958.
31. Sobsey MD, Shields PA, Hauchman FS, Davies AL, Rullman VA, Bosch A. 1988. Survival and persistence of hepatitis A virus in environmental samples. *In* Zuckerman AJ (ed), *Viral Hepatitis and Liver Disease*, pp. 121–124, Alan R. Liss, New York.
32. Coulepis A, Anderson B, Gust I. 1987. Hepatitis A. *Adv Virus Res* 32:129–169.
33. Polish LB, Robertson BH, Khanna B, Krawczynski K, Spelbring J, Olson F, Shapiro CN. 1999. Excretion of hepatitis A virus (HAV) in adults: Comparison of immunologic and molecular detection methods and relationship between HAV positivity and infectivity in tamarins. *J Clin Microbiol* 37:3615–3617.
34. Purcell RH, Wong DC, Shapiro M. 2002. Relative infectivity of hepatitis A virus by the oral and intravenous routes in 2 species of nonhuman primates. *J Infect Dis* 185:1668–1671.
35. Istre GR, Hopkins RS. 1985. An outbreak of foodborne hepatitis A showing a relationship between dose and incubation period. *Am J Public Health* 75:280–281.
36. Cuthbert JA. 2001. Hepatitis A: Old and new. *Clin Microbiol Rev* 14:38–58.
37. Blank CA, Anderson DA, Beard M, Lemon SM. 2000. Infection of polarized cultures of human intestinal epithelial cells with hepatitis A virus: Vectorial release of progeny virions through apical cellular membranes. *J Virol* 74:6476–6484.
38. Asher LVS, Binn LN, Mensing TL, Marchwicki RH, Vassell RA, Young GD. 1995. Pathogenesis of hepatitis A in orally inoculated owl monkeys (*Aotus trivirgatus*). *J Med Virol* 47:260–268.
39. Hollinger FB. 1991. Hepatitis viruses. *In* Balows A, Hausler WJJ, and Herrmann KL (eds), *Manual of Clinical Microbiology*, vol 959, pp. 959–983. American Society of Microbiology Press, Washington, DC.
40. Yotsuyanagi H, Koike K, Yasuda K, Moriya K, Shintani Y, Fujie H, Kurokawa K, Iino S. 1996. Prolonged fecal excretion of hepatitis A virus in adult patients with hepatitis A as determined by polymerase chain reaction. *Hepatology* 24:10–13.
41. Purcell RH, Feinstone SM, Ticehurst JR. 1984. Hepatitis A virus. *In* Vyas GN, Dienstag JL, and Hoofnagle JH (eds), *Viral Hepatitis and Liver Disease*, pp. 9–22. Grune and Stratton, Orlando, FL.
42. Koff RS. 1995. Seroepidemiology of hepatitis A in the United States. *J Infect Dis* 171:S19–S23.
43. Koff RS. 1998. Hepatitis A. *Lancet* 351:1643–1649.
44. Centers for Disease Control and Prevention (CDC). 2008. CDC ABC factsheet. http://www.cdc.gov/hepatitis/Resources/Professionals/PDFs/ABCTable_BW.pdf. Accessed April 29, 2015.
45. Nainan OV, Xia G, Vaughan G, Margolis HS. 2006. Diagnosis of hepatitis A virus infection: A molecular approach. *Clin Microbiol Rev* 19:63–79.
46. Lippy EC, Waltrip SC. 1984. Waterborne disease outbreaks—1946–1980: A thirty-five-year perspective. *J Am Water Works Ass* 76:60–67.
47. Fiore AE. 2004. Hepatitis A transmitted by food. *Clin Infect Dis* 38:705–715.
48. Mbithi JN, Springthorpe VS, Boulet JR, Sattar SA. 1992. Survival of hepatitis A virus on human hands and its transfer on contact with animate and inanimate surfaces. *J Clin Microbiol* 30:757–763.
49. Bidawid S, Farber J, Sattar S. 2000. Contamination of foods by food handlers: Experiments on hepatitis A virus transfer to food and its interruption. *Appl Environ Microbiol* 66:2759–2763.
50. Hooper RR, Juels CW, Routenberg JA, Harrison WO, Kilpatrick ME, Kendra SJ, Dienstag JL. 1977. An outbreak of type A viral hepatitis at the Naval Training Center, San Diego: Epidemiologic evaluation. *Am J Epidemiol* 105:148–155.
51. Centers for Disease Control and Prevention (CDC). 1993. Foodborne hepatitis A: Missouri, Wisconsin, and Alaska, 1990–1992. *MMWR* 42:526–534.
52. Catton MG, Locarnini SA. 2005. Epidemiology. *In* Thomas H, Lemon S, and Zuckerman A. (eds), *Viral Hepatitis*, 3rd edn, vol 92, pp. 92–108. Blackwell Publishing, Malden, MA.
53. Catton MG, Locarnini SA. 1998. Epidemiology. *In* Zuckerman AJ, and Thomas HC. (eds), *Viral Hepatitis*, 2nd edn, vol 29, pp. 29–41. Churchill Livingstone, London.
54. Papaevangelou GJ. 1994. Global epidemiology of hepatitis A. *In* Gerety RJ (ed), *Hepatitis A*, pp. 101–132. Academic Press, Orlando, FL.
55. Centers for Disease Control and Prevention (CDC). 1999. Prevention of hepatitis A through active or passive immunization: Recommendations of the advisory committee on immunization practices (ACIP). *MMWR* 48(RR-12):1–37.

56. Centers for Disease Control and Prevention (CDC). 1999. Summary of Notifiable Diseases, United States. *MMWR* 48(53):1–104.
57. Gravelle CR, Hornbeck CL, Maynard JE, Schable CA, Cook EH, Bradley DW. 1975. Hepatitis A: Report of a common-source outbreak with recovery of a possible etiologic agent. II. Laboratory studies. *J Infect Dis* 131:167–171.
58. Hutin YJ, Pool V, Cramer EH, Nainan OV, Weth J, Williams IT, Goldstein ST, Gensheimer KF, Bell BP, Shapiro CN. 1999. A multistate, foodborne outbreak of hepatitis A. *N Engl J Med* 340:595–602.
59. De Serres G, Cromeans TL, Levesque B, Brassard N, Barthe C, Dionne M, Prud'homme H, Paradis D, Shapiro CN, Nainan OV. 1999. Molecular confirmation of hepatitis A virus from well water: Epidemiology and public health implications. *J Infect Dis* 179:37–43.
60. Tallon LA, Love DC, Moore ZS, Sobsey MD. 2008. Recovery and sequence analysis of hepatitis A virus from springwater implicated in an outbreak of acute viral hepatitis. *Appl Environ Microbiol* 74:6158–6160.
61. Calder L, Simmons G, Thornley C, Taylor P, Pritchard K, Greening G, Bishop J. 2003. An outbreak of hepatitis A associated with consumption of raw blueberries. *Epidemiol Infect* 131:745–751.
62. Sánchez G, Pintó RM, Vanaclocha H, Bosch A. 2002. Molecular characterization of hepatitis A virus isolates from a transcontinental shellfish-borne outbreak. *J Clin Microbiol* 40:4148–4155.
63. Chironna M, Lopalco P, Prato R, Germinario C, Barbuti S, Quarto M. 2004. Outbreak of infection with hepatitis A virus (HAV) associated with a foodhandler and confirmed by sequence analysis reveals a new HAV genotype IB variant. *J Clin Microbiol* 42:2825–2828.
64. Purcell RH, Mannucci PM, Gdovin S, Gringer A, Colombo M, Melec A, Schinaia N, Ciavarella N, Emerson SU. 1994. Virology of the hepatitis A epidemic in Italy. *Vox Sang* 67:2–7.
65. Chudy M, Budek I, Keller-Stanislawski B, McCaustland KA, Neidhold S, Robertson BH, Nübling CM, Seitz R, Löwer J. 1999. A new cluster of hepatitis A infection in hemophiliacs traced to a contaminated plasma pool. *J Med Virol* 57:91–99.
66. Centers for Disease Control and Prevention (CDC). 2007. Surveillance for acute viral hepatitis—United States 2005. *MMWR* 56(SS03):1–24.
67. Centers for Disease Control and Prevention (CDC). 2008. Surveillance for acute viral hepatitis—United States 2006. *MMWR* 57(SS02):1–24.
68. Heng B, Goh K, Doraisingham S, Quek G. 1994. Prevalence of hepatitis A virus infection among sewage workers in Singapore. *Epidemiol Infect* 113:121–128.
69. Centers for Disease Control and Prevention (CDC). 2013. Viral hepatitis surveillance—United States, 2013. http://www.cdc.gov/hepatitis/statistics/2013surveillance/pdfs/2013hepsurveillancerpt.pdf. Accessed February 2016.
70. Barker J, Stevens D, Bloomfield S. 2001. Spread and prevention of some common viral infections in community facilities and domestic homes. *J Appl Microbiol* 91:7–21.
71. Mbithi JN, Springthorpe VS, Sattar SA. 1993. Comparative in vivo efficiencies of hand-washing agents against hepatitis A virus (HM-175) and poliovirus type 1 (Sabin). *Appl Environ Microbiol* 59:3463–3469.
72. Casteel M, Schmidt C, Sobsey M. 2009. Chlorine inactivation of coliphage MS2 on strawberries by industrial-scale water washing units. *J Water Health* 7:244–250.
73. Carl M, Francis DP, Maynard JE. 1983. Food-borne hepatitis A: Recommendations for control. *J Infect Dis* 148:1133–1135.
74. Winokur PL, Stapleton JT. 1992. Immunoglobulin prophylaxis for hepatitis A. *Clin Infect Dis* 14:580–586.
75. Bell BP, Kruszon-Moran D, Shapiro CN, Lambert SB, McQuillan GM, Margolis HS. 2005. Hepatitis A virus infection in the United States: Serologic results from the Third National Health and Nutrition Examination Survey. *Vaccine* 23:5798–5806.
76. Franco E, Giambi C, Ialacci R, Coppola RC, Zanetti AR. 2003. Risk groups for hepatitis A virus infection. *Vaccine* 21:2224–2233.
77. Ochnio JJ, Scheifele DW, Ho M, Mitchell LA. 1997. New, ultrasensitive enzyme immunoassay for detecting vaccine and disease-induced hepatitis A virus-specific immunoglobulin G in saliva. *J Clin Microbiol* 35:98–101.
78. Brown EA, Stapleton JT. 1991. Hepatitis A virus. *In* Murray PR, Baron EJ, Jorgenson JH, and Pfaller MA (eds), *Manual of Clinical Microbiology*, 8th edn. ASM Press, Washington, DC.
79. United States Environmental Protection Agency (USEPA). 2001. Manual of Methods for Virology. EPA/600/4-84/013 (N14).
80. Sobsey MD, Oglesbee SE, Wait DA. 1985. Evaluation of methods for concentrating hepatitis A virus from drinking water. *Appl Environ Microbiol* 50:1457–1463.

81. U.S Food and Drug Administration (FDA). 2003. Food Sampling/Preparation of Sample Homogenate, *on* U.S Food and Drug Administration (FDA). http://www.fda.gov/food/foodscienceresearch/laboratorymethods/ucm063335.htm. Accessed November, 2015.
82. The International Commission on Microbiological Specifications for Foods (ICMSF). 1986. *Microorganisms in Foods 2: Sampling for Microbiological Analysis: Principles and Specific Applications*, 2nd edn. University of Toronto Press, Buffalo, NY.
83. Boom R, Sol C, Salimans M, Jansen C, Wertheim-van Dillen P, Van der Noordaa J. 1990. Rapid and simple method for purification of nucleic acids. *J Clin Microbiol* 28:495–503.
84. Sambrook J, Russell DW. 2001. *Molecular Cloning: A Laboratory Manual*, 3rd edn. CSHL Press, Cold Spring Harbor, NY.
85. Love DC, Casteel MJ, Meschke JS, Sobsey MD. 2008. Methods for recovery of hepatitis A virus (HAV) and other viruses from processed foods and detection of HAV by nested RT-PCR and TaqMan RT-PCR. *Int J Food Microbiol* 126:221–226.
86. Jothikumar N, Cromeans T, Sobsey M, Robertson B. 2005. Development and evaluation of a broadly reactive TaqMan assay for rapid detection of hepatitis A virus. *Appl Environ Microbiol* 71:3359–3363.
87. Cliver DO, Matsui SM, Casteel, M. 2006. Infections with viruses and prions. *In* Riemann HP, and Cliver DO (eds), *Foodborne Infections and Intoxications*, 3rd edn, pp. 367–448. Academic Press (Elsevier), London.
88. Sánchez G, Bosch A, Pintó R. 2007. Hepatitis A virus detection in food: Current and future prospects. *Lett Appl Microbiol* 45:1–5.
89. Amon JJ, Devasia R, Xia G, Nainan OV, Hall S, Lawson B, Wolthuis JS, MacDonald PD, Shepard CW, Williams IT. 2005. Molecular epidemiology of foodborne hepatitis A outbreaks in the United States, 2003. *J Infect Dis* 192:1323–1330.
90. Sanger F, Nicklen S, Coulson AR. 1977. DNA sequencing with chain-terminating inhibitors. *Proc Natl Acad Sci USA* 74:5463–5467.
91. Daemer RJ, Feinstone SM, Gust I, Purcell RH. 1981. Propagation of human hepatitis A virus in African green monkey kidney cell culture: Primary isolation and serial passage. *Infect Immun* 32:388–393.
92. Reynolds KA, Gerba CP, Abbaszadegan M, Pepper IL. 2001. ICC/PCR detection of enteroviruses and hepatitis A virus in environmental samples. *Can J Microbiol* 47:153–157.
93. Wasley A, Miller JT, Finelli L, Centers for Disease C, Prevention. 2007. Surveillance for acute viral hepatitis–United States, 2005. *MMWR* 56:1–24.
94. Centers for Disease Control and Prevention (CDC). 2009. Surveillance for acute viral hepatitis—United States, 2007. *MMWR* 58(SS03):1–27.
95. Centers for Disease Control and Prevention (CDC). 2009. Hepatitis A vaccination coverage among children aged 24–35 months—United States, 2006 and 2007. *MMWR* 58(25):689–694.
96. Lu PJ, Euler GL, Hennessey KA, Weinbaum CM. 2009. Hepatitis A vaccination coverage among adults aged 18–49 years in the United States. *Vaccine* 27:1301–1305.
97. Berge JJ, Drennan DP, Jacobs RJ, Jakins A, Meyerhoff AS, Stubblefield W, Weinberg M. 2000. The cost of hepatitis A infections in American adolescents and adults in 1997. *Hepatology* 31:469–473.
98. Centers for Disease Control and Protection (CDC). OutbreakNet team: Foodborne disease surveillance and outbreak investigation toolkit. http://www.cdc.gov/foodsafety/outbreaks/surveillance-reporting/investigation-toolkit.html. Accessed July 2016.
99. Lemon S, Shapiro C. 1994. The value of immunization against hepatitis A. *Infect Agents Dis* 3:38–49.

8 Hepatitis E Virus, an Etiological Agent of Foodborne and Waterborne Hepatitis

Natalie E. Netzler, Andrew G. Kelly, and Peter A. White

CONTENTS

INTRODUCTION

Hepatitis E virus (HEV) is a major etiological agent responsible for acute liver disease, causing widespread epidemics and sporadic infections worldwide. Based on seroprevalence, it is estimated that one-third of the global population has been infected with HEV at some stage in their lifetime [1].

HEV imposes a significant health burden on society, with annual estimates of 20 million infections, over 3 million reported acute cases, 3,000 stillbirths, and 50,000–70,000 HEV-related deaths globally [2,3]. However, it is plausible that these figures are an underestimation, given that the virus is often asymptomatic and usually self-limiting, and that HEV epidemics strike in the geographical areas least likely to have effective diagnostic testing available [4,5].

HEV is a member of the *Hepeviridae* family, which is divided into two genera: *Orthohepevirus*, which infects humans, mammals, and birds and *Piscihepevirus*, which infects fish [6]. Within *Orthohepevirus* A, the subgroup of *Orthohepevirus* that infects humans, there are currently seven HEV genotypes, five of which are known to infect humans [6,7], and each has a different epidemiological profile [7]. Genotype (G)I and GII only infect humans, mostly in the developing world, while GIII and GIV infect a range of animal hosts, as well as humans. GV and GVI infect boars, while GVII infects camels [8] but has also been detected in a chronic hepatitis patient [9]. Viruses from GIII, GIV, and GVII are transmitted to humans via zoonosis through animal contact and meat ingestion [9–11]. Due to the recent increase in fully sequenced HEV genomes, additional mammalian HEV genotypes from a wide number of species have been described in the literature, resulting in the new classification system recently adopted in the 2014 International Committee on Taxonomy of Viruses (ICTV) report [6,11,12].

Acute HEV infections can vary in severity, with symptomatic infections most common in young adults aged between 15 and 40 years [2]. Symptoms of HEV infection include anorexia, vomiting, nausea, arthralgia, fever, hepatomegaly, and jaundice. This is often accompanied by dark urine, pale stools, and upper right quadrant pain [2]. HEV does not cause chronic hepatitis unless the patient is immunocompromised or immunosuppressed. However, HEV infections can be particularly severe in pregnant women, especially within the third trimester, with mortality rates of 27% [13], compared with between 0.07% and 0.6% in the general population [14].

Until recently, HEV was thought to be solely transmissible through the fecal–oral route via contaminated water, mostly in developing countries where poor sanitation and a lack of wastewater infrastructure foster the transmission of infectious diseases. However, in the last 15 years, there have been increasing reports of patients presenting with acute HEV infections in many developed nations, despite a lack of recent travel history. These autochthonous incidences were first directly linked to the consumption of undercooked pork products in 2003 [15]. Since then, it has been recognized that HEV can be transmitted zoonotically to humans, particularly from swine, a major animal HEV reservoir (reviewed in [16,17]). Following the evidence of foodborne HEV transmission, further studies identified specific cases of HEV infection linked to the consumption of game meats, such as Japanese wild boar [18] and sika deer [19]. It is now widely accepted that HEV can be spread by foodborne transmission, especially from undercooked meats, shellfish, and by consumption of contaminated fresh and unprocessed produce.

This chapter reviews HEV and its transmission via consumption of contaminated food and water.

DIAGNOSIS OF HEV INFECTIONS

The specific diagnosis of HEV infection relies on the serological detection of either anti-HEV antibodies and/or HEV RNA in the stool or serum [20]. Recent HEV infections can be detected by the presence of immunoglobulin M (IgM) anti-HEV antibodies in the first few months postinfection (Figure 8.1), while IgG anti-HEV antibodies can indicate a recent or past exposure (reviewed in [21]). The detection of HEV-RNA is a marker of a current infection (Figure 8.1), whether acute, asymptomatic, or chronic [21,22]. Following HEV infection, RNA is shed in the stool and is detectable for up to 28 days after the onset of symptoms [23]. While anti-HEV diagnostic assays are widely available, significant inter-assay discrepancies have been reported with variable results between methods [21]. As a result, there is still a need for the development of improved HEV detection assays. Diagnosis in immunocompromised patients requires a combination of both serological and nucleic acid detection assays (reviewed in [24]).

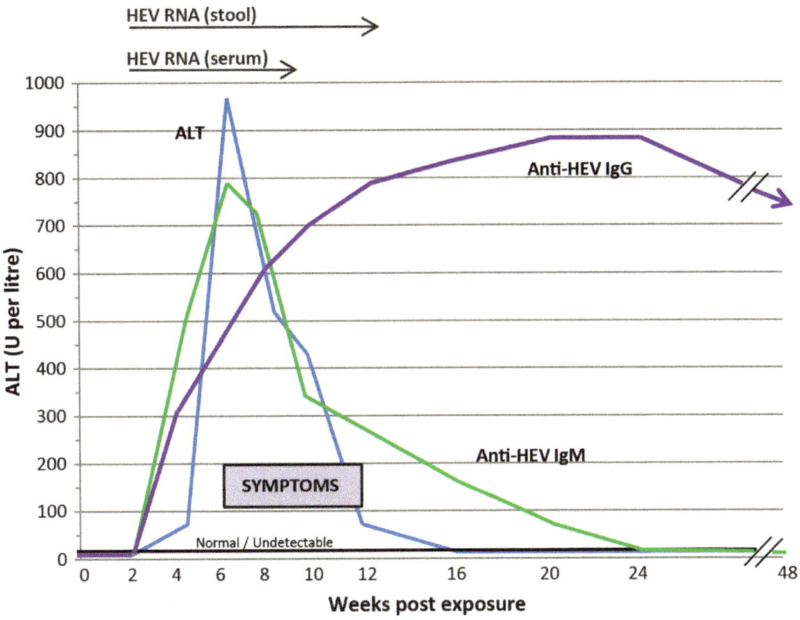

FIGURE 8.1 Course of acute HEV infection in immunocompetent humans. HEV RNA is detectable in human stool and serum within the incubation period (mean of 40 days) before IgM and then IgG anti-HEV antibodies are detectable. Symptoms of HEV infection include malaise, nausea, and jaundice, followed by elevated levels of serum alanine aminotransferase (ALT), an enzyme marker of liver damage. Although HEV RNA is no longer detectable in serum after recovery, it can persist for several weeks in the patient's stool. (Adapted from a review by Hoofnagle, J.H. et al. *N Engl J Med*, 367, 1237–1244, 2012.)

Global standardization of HEV diagnostic assays would be useful, because different HEV genotype antigens are used in different assays, creating sensitivity discrepancies. In particular, HEV GIII is often detected more efficiently than other genotypes. This has affected seroprevalence studies from different regions, so epidemiological data cannot always be compared directly [26,27].

CLINICAL MANIFESTATIONS OF HEV INFECTIONS

HEV infections manifest with similar symptoms to other forms of acute viral hepatitis and range in severity. Infections can sometimes be asymptomatic or anicteric, associated with mild biochemical abnormalities such as elevated alanine aminotransferase (ALT) and bilirubin levels [28]. However, in some cases, infections can be very severe and can lead to fulminant hepatic failure, increased morbidity, and even mortality [29,30].

Symptoms of acute infection usually include fever, nausea, malaise, joint aches, weight loss, dark urine, and jaundice. HEV has a long incubation period of 3–8 weeks with a mean of 40 days (reviewed in [2]). Following a brief prodromal period before symptoms arise, acute cases generally last 4–6 weeks before symptoms resolve and the virus is cleared (Figure 8.1). It is understood that different genotypes also have different levels of infection severity [31,32].

Mortality rates range between 0.07% and 0.6% within the general population (reviewed in [14]), but are significantly higher in pregnant women, with spontaneous abortions and stillbirths common in those that survive [29,33]. It is still not fully understood why disease severity and mortality rates are so high in pregnant women, but several theories have been proposed, including malnourishment or altered hormone and immunity levels.

HEV occasionally presents with extrahepatic manifestations, which although rare, can be observed in both acute and chronic patients. These multiorgan symptoms include neurological disorders such as Guillain Barré syndrome [34], myasthenia gravis [35], peripheral neuropathy, pseudotumor cerebri, meningoencephalitis, and cranial nerve palsies, among others [30]. Other nonhepatic symptoms include acute pancreatitis [30], rashes, kidney damage, and hematological disorders [36] such as aplastic anemia [37], thrombocytopenia, and hemolysis [30].

Shedding of HEV in the stool starts before the onset of symptoms, with up to 10^8 HEV genome copies shed per milligram of feces over several days before levels decline [38]. This is also observed in animals, with detectable HEV RNA found in swine feces even when the infection is asymptomatic, which has implications for environmental transmission and contamination of crops by runoff from farms and abattoirs [39].

HEV VACCINES

Infection studies in humans and animals have shown that HEV invokes the production of anti-HEV antibodies, which give protective immunity against subsequent infections [40]. HEV genotypes represent one serotype, with anti-HEV antibodies demonstrating cross-reactivity across different HEV genotypes [41]. These properties indicate that the development of a successful vaccine is achievable. However, the development of widely available HEV vaccines has been hindered by the lack of knowledge on the viral replication cycle, host cell targets, and viral entry and egress. In addition, HEV has proved challenging to cultivate, so a cell culture method has not yet been developed that can yield sufficient HEV to produce inactivated or live-attenuated vaccines [42]. Another major challenge has been a lack of interest in the virus from higher-income nations with the ability to fund vaccine research. HEV usually causes epidemics in developing countries that generally do not have the funds for this type of research [43].

Several recombinant vaccine candidates have been developed to induce neutralizing antibodies that target epitopes of the open reading frame 2 (ORF2) encoded capsid protein, as reviewed by Amini-Bavil-Olyaee et al. [44]. However, only one vaccine is currently in use in China, and was demonstrated to be effective following clinical trials [45]. This vaccine, HEV 239 (marketed as Hecolin®), was developed as a joint venture between Xiamen University and Xiamen Innovax, a biotechnology company. The HEV 239 antigen is a recombinant HEV capsid protein fragment, expressed in *Escherichia coli.* This fragment comprises capsid amino acids 368–606, extending across the capsid neutralizing epitope (amino acids 458–607) of a GI Chinese HEV strain (DDBJ accession number D11092) [46,47]. The recombinant protein self-assembles into 23 nm virus-like particles (VLPs) that promote a strong antibody response [46], and was shown to provide protective immunity against subsequent HEV challenge in rhesus monkeys [48]. Phase II randomized, controlled HEV 239 trials in humans were completed to test vaccine safety and immunogenicity [49], and were followed by phase III testing. The phase III trial was a randomized, double-blind, placebo-controlled trial in 11,165 participants aged between 16 and 65 years [50]. Half of the study group was administered three doses of the HEV 239 vaccine and the other half was administered three doses of a placebo. The trial results showed that the vaccine was 100% efficacious after the third dose, with minimal side effects reported [50]. HEV 239 is currently undergoing a post-authorization phase IV clinical trial to test efficacy in adults over 65 years [51].

A recent, extended follow-up study was published assessing the long-term efficacy of the HEV 239 vaccine within the Chinese population. In this double-blinded study of 112,604 participants aged 16–65, half of the study group received the hepatitis B virus (HBV) vaccine as a control and half received the HEV 239 vaccine to evaluate ongoing protection [52]. HEV 239 was found to provide immunogenicity for up to 4.5 years in almost 90% of those vaccinated [52]. The HEV 239 vaccine was approved for use in China in 2011, but is not currently available elsewhere.

Another HEV vaccine candidate, the rHEV vaccine, reached phase II clinical trials in Nepal and was reported to have an efficacy of 85.7% protection with a two-dose course or 95.5% efficacy after a three-dose course. This vaccine candidate was jointly developed by the U.S. Army,

GlaxoSmithKline, and researchers in Thailand and Nepal [53]. The vaccine comprised a truncated 56 kDa ORF2 encoded capsid protein, expressed in baculovirus. The recombinant protein spanned amino acids 112–607 [44] and self-assembled into VLPs. This vaccine candidate has been shown to promote a serum immune response (IgM, IgA, and IgG) and an intestinal response (IgA) in mice 2–8 weeks postimmunization [54]. Safety and efficacy were tested in 1794 soldiers in the Nepalese army. The rHEV test vaccine cohort included 898 soldiers, with 896 soldiers in the placebo cohort. Both groups were followed for a median period of 804 days postvaccination [53]. The trial reported a vaccine efficacy of 95.5%, with 69 individuals developing HEV, 66 of whom were within the placebo cohort [53]. However, these trials were considered controversial and further trials were therefore abandoned [53,55].

An early stage candidate for a HEV vaccine is a bivalent vaccine currently under development for both norovirus (NoV) and HEV [56]. The vaccine candidate uses a polyvalent complex platform with fusion of the dimeric protruding capsid domains of both NoV and HEV, linked to dimeric glutathione-S-transferase (GST). *In vivo* murine studies have been used to show that the fusion protein elicits a significantly higher immune response than that observed with simultaneous delivery of the two unfused proteins [56]. While this discovery is encouraging, this vaccine candidate is still in the early developmental stages.

Despite these examples of HEV vaccine research and the development of an efficacious HEV vaccine, HEV 239 is currently only approved for use within China. Without a readily available, worldwide HEV vaccine, HEV infections will continue to place several populations at risk, including pregnant women, throughout the developing world. The manufacturers of HEV 239 are reportedly in discussions with the World Health Organization (WHO) to make the vaccine readily available outside China through charitable agencies [43]. A memorandum of understanding was signed in June 2015 between the European Vaccine Initiative, the International Vaccine Institute of Korea, and Xiamen Innovax, with the aim of bringing Hecolin® to WHO prequalification standards so that it can be made more widely available [57].

TREATMENT FOR HEV INFECTION

In general, only supportive care can be offered to acute HEV patients, as there are currently no antiviral treatments developed specifically for this virus. HEV causes chronic infections in immunosuppressed individuals, including HIV patients and solid organ transplant recipients on immunosuppressive medication (reviewed in [58]). In these cases, ribavirin or pegylated interferon can be administered (Figure 8.2) with variable success [59–61]. Alternatively, reduced immunosuppression in organ transplant recipient patients can sometimes allow the immune system time to clear the virus [62,63]. The possible clinical courses of HEV infection are shown in Figure 8.2.

HISTORY OF HEV HEPATITIS OUTBREAKS

The most recent common ancestor of contemporary HEV genotypes evolved some 536–1344 years ago, according to one phylogenetic evolutionary study, before diverging into GI and GII (anthropotropic) and GIII–GVII (ezoonotic) genotypes [64]. A series of successive animal host adaptation steps are thought to have occurred in the lead-up to the infection of a human host, and increased population growth and travel likely contributed to the spread of HEV [64].

It has been postulated that as far back as the Middle Ages, HEV would have caused some of the documented jaundice outbreaks (reviewed in [65,66]). Most of the population in those times would have had early childhood hepatitis A virus (HAV) infections, giving them lifelong HAV protective immunity. As such, it seems reasonable that several large-scale reported hepatitis outbreaks in previous times could have been caused by HEV. This is supported by the notable rate of infections in adults rather than in children, and poor levels of sanitation common in earlier times that allowed easy transmission of HEV [1,65].

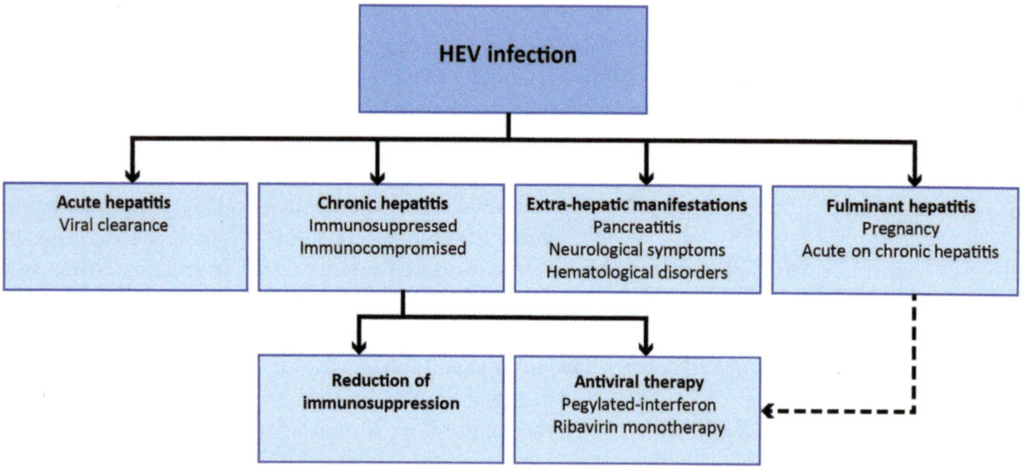

FIGURE 8.2 The possible clinical courses of HEV infection in humans. The known clinical courses of HEV infection in humans are shown with solid black arrows, while one possible course of infection is shown with a hashed arrow. Acute hepatitis is self-limiting and is cleared naturally. Chronic hepatitis is often associated with HEV genotype III. Chronicity is associated with immunocompromised patients, including HIV patients and solid organ transplant recipients. Immunosuppressed patients may respond to a reduction in posttransplant immunosuppression to allow viral clearance by host immunity and treatments include pegylated-interferon and ribavirin monotherapy. Extrahepatic manifestations can include pancreatitis, neurological symptoms, and hematological disorders. (Adapted from a review by Kamar, N. et al. *Clin Microbiol Rev*, 27, 116–138, 2014.)

One of the earliest large-scale reported outbreaks of HEV was retrospectively identified from a 1955 hepatitis outbreak in India. The hepatitis epidemic was reported in Delhi over the winter of 1955–1956, when monsoons caused severe flooding in the city [67]. The flooding carried raw sewage into the city's potable water supply at the same time as the city's chlorination treatment plant broke down. At the time, the disease outbreak was thought to be a HAV epidemic; however, mortality rates in pregnant women were unusually high and the large number of icteric hepatitis cases did not fit with HAV epidemiology, given that most of the population was likely to be immune to HAV [67]. Archived patient samples were tested several years later and the etiological agent was identified as HEV [65,68].

Another large HEV outbreak was reported in 1978 in the Kashmir Valley. This outbreak caused 52,000 cases of icteric hepatitis and 1,700 deaths, affecting 200 villages. The outbreak was caused by fecal contamination of open drinking water sources [69].

HEV has historically been dismissed as an issue relevant only in developing countries, with widespread outbreaks resulting from poor sanitation. However, in recent years, HEV recognition has increased within developed countries, with growing evidence of zoonotic transmission of the virus from food sources and animal reservoirs as well as transmission through fecal contamination of water and crops. As research has intensified, HEV has now been recognized as a significant health burden and a global threat [70].

DISCOVERY OF HEV

The discovery and identification of HEV occurred in 1983, following an outbreak of acute viral hepatitis during the occupation of Afghanistan by Soviet troops in 1980 [71]. A fecal extract from one of the infected troops was filtered and self-administered by Mikhail Balayan, a zealous pioneer of early HEV research. Balayan then visualized the HEV virion from his own stool sample using immune electron microscopy (EM), after developing hepatitis [72].

Before HEV-specific diagnostic techniques were developed, the virus was diagnosed only by eliminating HAV and HBV infection [73]. Before the discovery of HEV in 1983, the virus was referred to as "enterically transmitted non-A, non-B hepatitis" (ET-NANBH) and after its discovery, the virus was attributed to many earlier outbreaks via retrospective studies, as described in the "History of HEV Hepatitis Outbreaks" section.

A partial genomic sequence of HEV was cloned in 1990 by Reyes et al. [74], after isolating cDNA from an infected cynomolgus macaque liver. This study provided evidence of the presence of a viral RNA-dependent RNA polymerase (RdRp) encoding region, by identifying a conserved polymerase motif within the isolated sequence. Shortly afterward, in 1991, the full length viral genome was sequenced using contiguous, overlapping cDNA clones of a Burmese HEV strain, (HEV[B]), isolated from the stool samples of four acutely ill Burmese patients [75]. This allowed further identification of the nonstructural and structural proteins and full characterization of the genome [76].

HEV EPIDEMIOLOGY

Throughout many parts of the developing world, HEV is hyperendemic, causing widespread epidemics and sporadic cases in regions of Africa, Central America, China, and Southeast Asia [69]. While it was originally thought that HEV cases in higher-income nations were solely the result of travel to HEV endemic regions [77], it has since been recognized that HEV infection in these countries can be transmitted via HEV contaminated meats and zoonotic infections [78,79].

The five HEV genotypes known to infect humans have different epidemiological profiles for transmission, disease severity, host species range, and geographic distribution (Table 8.1). HEV GI and GII only infect humans via the fecal–oral route, usually from contaminated water, and cause outbreaks in the developing world. GI and GII viruses usually infect young adults [80] and are frequently subclinical. GI is found in many parts of Africa and Asia, while GII has been mostly found in Mexico and, more recently, in Africa, where the seroprevalence of anti-HEV antibodies within populations has been reported to be 30%–80% (reviewed in [25]). The host range for ezoonotic genotypes varies considerably more than for anthropotropic genotypes. HEV GIII infects a wide host range, including humans, swine, rabbits, boar, and deer; GIV infects humans and swine; GV and GVI infect wild boar; and GVII infects humans and camels [6]. GIII and GVII have been associated with chronic hepatitis in human patients, although GIII and GIV HEV usually cause sporadic cases of hepatitis in humans in developed countries [9,25]. Human HEV infections with GIII, GIV, and GVII strains are caused via zoonotic transfer from animal exposure, environmental transmission, or from ingestion of contaminated raw food and animal products [9,25].

In developed countries, symptomatic infections generally present in older males (mean age of 60) with a ratio of three males infected to each female [58]. Increased HEV disease in this group could be attributed to undiagnosed but preexisting liver disease [58]. A summary of HEV epidemiology is shown in Table 8.1.

HEV CLASSIFICATION

HEV is a single-stranded, nonenveloped RNA virus, currently classified within the *Hepeviridae* family [82]. HEV was originally classified within the *Caliciviridae* family [76] due to its genetic organization and morphological similarities visualized by EM, but in 2004 it was reclassified to the *Hepeviridae* family (reviewed in [1]). Early *in silico* studies were used to identify conserved functional domains within the HEV genome sequence [83]. These studies noted that the nonstructural proteins had the closest similarity to rubella virus and to the plant virus, beet necrotic yellow vein virus, while the genome organization of encoded structural proteins had the closest similarity to animal enteric caliciviruses [83].

In an attempt to clarify the evolutionary history of the *Hepeviridae*, an investigation revealed that the ORF1 region of the HEV genome falls within the "alpha-like" supergroup of viruses, while

TABLE 8.1
Geographical Distribution and Characteristics of Human HEV Infections

	HEV Genotype				
Characteristics	**Genotype I**	**Genotype II**	**Genotype III**	**Genotype IV**	**Genotype VII**
Geographical distribution	Africa and Asia	Mexico, Africa	Developed countries (autochthonous)	China, Japan, Taiwan (autochthonous)	Middle East
Transmission	Water-borne, fecal–oral, and person-to-person	Water-borne and fecal–oral	Foodborne and zoonotic	Foodborne and zoonotic	Foodborne and zoonotic
High risk demographic	Adolescents and young adults	Adolescents and young adults	Older adults (>40 years), males, and immunocompromised	Older adults (>40 years), males, and immunocompromised	Immunosuppressed and immunocompromised
Mortality	High risk in pregnant women	High risk in pregnant women	Usually low, can be higher in older adults	Usually low, can be higher in older adults	Not determined
Chronicity	No	No	Yes	No	Yes
Pattern of occurrence	Hyper-endemic	Smaller scale outbreaks	Sporadic	Sporadic	Sporadic

Source: Reviewed in Hoofnagle, J.H. et al. *N Engl J Med,* 367, 1237–1244, 2012; Khuroo, M.S. *Virus Res,* 161, 3–14, 2011; Centers for Disease Control and Prevention (CDC). Viral Hepatitis: Hepatitis E Information—Hepatitis E FAQs for Health Professionals, 2015. http://www.cdc.gov/hepatitis/HEV/HEVfaq.htm. Accessed November, 2015 [25,69,81].

the capsid encoding region falls within the "Picorna-like" supergroup [84]. These results strongly indicate an ancient recombination event that occurred at the junction of the nonstructural and structural-encoding regions, between a plant-feeding insect virus (*Alphatetraviridae*-like) and an animal virus (*Astroviridae*-like), which likely led to the emergence of the *Hepeviridae* family [84].

In 2014, a comprehensive review of *Hepeviridae* taxonomy was undertaken and two new genera were created, *Orthohepevirus* and *Piscihepevirus* [6]. HEV isolates within these genera are split further into *Orthohepevirus* A, for isolates of human, porcine, camel, deer, and rabbit HEV; *Orthohepevirus* B, for avian HEV; *Orthohepevirus* C, for isolates from rat, bandicoot, ferret, and shrew HEV; *Orthohepevirus* D, for bat HEV; and *Piscihepevirus*, for cutthroat trout virus (CTV) [6,82,85].

Of the seven *Orthohepevirus* A genotypes (GI–GVII), only viruses from GIII, GIV, and GVII [9] are confirmed to be transmissible zoonotically to humans (reviewed in [16]). GIII is highly prevalent worldwide in swine, which are recognized as a GIII HEV reservoir (reviewed in [17]). GIV is found in swine and humans, mainly within Asian countries [86]. The epidemiology of transmission for HEV GIII and GIV is reminiscent of influenza, where swine strains can be transmitted zoonotically to humans. GVII is found in camelids [8] and has been detected as the cause of hepatitis in a liver transplant patient in the Middle East [9]. A phylogenetic tree showing the common ancestry of HEV full length genomes is shown in Figure 8.3.

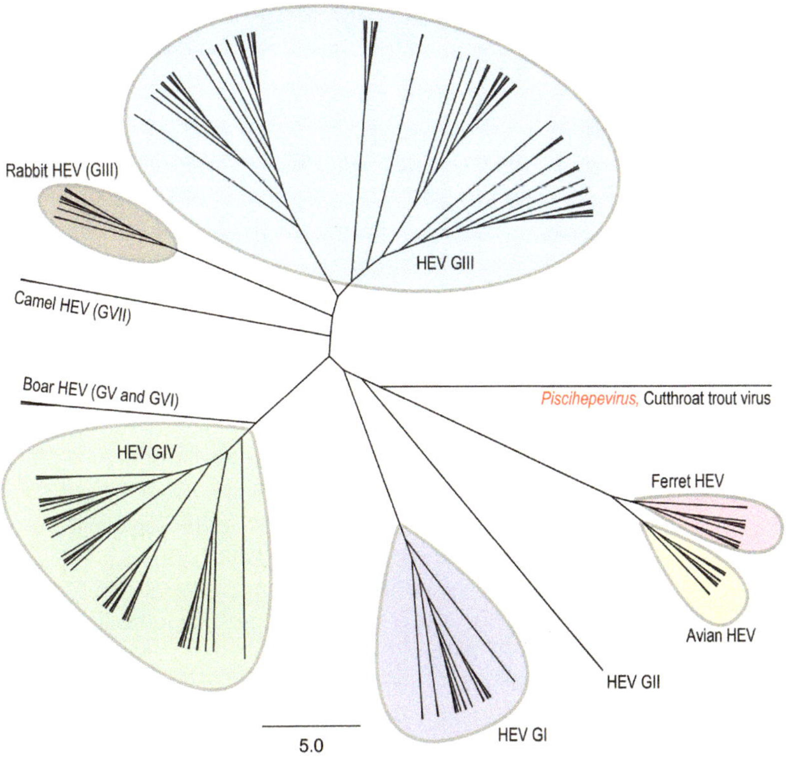

FIGURE 8.3 HEV common ancestry: An unrooted phylogenetic tree of 173 *Hepeviridae* full-leth genomes. The *Hepeviridae* genomes were aligned by multiple alignment using the fast Fourier transform (MAFFT) method, from Katoh, K. et al. *Nucleic Acids Res,* 30, 3059–3066, 2002 and the maximum-likelihood tree was inferred using FastTree 2.1.5 from Price, M.N. et al. *PLoS One,* 5, e9490, 2010. All genomes shown are from the *Orthohepevirus* genus, except for cutthroat trout virus, which is from the *Piscihepevirus* genus (shown in red). The tree shows common ancestry, but does not indicate evolutionary distance. The scale bar indicates the number of nucleotide substitutions per site.

Animal Hosts and Reservoirs for HEV

HEV is unusual compared with other hepatitis viruses, because it has known animal reservoirs (reviewed in [17]). For example, swine is a well-recognized HEV reservoir, and zoonotic transmission of HEV from pigs to humans is relatively common [16,89]. There is a growing list of nonhuman HEV reservoirs as novel strains are identified within different animal species, with several species of rats recently identified as additional HEV hosts [16].

HEV has been detected in many animal species [90] with a high degree of genetic identity (Table 8.2) to the HEV genotypes that only infect humans (GI and GII). Table 8.2 shows the percentage of nucleotide identity between 13 HEV genomes isolated from different host species. HEV has been identified in domesticated swine [91], deer [19], moose [92], Japanese wild boar [18], bats [93], camels [8], chickens [94,95], mongooses [96], rabbits [97], domesticated pet birds [98], rats [99], and various primates, including chimpanzees, tamarins cynomolgus macaques, and rhesus, owl, and African green monkeys [40,72,75,100]. In addition, anti-HEV IgG antibodies have been identified in a number of other animals, although no HEV RNA has been isolated from these animals to date. These seropositive animals include cattle, dogs, horses, rodents, and bison [90].

Avian HEV (*Orthohepevirus* B)

A more divergent HEV virus, classified as *Orthohepevirus* B, has been detected in birds [6]. The "big liver and spleen" (BLS) disease found in chickens was first reported in 1980 in Australia, although the etiological agent responsible was only recognized as avian HEV in 1999 [101] and reclassified as *Orthohepevirus* B in 2014 [6]. BLS disease is associated with lower egg production and a marginal increase in mortality in infected birds [101,102]. A similar disease was reported throughout North America and called hepatitis-splenomegaly (HS) syndrome, and in 2001 it was associated with an infection by a novel HEV strain [103]. The virus was shown to have 80% nucleotide identity to the *Orthohepevirus* B strain isolated from HEV-infected Australian chickens and 50%–60% identity to the antigenic capsid region of human HEV strains [103]. While *Orthohepevirus* B is prevalent in chicken flocks, it is not currently thought to be transmissible to humans [79].

HEV Genome Structure and Protein Function

HEV has a positive sense, single-stranded RNA genome, contained within a small, nonenveloped, icosahedral virion that is 27–34 nm in size [104]. The HEV genome size is around 7.2 kb, depending on the genotype, with the nonstructural protein coding regions at the 5′ end and the structural protein coding regions at the 3′ end of the genome [105]. The genome includes a 5′ 7-methylguanosine (7 mG) cap and a polyadenylated (polyA) tail at the 3' end, which is essential for replication [106]. There are short 5′ and 3′ untranslated regions (UTRs) that form stem loops and hairpin structures, likely to be required for replication [107,108].

The viral genome (Figure 8.4) contains three partially overlapping ORFs; the first encodes a large nonstructural polypeptide of 1693 amino acids (ORF1) and the remaining ORFs (ORF2 and ORF3) encode two structural proteins [76]. Thus, the genome demonstrates similarities in organization to viruses of the *Caliciviridae* family, as previously mentioned [76].

The nonstructural polyprotein has several functional domains in common with other RNA viruses [83], shown in Figure 8.4. These include a viral methyltransferase (MT) domain, a "Y" protein domain of unknown function, a putative papain-like cysteine protease (Pro) domain, a helicase (Hel) domain, and an RNA-dependent RNA polymerase (RdRp) domain [42,83]. The eight conserved RdRp motifs present in many other positive-sense RNA viruses (motif I-VIII) have been identified in the HEV RdRp domain, including the canonical "GDD" motif within the active

TABLE 8.2
Similarity Matrix of HEV Genomes of Various HEV Strains (% identical nucleotides)

Similarity Matrix		Percentage Nucleotide Identity [%]												
HEV Genome	Accession Numbers	AB074915	AB236320	AF455784	AB740220	KJ496143	M74506	NC001434	KF951328	AB890001	JX120573	NC018382	NC023425	NC015521
HEV GIV	AB074915	—	75.3	75.1	73.3	74.8	73.3	74.2	58.5	51.1	51.1	46.9	44.5	36.6
Mongoose HEV	AB236320		—	81.7	77.6	75.7	73.2	74.2	59.0	51.6	51.1	46.8	44.9	35.7
HEV GIII	AF455784			—	77.5	75.3	73.5	74.0	58.9	51.4	50.6	46.8	45.0	35.9
Rabbit HEV	AB740220				—	73.9	71.7	72.3	57.9	51.4	51.4	46.7	44.3	36.5
Camel HEV	KJ496143					—	73.4	74.1	58.4	51.0	50.8	46.6	44.6	35.9
HEV GII	M74506						—	75.3	58.9	50.5	50.8	46.6	45.1	35.6
HEV GI	NC001434							—	59.4	51.2	51.5	47.1	45.0	36.8
Moose HEV	KF951328								—	51.8	52.4	47.2	46.5	35.7
Ferret HEV	AB890001									—	68.3	47.0	47.0	35.1
Rat HEV	JX120573										—	46.6	45.9	34.8
Bat hepevirus	NC018382											—	47.0	34.7
Avian HEV	NC023425												—	33.6
Cutthroat trout virus	NC015521													—

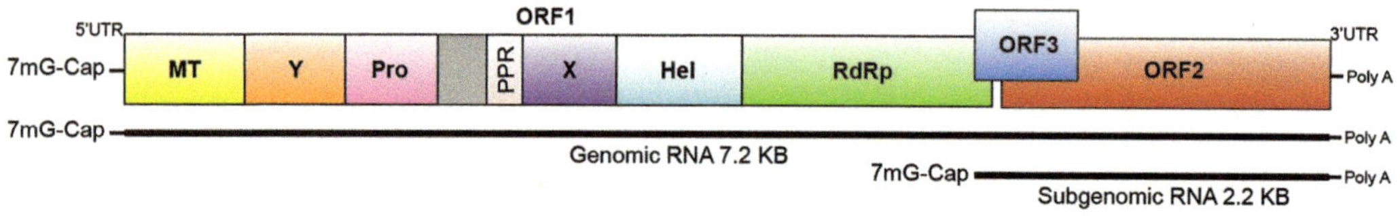

FIGURE 8.4 HEV genome organization. The HEV genome is approximately 7.2 kb long with three ORFs, flanked by a 7 mG-cap, 5′UTR, 3′UTR, and 3′ polyadenylated tail. ORF1 encodes the nonstructural proteins, which include methyltransferase (MT), Y protein (Y), protease (Pro), polyproline region (PPR), X domain (X), helicase (Hel), and the RNA-dependent RNA polymerase (RdRp). ORF2 encodes the capsid protein, while ORF3 is thought to mediate the host cell environment for virion egress. (Modified and adapted from review by Kamar, N. et al. *Clin Microbiol Rev*, 27, 116–138, 2014.)

site [109]. The nonstructural ORF1 region also includes an "X" or macro domain, of unknown function, and a hypervariable region (HVR), which has been shown to have a relationship with replication efficiency and infectivity [110]. The HVR overlaps a putative polyproline "hinge." The polyproline region (PPR) is intrinsically unstructured and is thought to allow multiple ligand binding events for the regulation of transcription and translation, although this function has not yet been fully elucidated [111].

HEV ORF1: Nonstructural Proteins

HEV ORF1 has been expressed as a ~186 kDa recombinant polyprotein in *E. coli,* insect cells, and mammalian cells, but it is still unclear how the protein is processed posttranslationally. Mixed experimental results give little clarity on how the larger polyprotein is processed into discrete protein products, with some evidence suggesting that there is no cleavage or processing of the polyprotein [112] and other studies giving evidence of proteolytic products that correlate to the proposed functional domains when expressed in baculovirus expression systems [113].

HEV ORF2: Capsid Protein

ORF2 encodes a 660 amino acid capsid protein, which assembles into an icosahedral virion. The ORF2 protein has three domains, the shell (S), middle (M), and protruding (P) regions [114]. The crystal structure of the HEV capsid protruding domain demonstrates that dimerization of the capsid protein is critical for host–cell interactions [115]. As the ORF2 protein includes neutralizing epitopes [116], it has been the main target for the development of the aforementioned HEV vaccine candidates.

HEV ORF3: Multifunctional Phosphoprotein

ORF3 encodes a multifunctional phosphoprotein of 114 amino acids, which has been shown to interact with the protein kinase signaling pathways [117]. It has been proposed that this protein modulates the host environment to facilitate virion egress [118]. The genome organization is detailed in Figure 8.4.

CELL ENTRY AND THE HEV REPLICATION CYCLE

Details on HEV replication and how the virus enters and egresses host cells are scarce due to the lack of an efficient cell culture system for HEV. However, heparin sulfate proteoglycans (HSPGs) on the host cell surface are thought to be involved in the binding of the capsid [119]. This was demonstrated in a study using a 56 kDa recombinant ORF2-encoded protein that could self-assemble into VLPs. When heparinase I enzyme was used to degrade heparin on the surface of Huh-7 cells (liver derived), the binding ability of the recombinant capsid protein was dramatically reduced, implicating HSPGs in HEV entry into host hepatocytes [119].

It is widely accepted that the HEV primary host cell target is the hepatocyte [120], although there is evidence of HEV replication in other nonhepatic tissues including the small intestine, lymph nodes, and the colon, as detected by negative strand (a marker of replication) specific reverse transcription polymerase chain reaction (RT-PCR) assays in swine [121].

Further details of the viral replication cycle have been mainly inferred by comparison with other RNA viruses [105] and studies in cell culture, or by using *in vivo* models such as chimpanzees, owl monkeys, and macaques [122]. Rats, chickens, rabbits, and pigs are also suggested small animal model candidates for further studies, given that HEV has now been isolated from natural infections in these species [123,124].

HEV CELL CULTURE SYSTEMS

HEV in Tissue Culture

HEV virion production has proved challenging because it has been difficult to establish robust *in vitro* HEV replication in cell lines. In a recent report from Japan, HEV was obtained from a patient with high viral RNA levels and successfully cultured in PLC/PRF/5 cells before being established in A549 cells [125]. HEV GIII and GIV have also been cultured in this cell line [126] and, in addition, HEV GIII has been recently cultured in human and porcine hepatoma-derived cell lines [127].

HEV replicons have also been created to study HEV replication within nonhuman primate cells, where ORF2 and ORF3 have been replaced with a marker gene encoding enhanced green fluorescent protein (EGFP) [128]. These studies showed that the 7 mG cap was required for infectivity, as only capped viral RNA could infect cells [128].

These examples illustrate that it is possible to culture HEV, but for the most part these systems have been slow to develop and are largely inefficient and difficult to replicate.

HEV TRANSMISSION

HEV causes hepatitis in humans through enteric transmission, which can be waterborne, foodborne, or zoonotic via animal contact (reviewed in [10]). It can also be transmitted vertically from person-to-person [129] and parenterally via blood transfusions [130–132] and solid organ transplants [133]; however, intrafamilial spread has been shown to be negligible [134]. Direct contact with infected animals and handling or consumption of contaminated undercooked meat are all possible routes of HEV transmission (reviewed in [16]). Direct evidence of HEV zoonoses has been demonstrated in cross-species HEV infection studies in humans, nonhuman primates, and swine [135,136]. In particular, the evidence of zoonoses between pigs, deer, wild boar, and humans is of great concern to public health.

There is evidence of HEV transmission to humans by vocational exposure on swine farms [137]. Additionally, abattoir effluent and swine farm runoff into water sources and onto pastures and use of slurry fertilizers and contaminated irrigation of crops are all methods for environmental transmission of HEV [39,138]. Studies have shown that swine farmers have a significantly higher HEV seroprevalence than the general population. In one study, the ratio of HEV infections was shown to be double for swine farmers compared with the general public, and the risks were shown to increase with drinking raw milk, years of exposure to piggeries, assisting sows at birth, and cleaning out barns [137]. In summary, HEV poses a potential health risk through zoonotic transfer from infected animal contact as well as through environmental transmission into potable water supplies and coastal waters that could contaminate shellfish beds or swimming areas [39].

FOODBORNE HEV TRANSMISSION

This section reviews the available body of evidence for direct foodborne HEV transmission to humans. Over the last decade, it has become increasingly evident that human HEV infections are a growing problem, as the global food network expands and diversifies. Mounting evidence has come to light indicating that foodborne transmission of HEV poses a significant risk to humans within developed countries, as additional transmission routes are recognized. There are four main routes for foodborne transmission of HEV to humans: (1) HEV GIII, GIV, and GVII are transmitted through consumption of contaminated, undercooked pork, boar, deer, and camel meats and other food products; (2) HEV GIII and GIV can also be transmitted to humans (usually swine farmers) via direct contact with animals farmed for human consumption [137]; (3) for GI and GII HEV infections, seafood consumption is a possible transmission route [39] and occurs when farmed shellfish are contaminated with enteric viruses (usually via sewage) [139,140]; and (4) there is mounting evidence for HEV transmission to humans from contaminated fresh produce such as berries and leafy vegetables [141–143], which are usually consumed raw.

Despite this growing recognition, there are only a few reported cases offering direct evidence of foodborne HEV transmission as the lengthy incubation period (mean of 40 days) often means source links are difficult to identify, because the HEV-contaminated food is no longer available for testing and further studies [39]. Of the available studies, RT-PCR is often used as the HEV detection method, so it should be noted that detection by RT-PCR only confirms the presence of the virus and does not necessarily indicate if the virus is infectious. While there are limited studies available on the environmental stability of HEV, it is known to stay viable through at least one freeze–thaw cycle [144]. This could have implications for HEV infection if the virus is able to remain viable in snap-frozen fresh produce.

EVIDENCE OF HEV IN FRESH PRODUCE SUPPLY CHAIN

There are usually no postharvest measures implemented in general crop production to kill pathogens [145]. As such, HEV is one of a number of viruses that can be detected on fresh produce for human consumption, along with norovirus, rotavirus, HAV, adenovirus, and astrovirus. Most fruit and many vegetables are intended to be eaten raw, and cooking to inactivate any foodborne enteric viruses is not usually feasible.

HEV has been isolated from fresh fruits such as berries and leafy green vegetables, which are grown in pastures and undergo little processing before they are consumed. For example, in one study, HEV was detected on strawberries 1 h after irrigation with polluted river water [141]. The HEV strain isolated from the berries had a sequence identity of 99% with a swine HEV strain detected on a pig farm in the same region of Quebec [141]. HEV has also been detected on frozen strawberries and raspberries at point-of-sale in Europe. Possible contamination routes were identified as handling, irrigation, or manure fertilization of the berries [142]. Animal-based manure fertilizer is a likely route for HEV transfer onto berries, given the detection of other enteric viruses in swine manure and irrigation water samples [142]. HEV has also been identified on fresh lettuce and in irrigation water applied to vegetable crops in Europe [143].

Changes in processing of fresh produce are required to help prevent the introduction of enteric viruses into the fresh food chain, specifically the avoidance of using swine manure fertilizer on crops or irrigation with contaminated water.

HEV INFECTIONS FROM MEAT CONSUMPTION

HEV Infection Risks from Pork

Swine are a well-recognized HEV reservoir [85], so the production and consumption of pork products carries a risk of HEV infection in humans. Indeed, thermal stability testing on HEV demonstrated that a small percentage of some HEV strains remained viable at 60°C, which has implications for consumers eating rare-cooked meats and rewarmed foods [144]. One of the first studies clearly linking transmission of HEV to humans from pork consumption was presented by Yazaki et al. in 2003 [15]. This study reported several cases of fulminant or sporadic acute hepatitis within 2–8 weeks of ingestion of undercooked pork liver in Japan. Testing of pig livers from 25 Japanese grocery stores in Hokkaido demonstrated that 2% of livers sold had detectable HEV RNA present in the meat [15]. Nucleotide sequence identity between the RNA isolated from the pig livers and the patients was between 98.5% and 100%, clearly implicating the pig livers as the source of human HEV infection [15]. This study further supports the accumulating evidence that HEV is transmissible to humans through consumption of undercooked meat.

Several studies across North America and Europe have also clearly linked the consumption of pork and pork products to human HEV infections. In Southern France, figatellu (plural: figatelli) sausage is a local delicacy of dried, cold-smoked pig liver, often eaten raw. Figatelli sausages from local supermarkets were tested using RT-PCR to detect HEV RNA levels, which confirmed that 58% (7 of 12) of samples were HEV RNA positive with an estimated 10^3–10^6 HEV genome copies per slice [146]. In the same case-controlled study, it was found that 54% (7 of 13) of individuals who consumed figatelli were serologically positive for HEV or had detectable HEV-RNA, compared with 0% (0 of 5) of individuals who did not consume figatelli [146]. Interestingly, HEV seroprevalence in blood donors is 52.5% in Southwest France [132] where figatelli is widely consumed, compared with just 3.2% in France overall [147]. Moreover, in the United States, 11% of pig livers sold in grocery stores tested positive for GIII HEV RNA using RT-PCR, which had the ability to detect HEV GI–GIV RNA. Two of three liver homogenates tested on swine resulted in HEV infections, demonstrating that the HEV was still infectious [148].

In Europe, commercially available pork livers sold in the Netherlands were tested for the presence of HEV RNA to establish routes for infection within the country. Of the livers tested, 6.5% (4 of 62) were positive for HEV using RT-PCR and Southern blot hybridization. Three of the four livers that were positive for HEV sequences were found to contain GIII viruses [149].

Similarly, in Britain, several steps within the pork food chain were monitored for HEV using RT-PCR, including samples and swabs from slaughterhouses, processing plants, and points-of-sale, with HEV RNA being detected at every step. It was revealed that 10% of all sausages sampled (6 out of 63) had detectable levels of HEV, as did 25% of surface swabs at the points-of-sale (2 of 8). As sausages are so widely consumed in Great Britain (over 212,746 tons from February 2011 to February 2012), this study raised several health concerns pertaining to HEV infections in the British public [150].

In summary, there is a risk of HEV infection, particularly GIII porcine viruses, from the consumption of pork and pork products, especially undercooked pig livers.

HEV Infection Risks from Venison

HEV infection has also been implicated from the consumption of venison. In 2003, a study linking HEV infections to the consumption of raw deer meat was published by Tei et al. [19]. Following the diagnosis of human HEV infections, frozen leftover venison meat was tested and shown to contain viruses with an identical nucleotide sequence to the HEV isolated from the infected individuals that had consumed deer meat. The patients' families that had not consumed the deer meat were not infected, further proving the zoonosis [19].

HEV Infection Risks from Offal and Wild Boar

Raw or undercooked offal (kidney and liver) and wild boar meat were found to be a statistically significant higher-risk food for HEV transmission in Germany, where HEV infection has been a notifiable disease since 2001. In one German study, 100% of patients (15 out of 15) with acute autochthonous HEV infection and without a recent travel history tested positive for HEV GIII or GIV genotypes, which are usually detected in swine. A bivariate assessment of risk factors and exposure of infected individuals was compiled, revealing that patients who ate offal were at a 40.9% chance of HEV infection compared with 18.5% of control individuals, demonstrating a significantly higher risk of HEV infection from eating offal [151].

HEV Infection Risks from Camel Meat and Milk Ingestion

In the Middle East and in regions of Africa, the consumption of camel meat is popular and often prized in celebratory dishes. In addition to the ingestion of camel meat, camel milk is also widely consumed, particularly by some Bedouin tribes. There is evidence of camel-to-human transmission of HEV. A HEV strain has been detected in a chronic hepatitis patient in the United Arab Emirates who regularly ingested camel meat and milk but did not consume pork, which is normally associated with foodborne HEV infections [9]. Phylogenetic studies of the HEV genome sequence from the patient revealed that the virus was a GVII, known to infect camels [9]. Therefore, there is a risk of zoonotic transmission of novel HEV isolates, and novel genotypes could be present in other animal species with the potential to cause human disease.

These studies all clearly link HEV infection to the consumption of raw or undercooked meats and animal products, highlighting the risk associated with meat consumption, in particular from pork products and game meats. As a result, HEV is now recognized as a zoonotic disease.

HEV INFECTIONS FROM FISH AND SHELLFISH CONSUMPTION

Despite the growing evidence for human HEV infections from fish and shellfish, several studies in shellfish from different regions where HEV is circulating in the human population did not detect any HEV RNA [152–154]. A recent systematic review of the scientific literature and ProMED reports of shellfish-borne viral outbreaks found that only 0.3% of infections were caused by HEV [155]. It is not known if this is because the assays used are not sensitive enough to detect the levels or type of HEV present, or if incidences of HEV contamination of shellfish beds are not common. Bivalve molluscan shellfish are a known high-risk food group for exposure to many enteric pathogens as their filter feeding can concentrate microorganisms, including protozoa, bacteria, and viruses, within the flesh [156].

In many countries, seafood is eaten raw (e.g., as sashimi) or straight from the shell. Even lightly steaming fish and shellfish might not be enough to inactivate HEV, given the ability of the virus to remain infectious at 60°C [144]. It has also been noted that HEV incidence is statistically higher in those who consume fish, relative to pork consumption, but no data regarding the cooking of the food were obtained, so conclusions could not be drawn as to the role of cooking in this analysis [157]. In summary, while there is very little documented evidence of HEV contamination in fish and shellfish, human HEV infection from sewage-contaminated seafood is a plausible transmission route.

CONCLUSIONS

While it has now been firmly established that HEV can be transmitted in food, there is still a long way to go to understanding the full scope of foodborne HEV transmission risks. HEV is a largely understudied virus, and several details of the viral replication cycle, host targets, and host range are still to be elucidated.

HEV transmission has clearly been linked to the consumption of undercooked meats, especially pork, which should be regarded as a high-risk food even in higher-income nations. Game meats, shellfish, and fresh, unprocessed produce also carry risks of HEV transmission. The testing of meat and fresh produce throughout the supply chain up to point-of-sale could ensure that introduction points of contamination are traced and controlled to prevent further human infection.

In addition to raising awareness of HEV transmission risks in consumers and producers, education should be promoted among produce growers, animal farmers, and abattoir workers on environmental transmission risks posed by runoff into rivers and coastlines, application of animal slurry, and contaminated crop irrigation. Decontamination and management of farm runoff is also required to further control the spread into drinking and swimming water sources. These measures would go some way toward the prevention of HEV infections as well as other enteric viruses that persist in the environment.

REFERENCES

1. Purcell RH, Emerson SU. 2008. Hepatitis E: An emerging awareness of an old disease. *J Hepatol* 48:494–503.
2. World Health Organization (WHO). 2015. Hepatitis E—Fact sheet no 280.
3. Rein DB, Stevens GA, Theaker J, Wittenborn JS, Wiersma ST. 2012. The global burden of hepatitis E virus genotypes 1 and 2 in 2005. *Hepatology* 55:988–997.
4. Pischke S, Wedemeyer H. 2013. Hepatitis E virus infection: Multiple faces of an underestimated problem. *J Hepatol* 58:1045–1046.
5. Pischke S, Behrendt P, Bock CT, Jilg W, Manns MP, Wedemeyer H. 2014. Hepatitis E in Germany: An under-reported infectious disease. *Dtsch Arztebl Int* 111:577–583.
6. Smith DB, Simmonds P, Jameel S, Emerson SU, Harrison TJ, Meng XJ, Okamoto H, Van der Poel WH, Purdy MA. 2014. Consensus proposals for classification of the family Hepeviridae. *J Gen Virol* 96:1191–1192.
7. Meng XJ, Anderson D, Arankalle VA, Emerson SU, Harrison TJ, Jameel S, Okamoto H. 2012. *Ninth Report of the International Committee on Taxonomy of Viruses*. Elsevier, Amsterdam, the Netherlands.
8. Woo PC, Lau SK, Teng JL, Tsang AK, Joseph M, Wong EY, Tang Y, et al. 2014. New hepatitis E virus genotype in camels, the Middle East. *Emerg Infect Dis* 20:1044–1048.
9. Lee GH, Tan BH, Chi-Yuan Teo E, Lim SG, Dan YY, Wee A, Aw PP, et al. 2016. Chronic infection with camelid hepatitis E virus in a liver transplant recipient who regularly consumes camel meat and milk. *Gastroenterology* 150:355–357 e353.
10. Meng XJ. 2010. Hepatitis E virus: Animal reservoirs and zoonotic risk. *Vet Microbiol* 140:256–265.
11. Smith DB, Purdy MA, Simmonds P. 2013. Genetic variability and the classification of hepatitis E virus. *J Virol* 87:4161–4169.
12. Purdy MA, Smith DB, Simmonds P, Emerson SU, Harrison TJ, Meng XJ, Okamoto H, Van der Poel WH, Jameel S. 2014. New classification scheme for hepeviridae. International Committee on Taxonomy of Viruses (ICTV). http://www.ictvonline.org/proposals/2014.008a-hV.A.v6.Hepeviridae.pdf, 15 June, 2016.
13. Kumar A, Beniwal M, Kar P, Sharma JB, Murthy NS. 2004. Hepatitis E in pregnancy. *Int J Gynaecol Obstet* 85:240–244.
14. Aggarwal R, Krawczynski K. 2000. Hepatitis E: An overview and recent advances in clinical and laboratory research. *J Gastroenterol Hepatol* 15:9–20.
15. Yazaki Y, Mizuo H, Takahashi M, Nishizawa T, Sasaki N, Gotanda Y, Okamoto H. 2003. Sporadic acute or fulminant hepatitis E in Hokkaido, Japan, may be food-borne, as suggested by the presence of hepatitis E virus in pig liver as food. *J Gen Virol* 84:2351–2357.
16. Pavio N, Meng XJ, Doceul V. 2015. Zoonotic origin of hepatitis E. *Curr Opin Virol* 10C:34–41.
17. Pavio N, Meng XJ, Renou C. 2010. Zoonotic hepatitis E: Animal reservoirs and emerging risks. *Vet Res* 41:46.
18. Hara Y, Terada Y, Yonemitsu K, Shimoda H, Noguchi K, Suzuki K, Maeda K. 2014. High prevalence of hepatitis E virus in wild boar (*Sus scrofa*) in Yamaguchi Prefecture, Japan. *J Wildl Dis* 50:378–383.
19. Tei S, Kitajima N, Takahashi K, Mishiro S. 2003. Zoonotic transmission of hepatitis E virus from deer to human beings. *Lancet* 362:371–373.
20. Huang S, Zhang X, Jiang H, Yan Q, Ai X, Wang Y, Cai J, et al. 2010. Profile of acute infectious markers in sporadic hepatitis E. *PLoS One* 5:e13560.
21. Aggarwal R. 2013. Diagnosis of hepatitis E. *Nat Rev Gastroenterol Hepatol* 10:24–33.
22. Baylis SA, Gartner T, Nick S, Ovemyr J, Blumel J. 2012. Occurrence of hepatitis E virus RNA in plasma donations from Sweden, Germany and the United States. *Vox Sang* 103:89–90.

23. Chandra NS, Sharma A, Malhotra B, Rai RR. 2010. Dynamics of HEV viremia, fecal shedding and its relationship with transaminases and antibody response in patients with sporadic acute hepatitis E. *Virol J* 7:213.
24. Dalton HR, Bendall R, Ijaz S, Banks M. 2008. Hepatitis E: An emerging infection in developed countries. *Lancet Infect Dis* 8:698–709.
25. Hoofnagle JH, Nelson KE, Purcell RH. 2012. Hepatitis E. *N Engl J Med* 367:1237–1244.
26. Baylis SA, Blümel J, Mizusawa S, Matsubayashi K, Sakata H, Okada Y, Nübling CM, Hanschmann KM, Group HCS. 2013. World Health Organization International Standard to harmonize assays for detection of hepatitis E virus RNA. *Emerg Infect Dis* 19:729–735.
27. Wenzel JJ, Preiss J, Schemmerer M, Huber B, Jilg W. 2013. Test performance characteristics of Anti-HEV IgG assays strongly influence hepatitis E seroprevalence estimates. *J Infect Dis* 207:497–500.
28. Clayson ET, Myint KS, Snitbhan R, Vaughn DW, Innis BL, Chan L, Cheung P, Shrestha MP. 1995. Viremia, fecal shedding, and IgM and IgG responses in patients with hepatitis E. *J Infect Dis* 172:927–933.
29. Khuroo MS, Teli MR, Skidmore S, Sofi MA, Khuroo MI. 1981. Incidence and severity of viral hepatitis in pregnancy. *Am J Med* 70:252–255.
30. Aggarwal R, Jameel S. 2011. Hepatitis E. *Hepatology* 54:2218–2226.
31. Navaneethan U, Al Mohajer M, Shata MT. 2008. Hepatitis E and pregnancy: Understanding the pathogenesis. *Liver Int* 28:1190–1199.
32. Baylis SA, Koc O, Nick S, Blümel J. 2012. Widespread distribution of hepatitis E virus in plasma fractionation pools. *Vox Sang* 102:182–183.
33. Patra S, Kumar A, Trivedi SS, Puri M, Sarin SK. 2007. Maternal and fetal outcomes in pregnant women with acute hepatitis E virus infection. *Ann Intern Med* 147:28–33.
34. van den Berg B, van der Eijk AA, Pas SD, Hunter JG, Madden RG, Tio-Gillen AP, Dalton HR, Jacobs BC. 2014. Guillain-Barré syndrome associated with preceding hepatitis E virus infection. *Neurology* 82:491–497.
35. Belbezier A, Deroux A, Sarrot-Reynauld F, Larrat S, Bouillet L. 2014. Myasthenia gravis associated with acute hepatitis E infection in immunocompetent woman. *Emerg Infect Dis* 20:908–910.
36. Kamar N, Rostaing L, Izopet J. 2013. Hepatitis E virus infection in immunosuppressed patients: Natural history and therapy. *Semin Liver Dis* 33:62–70.
37. Amarapurkar DN, Amarapurkar AD. 2002. Extrahepatic manifestations of viral hepatitis. *Ann Hepatol* 1:192–195.
38. Li X, Kamili S, Krawczynski K. 2006. Quantitative detection of hepatitis E virus RNA and dynamics of viral replication in experimental infection. *J Viral Hepat* 13:835–839.
39. Van der Poel WH. 2014. Food and environmental routes of Hepatitis E virus transmission. *Curr Opin Virol* 4:91–96.
40. Tsarev SA, Tsareva TS, Emerson SU, Govindarajan S, Shapiro M, Gerin JL, Purcell RH. 1994. Successful passive and active immunization of cynomolgus monkeys against hepatitis E. *Proc Natl Acad Sci USA* 91:10198–10202.
41. Tam AW, White R, Yarbough PO, Murphy BJ, McAtee CP, Lanford RE, Fuerst TR. 1997. *In vitro* infection and replication of hepatitis E virus in primary cynomolgus macaque hepatocytes. *Virology* 238:94–102.
42. Chandra V, Taneja S, Kalia M, Jameel S. 2008. Molecular biology and pathogenesis of hepatitis E virus. *J Biosci* 33:451–464.
43. Park SB. 2012. Hepatitis E vaccine debuts. *Nature* 491:21–22.
44. Amini-Bavil-Olyaee ST, Trautwein C, Tacke F. 2009. Hepatitis E vaccine: Current status and future prospects. *Future Virol* 4.2:143–154.
45. Yang L. 2012. China approves hepatitis E vaccine, *on* Xinhua News Agency, Beijing. http://news.xinhuanet.com/english/china/2012-01/11/c_131355108.htm. Accessed November, 2015.
46. Zhang JZ, Ng MH, Xia NS, Lau SH, Che XY, Chau TN, Lai ST, Im SW. 2001. Conformational antigenic determinants generated by interactions between a bacterially expressed recombinant peptide of the hepatitis E virus structural protein. *J Med Virol* 64:125–132.
47. Kamili S. 2011. Toward the development of a hepatitis E vaccine. *Virus Res* 161:93–100.
48. Li SW, Zhang J, Li YM, Ou SH, Huang GY, He ZQ, Ge SX, et al. 2005. A bacterially expressed particulate hepatitis E vaccine: Antigenicity, immunogenicity and protectivity on primates. *Vaccine* 23:2893–2901.
49. Zhang J, Liu CB, Li RC, Li YM, Zheng YJ, Li YP, Luo D, et al. 2009. Randomized-controlled phase II clinical trial of a bacterially expressed recombinant hepatitis E vaccine. *Vaccine* 27:1869–1874.

50. Zhu FC, Zhang J, Zhang XF, Zhou C, Wang ZZ, Huang SJ, Wang H, et al. 2010. Efficacy and safety of a recombinant hepatitis E vaccine in healthy adults: A large-scale, randomised, double-blind placebo-controlled, phase 3 trial. *Lancet* 376:895–902.
51. Clinicaltrials.gov. 2014. Phase IV clinical trial of recombinant hepatitis E vaccine (Hecolin®) ClinicalTrials.gov Identifier: NCT02189603. https://clinicaltrials.gov/ct2/show/NCT02189603?term=h. Accessed November, 2015.
52. Zhang J, Zhang XF, Huang SJ, Wu T, Hu YM, Wang ZZ, Wang H, et al. 2015. Long-term efficacy of a hepatitis E vaccine. *N Engl J Med* 372:914–922.
53. Shrestha MP, Scott RM, Joshi DM, Mammen MP, Jr., Thapa GB, Thapa N, Myint KS, et al. 2007. Safety and efficacy of a recombinant hepatitis E vaccine. *N Engl J Med* 356:895–903.
54. Li T, Takeda N, Miyamura T. 2001. Oral administration of hepatitis E virus-like particles induces a systemic and mucosal immune response in mice. *Vaccine* 19:3476–3484.
55. Andrews J. 2005. U.S. Military sponsored vaccine trials and la resistance in Nepal. *Am J Bioeth* 5:W1–W3.
56. Wang L, Cao D, Wei C, Meng XJ, Jiang X, Tan M. 2014. A dual vaccine candidate against norovirus and hepatitis E virus. *Vaccine* 32:445–452.
57. International Vaccine Institute and Xiamen Innovax Biotech Company Co. L. 2015. *Memorandum of Understanding*. International Vaccine Institute. EU Vaccine website: http://www.euvaccine.eu/sites/default/files/uploads/docs/MoU/EVI_MOU_IVI-INNOVAX-EVI-Signed_Final_150612.pdf. Accessed November, 2015.
58. Kamar N, Dalton HR, Abravanel F, Izopet J. 2014. Hepatitis E virus infection. *Clin Microbiol Rev* 27:116–138.
59. Alric L, Bonnet D, Laurent G, Kamar N, Izopet J. 2010. Chronic hepatitis E virus infection: Successful virologic response to pegylated interferon-alpha therapy. *Ann Intern Med* 153:135–136.
60. Kamar N, Rostaing L, Abravanel F, Garrouste C, Lhomme S, Esposito L, Basse G, et al. 2010. Ribavirin therapy inhibits viral replication on patients with chronic hepatitis E virus infection. *Gastroenterol J* 139:1612–1618.
61. Gerolami R, Borentain P, Raissouni F, Motte A, Solas C, Colson P. 2011. Treatment of severe acute hepatitis E by ribavirin. *J Clin Virol* 52:60–62.
62. Kamar N, Garrouste C, Haagsma EB, Garrigue V, Pischke S, Chauvet C, Dumortier J, et al. 2011. Factors associated with chronic hepatitis in patients with hepatitis E virus infection who have received solid organ transplants. *Gastroenterol J* 140:1481–1489.
63. Kamar N, Abravanel F, Selves J, Garrouste C, Esposito L, Lavayssiere L, Cointault O, et al. 2010. Influence of immunosuppressive therapy on the natural history of genotype 3 hepatitis-E virus infection after organ transplantation. *Transplantation* 89:353–360.
64. Purdy MA, Khudyakov YE. 2010. Evolutionary history and population dynamics of hepatitis E virus. *PLoS One* 5:e14376.
65. Teo CG. 2012. Fatal outbreaks of jaundice in pregnancy and the epidemic history of hepatitis E. *Epidemiol Infect* 140:767–787.
66. Aggarwal R. 2011. Hepatitis E: Historical, contemporary and future perspectives. *J Gastroenterol Hepatol* 26 Suppl 1:72–82.
67. Gupta DN, Smetana HF. 1957. The histopathology of viral hepatitis as seen in the Delhi epidemic (1955–1956). *Indian J Med Res* 45:101–113.
68. Wong DC, Purcell RH, Sreenivasan MA, Prasad SR, Pavri KM. 1980. Epidemic and endemic hepatitis in India: Evidence for a non-A, non-B hepatitis virus aetiology. *Lancet* 2:876–879.
69. Khuroo MS. 2011. Discovery of hepatitis E: The epidemic non-A, non-B hepatitis 30 years down the memory lane. *Virus Res* 161:3–14.
70. Dalton HR, Pas SD, Madden RG, van der Eijk AA. 2014. Hepatitis E virus: Current concepts and future perspectives. *Curr Infect Dis Rep* 16:399.
71. Labrique A, Nelson K. 2010. Hepatitis E virus infections among US military personnel deployed to Afghanistan. *J Infect Dis* 202:1297–1299.
72. Balayan MS, Andjaparidze AG, Savinskaya SS, Ketiladze ES, Braginsky DM, Savinov AP, Poleschuk VF. 1983. Evidence for a virus in non-A, non-B hepatitis transmitted via the fecal-oral route. *Intervirology* 20:23–31.
73. Labrique AB, Thomas DL, Stoszek SK, Nelson KE. 1999. Hepatitis E: An emerging infectious disease. *Epidemiol Rev* 21:162–179.
74. Reyes GR, Purdy MA, Kim JP, Luk KC, Young LM, Fry KE, Bradley DW. 1990. Isolation of a cDNA from the virus responsible for enterically transmitted non-A, non-B hepatitis. *Science* 247:1335–1339.

75. Bradley DW, Balayan MS. 1988. Virus of enterically transmitted non-A, non-B hepatitis. *Lancet* 1:819.
76. Tam AW, Smith MM, Guerra ME, Huang CC, Bradley DW, Fry KE, Reyes GR. 1991. Hepatitis E virus (HEV): Molecular cloning and sequencing of the full-length viral genome. *Virology* 185:120–131.
77. Centers for Disease Control and Prevention (CDC). 1993. *Hepatitis E among U.S. Travelers, 1989–1992*. Centers for Disease Control and Prevention (CDC), Atlanta, GA.
78. Takahashi M, Nishizawa T, Yoshikawa A, Sato S, Isoda N, Ido K, Sugano K, Okamoto H. 2002. Identification of two distinct genotypes of hepatitis E virus in a Japanese patient with acute hepatitis who had not travelled abroad. *J Gen Virol* 83:1931–1940.
79. Meng XJ. 2013. Zoonotic and foodborne transmission of hepatitis E virus. *Semin Liver Dis* 33:41–49.
80. Khuroo MS, Duermeyer W, Zargar SA, Ahanger MA, Shah MA. 1983. Acute sporadic non-A, non-B hepatitis in India. *Am J Epidemiol* 118:360–364.
81. Centers for Disease Control and Prevention (CDC). 2015. Viral Hepatitis: Hepatitis E Information—Hepatitis E FAQs for Health Professionals. http://www.cdc.gov/hepatitis/HEV/HEVfaq.htm. Accessed November, 2015.
82. International Committee on Taxonomy of Viruses (ICTV). 2014: EC 46, Montreal, Canada, July 2014, Email ratification 2015 (MSL #29), *on* ICTV. http://ictvonline.org/proposals/2014.008a-hV.A.v6.Hepeviridae.pdf. Accessed 6 October, 2015.
83. Koonin EV, Gorbalenya AE, Purdy MA, Rozanov MN, Reyes GR, Bradley DW. 1992. Computer-assisted assignment of functional domains in the nonstructural polyprotein of hepatitis E virus: Delineation of an additional group of positive-strand RNA plant and animal viruses. *Proc Natl Acad Sci USA* 89:8259–8263.
84. Kelly AG, Netzler NE, White PA. 2016. Ancient recombination events and the origins of hepatitis E virus. *Submitted, under review.*
85. Meng XJ. 2011. From barnyard to food table: The omnipresence of hepatitis E virus and risk for zoonotic infection and food safety. *Virus Res* 161:23–30.
86. Zhang W, Shen Q, Mou J, Gong G, Yang Z, Cui L, Zhu J, Ju G, Hua X. 2008. Hepatitis E virus infection among domestic animals in eastern China. *Zoonoses Public Health* 55:291–298.
87. Katoh K, Misawa K, Kuma K, Miyata T. 2002. MAFFT: A novel method for rapid multiple sequence alignment based on fast Fourier transform. *Nucleic Acids Res* 30:3059–3066.
88. Price MN, Dehal PS, Arkin AP. 2010. FastTree 2—approximately maximum-likelihood trees for large alignments. *PLoS One* 5:e9490.
89. Meng XJ, Purcell RH, Halbur PG, Lehman JR, Webb DM, Tsareva TS, Haynes JS, Thacker BJ, Emerson SU. 1997. A novel virus in swine is closely related to the human hepatitis E virus. *Proc Natl Acad Sci USA* 94:9860–9865.
90. Dong C, Meng J, Dai X, Liang JH, Feagins AR, Meng XJ, Belfiore NM, et al. 2011. Restricted enzooticity of hepatitis E virus genotypes 1 to 4 in the United States. *J Clin Microbiol* 49:4164–4172.
91. Clayson ET, Innis BL, Myint KS, Narupiti S, Vaughn DW, Giri S, Ranabhat P, Shrestha MP. 1995. Detection of hepatitis E virus infections among domestic swine in the Kathmandu Valley of Nepal. *Am J Trop Med Hyg* 53:228–232.
92. Lin J, Karlsson M, Olofson A-S, Belák S, Malmsten J, Dalin A-M, Widén F, Norder H. 2015. High prevalence of hepatitis E virus in Swedish moose—A phylogenetic characterization and comparison of the virus from different regions. *PLoS ONE* 10:e0122102.
93. Drexler JF, Seelen A, Corman VM, Fumie Tateno A, Cottontail V, Melim Zerbinati R, Gloza-Rausch F, et al. 2012. Bats worldwide carry hepatitis E virus-related viruses that form a putative novel genus within the family Hepeviridae. *J Virol* 86:9134–9147.
94. Bilic I, Jaskulska B, Basic A, Morrow CJ, Hess M. 2009. Sequence analysis and comparison of avian hepatitis E viruses from Australia and Europe indicate the existence of different genotypes. *J Gen Virol* 90:863–873.
95. Marek A, Bilic I, Prokofieva I, Hess M. 2010. Phylogenetic analysis of avian hepatitis E virus samples from European and Australian chicken flocks supports the existence of a different genus within the Hepeviridae comprising at least three different genotypes. *Vet Microbiol* 145:54–61.
96. Nidaira M, Takahashi K, Ogura G, Taira K, Okano S, Kudaka J, Itokazu K, Mishiro S, Nakamura M. 2012. Detection and phylogenetic analysis of hepatitis E viruses from mongooses in Okinawa, Japan. *J Vet Med Sci* 74:1665–1668.
97. Zhao C, Ma Z, Harrison TJ, Feng R, Zhang C, Qiao Z, Fan J, et al. 2009. A novel genotype of hepatitis E virus prevalent among farmed rabbits in China. *J Med Virol* 81:1371–1379.

98. Cong W, Meng QF, Shan XF, Sun WW, Qin SY, Zhang XX, Huang SY, Qian AD. 2014. Seroprevalence and risk factors associated with hepatitis E virus infection in three species of pet birds in northwest China. *Sci World J* 2014:296285.
99. Kabrane-Lazizi Y, Fine JB, Elm J, Glass GE, Higa H, Diwan A, Gibbs CJ, Jr., Meng XJ, Emerson SU, Purcell RH. 1999. Evidence for widespread infection of wild rats with hepatitis E virus in the United States. *Am J Trop Med Hyg* 61:331–335.
100. Arankalle VA, Goverdhan MK, Banerjee K. 1994. Antibodies against hepatitis E virus in Old World monkeys. *J Viral Hepat* 1:125–129.
101. Payne CJ, Ellis TM, Plant SL, Gregory AR, Wilcox GE. 1999. Sequence data suggests big liver and spleen disease virus (BLSV) is genetically related to hepatitis E virus. *Vet Microbiol* 68:119–125.
102. Handlinger JH, Williams W. 1988. An egg drop associated with splenomegaly in broiler breeders. *Avian Dis* 32:773–778.
103. Haqshenas G, Shivaprasad HL, Woolcock PR, Read DH, Meng XJ. 2001. Genetic identification and characterization of a novel virus related to human hepatitis E virus from chickens with hepatitis-splenomegaly syndrome in the United States. *J Gen Virol* 82:2449–2462.
104. Krawczynski K, Aggarwal R, Kamili S. 2000. Hepatitis E. *Infect Dis Clin North Am* 14:669–687.
105. Reyes GR, Huang CC, Tam AW, Purdy MA. 1993. Molecular organization and replication of hepatitis E virus (HEV). *Arch Virol* Suppl 7:15–25.
106. Agrawal S, Gupta D, Panda SK. 2001. The 3' end of hepatitis E virus (HEV) genome binds specifically to the viral RNA-dependent RNA polymerase (RdRp). *Virology* 282:87–101.
107. Purdy MA, Tam AW, Huang C, Yarbough P, Reyes G. 1993. Hepatitis E virus: A non-enveloped member of the "alpha-like" RNA virus supergroup. *Sem Virol* 4:319–326.
108. Ahmad I, Holla RP, Jameel S. 2011. Molecular virology of hepatitis E virus. *Virus Res* 161:47–58.
109. Koonin EV. 1991. The phylogeny of RNA-dependent RNA polymerases of positive-strand RNA viruses. *J Gen Virol* 72 (Pt 9):2197–2206.
110. Pudupakam RS, Kenney SP, Córdoba L, Huang YW, Dryman BA, Leroith T, Pierson FW, Meng XJ. 2011. Mutational analysis of the hypervariable region of hepatitis E virus reveals its involvement in the efficiency of viral RNA replication. *J Virol* 85:10031–10040.
111. Purdy MA, Lara J, Khudyakov YE. 2012. The hepatitis E virus polyproline region is involved in viral adaptation. *PLoS One* 7:e35974.
112. Ansari IH, Nanda SK, Durgapal H, Agrawal S, Mohanty SK, Gupta D, Jameel S, Panda SK. 2000. Cloning, sequencing, and expression of the hepatitis E virus (HEV) nonstructural open reading frame 1 (ORF1). *J Med Virol* 60:275–283.
113. Sehgal D, Thomas S, Chakraborty M, Jameel S. 2006. Expression and processing of the hepatitis E virus ORF1 nonstructural polyprotein. *Virol J* 3:38.
114. Yamashita T, Mori Y, Miyazaki N, Cheng RH, Yoshimura M, Unno H, Shima R, et al. 2009. Biological and immunological characteristics of hepatitis E virus-like particles based on the crystal structure. *Proc Natl Acad Sci USA* 106:12986–12991.
115. Li S, Tang X, Seetharaman J, Yang C, Gu Y, Zhang J, Du H, et al. 2009. Dimerization of hepatitis E virus capsid protein E2s domain is essential for virus–host interaction. *PLoS Pathog* 5:e1000537.
116. Zhou YH, Purcell RH, Emerson SU. 2005. A truncated ORF2 protein contains the most immunogenic site on ORF2: Antibody responses to non-vaccine sequences following challenge of vaccinated and non-vaccinated macaques with hepatitis E virus. *Vaccine* 23:3157–3165.
117. Yamada K, Takahashi M, Hoshino Y, Takahashi H, Ichiyama K, Nagashima S, Tanaka T, Okamoto H. 2009. ORF3 protein of hepatitis E virus is essential for virion release from infected cells. *J Gen Virol* 90:1880–1891.
118. Chandra V, Kar-Roy A, Kumari S, Mayor S, Jameel S. 2008. The hepatitis E virus ORF3 protein modulates epidermal growth factor receptor trafficking, STAT3 translocation, and the acute-phase response. *J Virol* 82:7100–7110.
119. Kalia M, Chandra V, Rahman SA, Sehgal D, Jameel S. 2009. Heparan sulfate proteoglycans are required for cellular binding of the hepatitis E virus ORF2 capsid protein and for viral infection. *J Virol* 83:12714–12724.
120. Kawai HF, Koji T, Iida F, Kaneko S, Kobayashi K, Nakane PK. 1999. Shift of hepatitis E virus RNA from hepatocytes to biliary epithelial cells during acute infection of rhesus monkey. *J Viral Hepat* 6:287–297.
121. Williams TP, Kasorndorkbua C, Halbur PG, Haqshenas G, Guenette DK, Toth TE, Meng XJ. 2001. Evidence of extrahepatic sites of replication of the hepatitis E virus in a swine model. *J Clin Microbiol* 39:3040–3046.

122. Purcell RH, Emerson SU. 2001. Animal models of hepatitis A and E. *ILAR J* 42:161–177.
123. Cheng X, Wang S, Dai X, Shi C, Wen Y, Zhu M, Zhan S, Meng J. 2012. Rabbit as a novel animal model for hepatitis E virus infection and vaccine evaluation. *PLoS One* 7:e51616.
124. Yugo DM, Cossaboom CM, Meng XJ. 2014. Naturally occurring animal models of human hepatitis E virus infection. *ILAR J* 55:187–199.
125. Tanaka T, Takahashi M, Kusano E, Okamoto H. 2007. Development and evaluation of an efficient cell-culture system for hepatitis E virus. *J Gen Virol* 88:903–911.
126. Shukla P, Nguyen HT, Torian U, Engle RE, Faulk K, Dalton HR, Bendall RP, Keane FE, Purcell RH, Emerson SU. 2011. Cross-species infections of cultured cells by hepatitis E virus and discovery of an infectious virus-host recombinant. *Proc Natl Acad Sci USA* 108:2438–2443.
127. Rogée S, Talbot N, Caperna T, Bouquet J, Barnaud E, Pavio N. 2013. New models of hepatitis E virus replication in human and porcine hepatocyte cell lines. *J Gen Virol* 94:549–558.
128. Emerson SU, Nguyen H, Graff J, Stephany DA, Brockington A, Purcell RH. 2004. *In vitro* replication of hepatitis E virus (HEV) genomes and of an HEV replicon expressing green fluorescent protein. *J Virol* 78:4838–4846.
129. Khuroo MS, Kamili S. 2009. Clinical course and duration of viremia in vertically transmitted hepatitis E virus (HEV) infection in babies born to HEV-infected mothers. *J Viral Hepat* 16:519–523.
130. Juhl D, Baylis SA, Blümel J, Görg S, Hennig H. 2014. Seroprevalence and incidence of hepatitis E virus infection in German blood donors. *Transfusion* 54:49–56.
131. Boxall E, Herborn A, Kochethu G, Pratt G, Adams D, Ijaz S, Teo CG. 2006. Transfusion-transmitted hepatitis E in a "nonhyperendemic" country. *Transfus Med* 16:79–83.
132. Mansuy JM, Bendall R, Legrand-Abravanel F, Sauné K, Miédouge M, Ellis V, Rech H, et al. 2011. Hepatitis E virus antibodies in blood donors, France. *Emerg Infect Dis* 17:2309–2312.
133. Schlosser B, Stein A, Neuhaus R, Pahl S, Ramez B, Kruger DH, Berg T, Hofmann J. 2012. Liver transplant from a donor with occult HEV infection induced chronic hepatitis and cirrhosis in the recipient. *J Hepatol* 56:500–502.
134. Arankalle VA, Chadha MS, Mehendale SM, Tungatkar SP. 2000. Epidemic hepatitis E: Serological evidence for lack of intrafamilial spread. *Indian J Gastroenterol* 19:24–28.
135. Meng XJ, Halbur PG, Shapiro MS, Govindarajan S, Bruna JD, Mushahwar IK, Purcell RH, Emerson SU. 1998. Genetic and experimental evidence for cross-species infection by swine hepatitis E virus. *J Virol* 72:9714–9721.
136. Arankalle VA, Chobe LP, Chadha MS. 2006. Type-IV Indian swine HEV infects rhesus monkeys. *J Viral Hepat* 13:742–745.
137. Drobeniuc J, Favorov MO, Shapiro CN, Bell BP, Mast EE, Dadu A, Culver D, Iarovoi P, Robertson BH, Margolis HS. 2001. Hepatitis E virus antibody prevalence among persons who work with swine. *J Infect Dis* 184:1594–1597.
138. Tyrrel SF, Quinton JN. 2003. Overland flow transport of pathogens from agricultural land receiving faecal wastes. *J Appl Microbiol* 94 Suppl:87S–93S.
139. La Rosa G, Muscillo M, Vennarucci VS, Garbuglia AR, La Scala P, Capobianchi MR. 2011. Hepatitis E virus in Italy: Molecular analysis of travel-related and autochthonous cases. *J Gen Virol* 92:1617–1626.
140. Shuval H. 2003. Estimating the global burden of thalassogenic diseases: Human infectious diseases caused by wastewater pollution of the marine environment. *J Water Health* 1:53–64.
141. Brassard J, Gagne MJ, Genereux M, Cote C. 2012. Detection of human food-borne and zoonotic viruses on irrigated, field-grown strawberries. *Appl Environ Microbiol* 78:3763–3766.
142. Maunula L, Kaupke A, Vasickova P, Soderberg K, Kozyra I, Lazic S, van der Poel WH, et al. 2013. Tracing enteric viruses in the European berry fruit supply chain. *Int J Food Microbiol* 167:177–185.
143. Kokkinos P, Kozyra I, Lazic S, Bouwknegt M, Rutjes S, Willems K, Moloney R, et al. 2012. Harmonised investigation of the occurrence of human enteric viruses in the leafy green vegetable supply chain in three European countries. *Food Environ Virol* 4:179–191.
144. Emerson SU, Arankalle VA, Purcell RH. 2005. Thermal stability of hepatitis E virus. *J Infect Dis* 192:930–933.
145. Lynch MF, Tauxe RV, Hedberg CW. 2009. The growing burden of foodborne outbreaks due to contaminated fresh produce: Risks and opportunities. *Epidemiol Infect* 137:307–315.
146. Colson P, Borentain P, Queyriaux B, Kaba M, Moal V, Gallian P, Heyries L, Raoult D, Gerolami R. 2010. Pig liver sausage as a source of hepatitis E virus transmission to humans. *J Infect Dis* 202:825–834.
147. Boutrouille A, Bakkali-Kassimi L, Crucière C, Pavio N. 2007. Prevalence of anti-hepatitis E virus antibodies in French blood donors. *J Clin Microbiol* 45:2009–2010.

148. Feagins AR, Opriessnig T, Guenette DK, Halbur PG, Meng XJ. 2007. Detection and characterization of infectious hepatitis E virus from commercial pig livers sold in local grocery stores in the USA. *J Gen Virol* 88:912–917.
149. Bouwknegt M, Lodder-Verschoor F, van der Poel WH, Rutjes SA, de Roda Husman AM. 2007. Hepatitis E virus RNA in commercial porcine livers in The Netherlands. *J Food Prot* 70:2889–2895.
150. Berto A, Martelli F, Grierson S, Banks M. 2012. Hepatitis E virus in pork food chain, United Kingdom, 2009–2010. *Emerg Infect Dis* 18:1358–1360.
151. Wichmann O, Schimanski S, Koch J, Kohler M, Rothe C, Plentz A, Jilg W, Stark K. 2008. Phylogenetic and case-control study on hepatitis E virus infection in Germany. *J Infect Dis* 198:1732–1741.
152. Krog JS, Larsen LE, Schultz AC. 2014. Enteric porcine viruses in farmed shellfish in Denmark. *Int J Food Microbiol* 186:105–109.
153. Grodzki M, Schaeffer J, Piquet JC, Le Saux JC, Chevé J, Ollivier J, Le Pendu J, Le Guyader FS. 2014. Bioaccumulation efficiency, tissue distribution, and environmental occurrence of hepatitis E virus in bivalve shellfish from France. *Appl Environ Microbiol* 80:4269–4276.
154. Namsai A, Louisirirotchanakul S, Wongchinda N, Siripanyaphinyo U, Virulhakul P, Puthavathana P, Myint KS, Gannarong M, Ittapong R. 2011. Surveillance of hepatitis A and E viruses contamination in shellfish in Thailand. *Lett Appl Microbiol* 53:608–613.
155. Bellou M, Kokkinos P, Vantarakis A. 2013. Shellfish-borne viral outbreaks: A systematic review. *Food Environ Virol* 5:13–23.
156. Robertson LJ. 2007. The potential for marine bivalve shellfish to act as transmission vehicles for outbreaks of protozoan infections in humans: A review. *Int J Food Microbiol* 120:201–216.
157. Faramawi MF, Johnson E, Chen S, Pannala PR. 2011. The incidence of hepatitis E virus infection in the general population of the USA. *Epidemiol Infect* 139:1145–1150.

9 Astroviruses

Eszter Kovács, Simona De Grazia, Vito Martella, and Krisztián Bányai

CONTENTS

INTRODUCTION

Astrovirus (AstV) is a member of the *Astroviridae* family. The name *astrovirus* derives from the star shape (*astron* in Greek) of the virion, as seen by electron microscopy (EM). In humans, AstV predominantly induces gastroenteritis and is a common cause of viral gastroenteritis in young children worldwide. In addition, AstV occasionally causes outbreaks in adult care facilities and within closed and semiclosed communities. Human AstV (HAstV) has been implicated in water- and foodborne gastroenteritis outbreaks, and has also been linked to the consumption of contaminated bivalve mollusks. These features indicate that the virus is an important foodborne pathogen, and that control and prevention of HAstV infection are of interest for public health.

HISTORY

HAstV was first identified using EM in 1975 in feces collected from hospitalized infants with diarrhea [1]. AstV virions have a starlike surface, a morphological feature that distinguishes them from picornaviruses and caliciviruses, also commonly found in human fecal specimens. In the same year, an outbreak of mild diarrhea and vomiting among infants in a maternity ward was also reported [2]. An investigation of the maternity ward outbreak followed. EM revealed virus particles of 29–30 nm in diameter, but without the distinguishing surface structures. Virus-specific immunologic reagents proved that these viruses were also AstV. The availability of commercial assays introduced during the 1980s led to the recognition of the medical importance of AstV infections, particularly gastroenteritis in humans.

CLASSIFICATION

Although AstV resembles other small, rounded RNA viruses in size, morphology, and genome organization [3], it has been classified into the distinct viral family, *Astroviridae*. The *Astroviridae* are further classified into two genera based on genetic and biological differences. These genera include *Mamastrovirus*, which encompasses human and mammalian AstV, and *Avastrovirus*, which includes avian AstV strains [4].

AstV species have long been defined on the basis of the host species infected. Taking advantage of improved diagnostic techniques, a number of novel AstVs have been discovered in recent years in both humans and animal species. These discoveries have revealed large genetic heterogeneity between viruses in this species, much greater than previously believed. Genetic relationships among various strains have been principally assessed using sequence data, usually from open reading frame 2 (ORF2), which encodes the viral capsid. This criterion has been implemented in the current classification found on the website of the International Committee on Taxonomy of Viruses (ICTV) [4], which proposes 19 species within the *Mamastrovirus* genus and three species in the *Avastrovirus* genus. Several recently identified animal AstV strains appear to be candidates for novel AstV species, although these have not yet been officially classified by the ICTV (Figure 9.1). In early studies of HAstV, marked antigenic differences were noted by the lack of cross-neutralization among some AstV isolates. These studies revealed two serotypes, type-1 and type-2, which could be defined antigenically. However, it was also noted that sera specific for both type-1 and type-2 HAstVs did not recognize some other AstV isolates [6]. Subsequent studies performed by immune EM, immunofluorescence, enzyme-linked immunosorbent assay (ELISA), and plaque neutralization assays revealed eight distinct serotypes [6–10]. Genetically, HAstV types 1–8 form a monophyletic group and are classified as a single unique species.

A number of unusual HAstVs discovered recently (MLB1, VA2 and HMO-A, VA-1, HMO-B, and HMO-C) are genetically more closely related to animal AstVs than to "typical" human AstVs [11–14], and all are now listed as a distinct AstV species. Epidemiological investigations suggest that these new HAstV species are distributed worldwide [6,15] and that human infections are not likely to be sporadic. Evidence for this has been demonstrated; antibodies specific for VA-1-like (HMO-B/HMO-C) AstV have been identified in nearly 65% of human sera in the United States [16]. Further work is clearly needed to determine the epidemiology of these new AstV strains.

MORPHOLOGY AND STRUCTURE

AstV is an icosahedral virus with a starlike surface structure and a diameter of 29–30 nm. However, the starlike surface structure is often not visible by EM due to sample preparation issues [1].

Ultrastructural analysis of type-1, type-2, and type-8 HAstVs grown in cell culture has revealed marked structural differences between mature and immature virions. Mature virus particles have icosahedral T = 3 symmetry and possess 30 dimeric spikes protruding from the surface, located on the icosahedral two-fold axes [17–19].

By amino acid (aa) sequence comparison of HAstVs, the capsid protein precursor can be divided into three domains, a highly conserved N-terminal domain (aa 1–424), a hypervariable domain [HVR (aa 425–688)], and a highly acidic C-terminal domain [20]. The mature, infectious virion contains three predominant protein species derived from the N-terminal domain (VP34) and the HVR (VP25 and VP27) following intra- and extracellular processing [6,21–23]. Neutralizing monoclonal antibodies have been mapped to residues in the HVR, suggesting that this part of the capsid is involved in virus binding to cellular receptors [23,24]. Structural analysis of a human type-8 AstV revealed that the AstV capsid spike is formed by dimerization of a polypeptide spanning the HVR, while the N-terminal-derived polypeptide forms the capsid shell [19,25]. In addition to proteins derived from ORF2, proteins encoded by ORF1a are also present in viral particles and associated with viral RNA [26].

GENOME STRUCTURE AND REPLICATION

AstV possesses a positive-sense, single-stranded, poly-adenylated RNA genome that varies in length from 6.4 to 7.3 kb [3]. The viral genome contains three ORFs. Two of these, ORF1a and ORF1b, are located at the 5′ end of the genome and encode the nonstructural proteins. The third ORF (ORF2) is located at the 3′ end and encodes the capsid proteins [3,27] (Figure 9.2).

During infection of susceptible host cells, a subgenomic RNA (sgRNA) of 2.4 kb is produced, which encompasses ORF2 [22,30]. Synthesis of genomic and subgenomic RNA is thought to be mediated by a full-length, negative-sense RNA template [3]. A motif upstream of the AstV ORF2 is believed to contain the promoter for the synthesis of subgenomic RNA [6]. This motif, ATTTGGAGNGGNGGACCNAAN$^{5-8}$<u>ATG</u>NC (where the ORF2 ATG start codon is underlined), is highly conserved among mammalian and avian AstVs. This sequence also shows identity with the 5′ end of the genomic RNA, providing further evidence for its role as an RdRp promoter [31].

ORF1a follows a short 5′ end untranslated region (UTR) and is the largest AstV ORF (2.5–3.4 kb), with the 3′ end partially overlapping the 5′ end of ORF1b. The overlapping region contains signals that are essential for transcription and translation of the viral RNA polymerase (encoded by ORF1b) [32,33] through a frameshift mechanism [34,35]. The polypeptide encoded by ORF1a contains helical transmembrane motifs followed by a viral serine protease motif [31]. ORF1b is 1.5–1.6 kb in length and overlaps with the 5′ end of ORF2 in mammalian AstVs. ORF2 encodes the capsid protein precursor and varies in size from 2.0 to 2.4 kb. A UTR of variable length (0.1–0.3 kb) lies at the 3′ end of the genome and contains highly conserved sequences [36].

EPIDEMIOLOGY

Transmission of AstV and Burden of Disease

The incidence of HAstV in children with gastroenteritis ranges from 2% to 9% [37], a rate sufficient to suggest that the control of HAstV gastroenteritis may be of interest in terms of public health.

AstV is most frequently transmitted person to person through the fecal–oral route. Virally contaminated water and food are also sources for AstV transmission, and can cause large outbreaks of disease in both children and adults [3]. AstV outbreaks also often affect parents, teachers, and

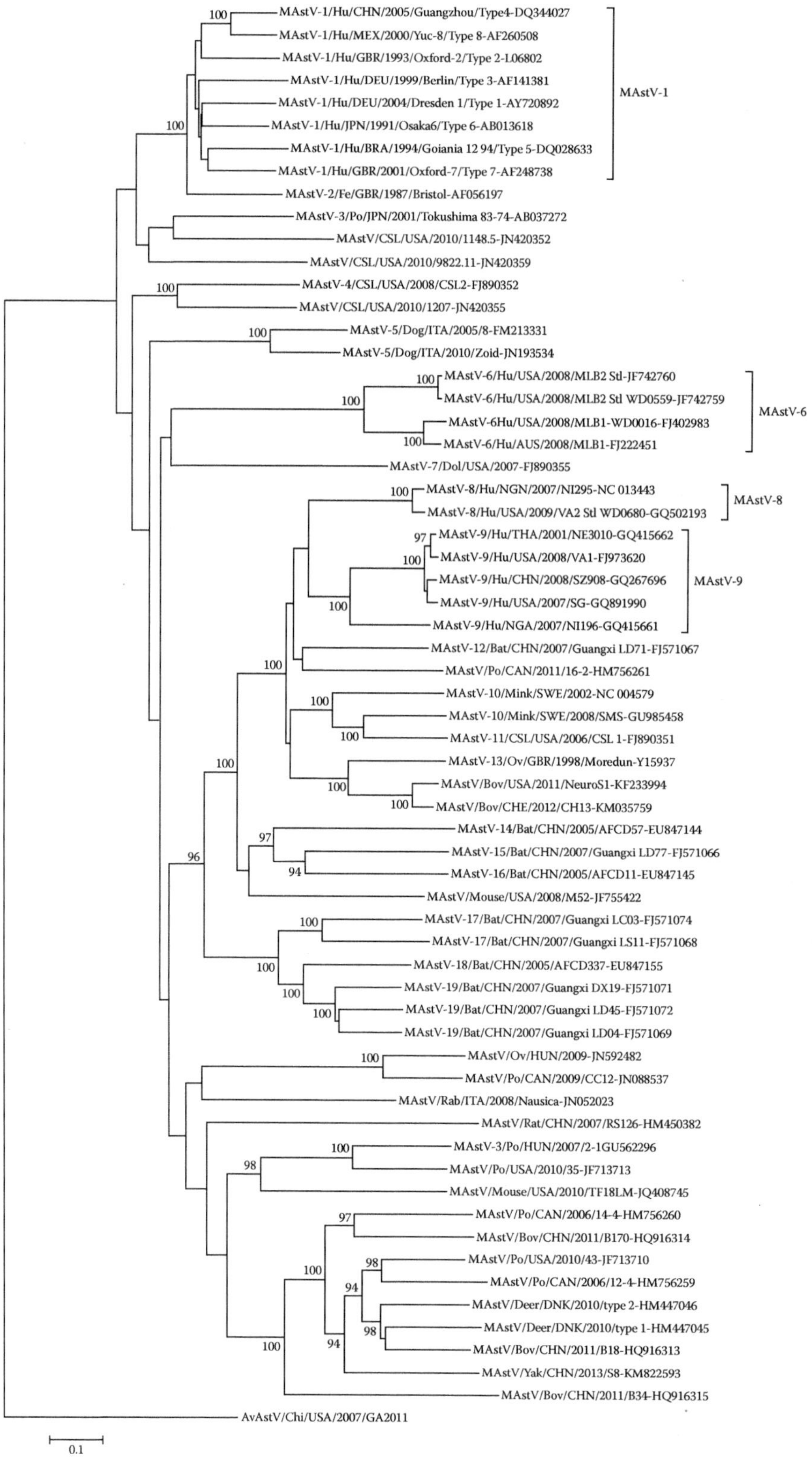

FIGURE 9.1 (CONTINUED)

FIGURE 9.1 (CONTINUED) Phylogenetic analysis of astrovirus ORF2 sequences among a selection of viruses from the Mamastrovirus genus. The analysis was inferred using the complete capsid coding gene (ORF2), eliminating all gap positions. Analysis was performed using the neighbor-joining method within the MEGA7 software package (from Kumar, S. et al. *Mol Biol Evol*, 9, 141, 2015). Accession numbers are indicated after astrovirus strain names. Square brackets are used to indicate human astroviruses. Abbreviations: AAstV, *Avastrovirus*; MAstV, *Mamastrovirus*; Bov, bovine; Chi, chicken; CSL, California sea lion; Dol, dolphin; Fe, feline; Hu, human; Ov, ovine; Po, porcine; Rab, rabbit. The scale bar indicates the number of nucleotide substitutions per site.

medical personnel in contact with infected children. AstVs are environmentally resistant and maintain their infectivity over time, even on inert surfaces. They remain stable at acidic pH and high temperature (60°C) for 5–10 min and are resistant to a variety of disinfectants and detergents [38]. Raw and treated sewage is considered an important source of AstV contamination because of its ability to persist in surface waters [1,39]. Indeed, AstV has been detected in various water sources, including drinking water, rivers, dams, wastewater, and effluents from water treatment plants [6,40–43]. The detection of identical HAstV strains from water and fecal samples collected from hospitalized patients clearly demonstrates the role of water in the transmission of HAstV [6]. HAstV is shed in stools for up to 2 weeks after the resolution of symptoms and for 3 months or more in immunocompromised patients [44].

Food is thought to play an important role in the transmission of HAstV, particularly through contaminated bivalve mollusks such as oysters. The ability of AstV to persist in marine water for extended periods may account for the high levels of contamination observed in shellfish. The AstV contamination peak appears to be seasonal, usually occurring in winter and following periods of heavy rain [45]. However, the peak in viral infections can vary from the winter months in temperate regions to late spring/early summer in tropical areas [6,46].

AstVs are generally considered strictly species-specific, but evidence for interspecies transmission and/or recombination between distantly related strains has been documented in cows and red deer [47], humans and marine mammals [48], and humans and ruminants or mink [14]. Direct interspecies transmission of animal (heterologous) strains coupled with genetic recombination may generate novel, viable viruses capable of spreading throughout the human population [11,13].

Molecular Epidemiology of AstV

HAstV is genetically and antigenically heterogeneous [10]. Sequence analysis of short fragments of ORF2 and reverse transcription polymerase chain reaction (RT-PCR) genotyping protocols with

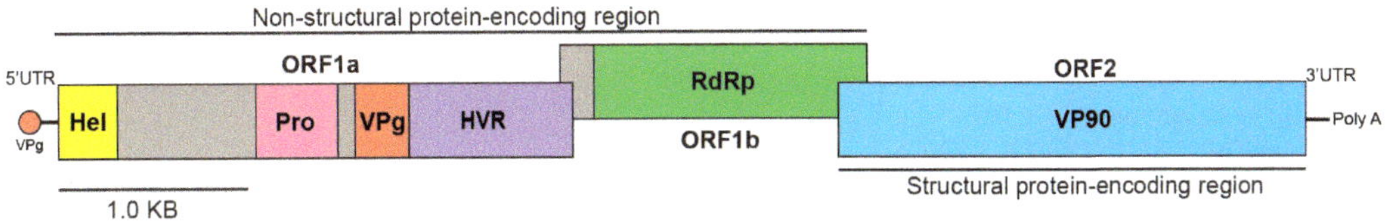

FIGURE 9.2 Genome organization of human astrovirus (HAstV). HAstV genomic RNA comprises three ORFs, ORF1a, ORF1b, and ORF2. The 5′ end of the genome features a covalently attached VPg and a UTR, and the 3′ end contains a UTR and a polyadenylated tail. Known functional domains are highlighted in color, with putative helicase (Hel), protease (Pro), VPg, and hypervariable region (HVR) domains on ORF1a, the RNA-dependent RNA polymerase (RdRp) domain on ORF1b, and the capsid protein (VP90) encoded by ORF2. The scale bar represents 1.0 kb. (Modified and adapted from Bosch, A. et al., *Clin Microbiol Rev*, 27, 1048–1074, 2014; SIB Swiss Institute of Bioinformatics: ViralZone. ViralZone online: Mamastrovirus *on* SIB Swiss Institute of Bioinformatics. http://viralzone.expasy.org/viralzone/all_by_species/281.html. Accessed 29 February 2016).

type-specific primers have been used for the genetic characterization of HAstV types 1–8. The results of nucleic acid–based typing were found to correlate with antigenic characterization when using type-specific polyclonal or monoclonal antibodies [49]. Irrespective of the methods utilized, typing surveys have shown that HAstV-1 is the most common type infecting children, followed by HAstV-2 to -5, whereas HAstV-6 to -8 have only been detected sporadically [20,50]. Interestingly, upon analysis of ORF2, discrete sequence variation has been observed within each HAstV type. At present, there is a lack of clearly defined criteria in the literature for the classification of HAstV lineages. Guix et al. [51] proposed that viruses belonging to different lineages within the same type should diverge by at least 7% at the nucleotide level, while Gabbay et al. [50] used a 5% nucleotide divergence cut-off value, coupled with a high bootstrap value for phylogenetic clusters, to define a new lineage. On the bases of these criteria, HAstV-1 has been divided into six lineages (HAstV-1a–1f), HAstV-2 into four lineages (HAstV-2a–2d), and HAstV-3 and HAStV-4 into two lineages (HAstV-3a–3b and HAstV-4a–4b, respectively) [50,52,53].

The incidence and distribution of HAstV sero/genotypes vary seasonally and geographically, but large nationwide epidemiological studies spanning long time periods are lacking in the literature. The cocirculation of multiple HAstV types and/or intratypic genetic variants has been reported in some countries. For example, surveillance studies conducted in Italy have demonstrated the emergence and/or reemergence of strains belonging to different HAstV lineages over time. In particular, HAstV-1d was first detected in 1999 and was replaced by HAstV-1b in 2000. This lineage reemerged in 2003–2005 and continued to circulate in 2008 along with HAstV-1a, which became predominant in 2010. During this survey, three additional genotypes (HAstV-2, -4, and -5) were also detected, but at low frequency [53]. However, non-genotype 1 HAstVs, such as HAstV-2, can become temporally predominant in some areas, as described in Chile, Spain, Colombia, and Brazil [37,53,54].

Changing Epidemiology of AstV

Until 2008, eight classical AstV types were known to cause human disease. However, our understanding of the epidemiology of AstV-associated gastroenteritis has changed markedly due to recent improvements in detection methods. As mentioned, several new AstV species have been discovered in human fecal samples from patients with diarrhea. These novel AstVs, tentatively called human, mink, and ovine-like AstVs (HMOAstVs), were identified and classified on the basis of their genetic relationship to classical HAstVs and some animal AstVs [6,11,55]. Since the detection of human AstV MLB1, a novel AstV, in a stool sample from a 3-year-old boy in Australia, evidence has suggested that MLB1-like strains circulate globally. However, a case-controlled study from India showed no association of MLB-1 with diarrhea [15]. Likewise AstV-VA1, discovered during a sporadic diarrheal disease outbreak in Virginia, has also been detected in different geographical areas [55]. The identification of multiple lineages of human AstVs (AstV-MLB and HMO-AstV), phylogenetically distinct from but closer to animal AstVs, likely suggests independent origins for each of these lineages [12,56].

IMMUNITY AND REPEATED INFECTIONS

The determinants of immunity to AstV are not well understood. After the first episode of infection, which usually affects children less than 2 years of age, AstV-specific antibodies appear and contribute to protection from further illness. However, humoral immunity decreases later in life, and older adults can again become susceptible to symptomatic infections. Protection from HAstV infection might be partially explained by passive immunity in infants or active immunity from previously acquired infections in children older than 2 years of age [45]. A decrease in the frequency of severe dehydration in children has been observed with subsequent symptomatic AstV infections, compared with children infected for the first time [57]. The circulation of different AstV types over time suggests a possible role of immune selection in the evolution of AstV [44]. Supporting this theory, phylogenetic analyses consistently show a replacement of different genotypes over time, with HAstV-1 being the most prevalent genotype. The identification of different lineages within HAstV-1 may suggest a lack of homotypic immunity to antigenic variants of the same serotype. However, supporting data are lacking, and it remains to be determined whether the humoral immune response is focused homotypically or heterotypically.

Immunoglobulin therapy in immunocompromised patients with persistent HAstV infection resulted in virus clearance and the elimination of diarrhea [44]. The normal mucosal immune response and the production of interferon gamma, tumor necrosis factor, and cytokines released by CD4+ T cells are thought to induce protection against repeated HAstV infections. Moreover, even in immunocompromised (e.g., HIV positive) patients, CD4+ T cells were found to have an important role in the control of viral replication. The control of viral replication in animals (i.e., turkeys infected with turkey AstV) was attributed to innate immune responses, although the role of adaptive immune response in viral clearance remains unclear [44].

ILLNESS AND PATHOGENESIS

HAstV infection is characterized by an incubation period of 3–4 days followed by mild, watery diarrhea that lasts for 2–4 days. A shorter incubation period (24–36 h) is usually observed during an outbreak of gastroenteritis. Clinical symptoms, apart from watery diarrhea and vomiting, also include headaches, fever, abdominal pain, and anorexia [58,59]. However, HAstV infections can be asymptomatic in both children and adults. It is virtually impossible to distinguish diarrhea caused by HAstV from that caused by other enteric viruses on the basis of clinical signs alone. Generally, HAstV diarrhea is milder than that caused by rotavirus, and it does not lead to significant dehydration or hospitalization [60]. The typical clinical features associated with HAstV infection are shown in Table 9.1. In children, the main clinical manifestations observed with HAstV infection are diarrhea and, less frequently, vomiting. HAstV infection causes moderate dehydration, often in association with elevated body temperature. Symptoms usually resolve within a few days. A range

TABLE 9.1
Clinical Features of Human Astrovirus Infection

Diarrhea (%)	72–100
Days of diarrhea (average)	2–7
Maximum number of stools/day	4–13
Abdominal pain (%)	50
Vomiting (%)	20–84
Days of vomiting (average)	1
Maximum incidents of vomiting/day	1
Fever (%)	20–25
Maximum temperature of fever (°C)	38.5
Dehydration (%)	10–30
Hospitalization (%)	6
Days of hospitalization (average)	2–6
Severity score[a]	5–17 points

Source: Adapted and modified from: Nguyen, T.A. et al., *J Med Virol,* 80, 298–305, 2008; Walter, J.E. and Mitchell, D.K., *Curr Opin Infect Dis,* 16, 247–253, 2003; Colomba, C. et al., *Eur J Clin Microbiol Infect Dis,* 25, 570–575, 2006.

[a] Twenty point scoring system (Ruuska, T. and Vesikari, T. *Scand J Infect Dis,* 22, 259–267, 1990).

of mixed infections involving AstV and rotavirus, adenovirus, or norovirus may also occur. No specific difference in clinical severity was evidenced when single and dual HAstV infections were compared [61].

Adult volunteer studies have facilitated the assessment of HAstV pathogenicity in humans. The administration of HAstV-positive fecal extracts to adult seronegative volunteers showed that the majority of volunteers developed asymptomatic infection or mild symptoms, such as vomiting and watery diarrhea, or abdominal discomfort without vomiting or diarrhea [62]. However, immunocompromised patients infected with HAstV often developed chronic diarrhea. In some animal species and, recently, in some human patients, extraintestinal localization and diseases (encephalitis, nephritis, and hepatitis) have been described in patients infected with HAstV [6,39,47,63–67]. Disseminated diseases and additional putative complications such as intussusception were also associated with AstV infections in studies from Hungary and Nigeria [68,69]. Mortality associated with AstV infections is extremely rare, but has been reported in immunodeficient patients [70].

Pathogenesis

Histopathologic studies demonstrate active replication of AstV within the small intestine. Viral replication mainly occurs in the mature epithelial cells near the villus tip of the jejune. Studies performed with ovine and bovine strains (OAstV and BAstV) have shown that AstV can also infect subepithelial macrophages (OAstV) and M cells (BAstV) of the small intestine [72].

Turkeys infected with turkey AstV (TAstV) have been largely used as the preferred animal model to study astroviral pathogenesis [73,74]. These experiments have demonstrated that infection induces severe diarrhea in the absence of significant changes in intestinal morphology, including a lack of change in villus height, width, or surface area. Furthermore, there is no evidence of inflammation and no increase in cell death following AstV infection [6,63,73,75]. Collectively, these observations indicate that AstV induces diarrhea through a mechanism independent of virus-mediated destruction of the epithelium. It is therefore thought that AstV primarily affects the physiological function of intestinal epithelial cells.

Recent *in vitro* studies examining the effect of AstV infection in human colonic adenocarcinoma (Caco-2) cells suggested that the AstV capsid protein alone can increase the permeability of cells in a monolayer, independently of viral replication. This increased permeability is one component of pathogenesis in viral diarrhea, and its mechanism is thought to act through actin rearrangement [76].

Investigation of electrophysiological changes associated with AstV infection *ex vivo* suggested that AstV may induce ultrastructural changes in the intestinal epithelium and rearrangement of F-actin. Nighot and coworkers [77] hypothesized that these actin changes disrupt the normal expression of transporter proteins in the apical membrane, specifically sodium/hydrogen exchanger 3 (NHE3), and lead to malabsorption of sodium ions (Na^+). This results in the failure to absorb water completely and induces fluid loss. Moreover, AstV induces apoptosis in cultured cells, suggesting that programmed cell death may also contribute to pathogenesis in some host species [78].

ASTROVIRUS DETECTION METHODS

Significant improvements in the detection of AstV have been achieved over the past three decades, and some of the methods have become commercially available as diagnostic kits to allow standardized laboratory diagnosis of AstV infections (Tables 9.2 and 9.3). In general, routine laboratory diagnosis is based on the detection of either viral antigen or genomic RNA.

MICROSCOPY AND VIRUS ISOLATION

HAstV was first identified using EM [1], and both EM and immune EM (IEM) were initially used as conventional diagnostic techniques for the detection of AstV in stool samples. IEM enhances

TABLE 9.2
Commercially Available Astrovirus Antigen Detection Tests

Product Name	Features	Producer	Country of Origin
NADAL® Astrovirus Test	Chromatographic immunoassay	Nal von Minden Ltd	Germany
Astrovirus Rapid Test	Chromatographic immunoassay	Creative Diagnostics	USA
Intermedical Astrovirus Device	Chromatographic immunoassay	INTERMEDICAL s.r.l	Italy
CerTest Astrovirus	Chromatographic immunoassay	CerTest Biotec	Spain
CerTest Rotavirus-Adenovirus + Astrovirus Combo Card	Chromatographic immunoassay; mulitplex assay detects rotavirus, adenovirus, and astrovirus	CerTest Biotec	Spain
ProSpecT™ Astrovirus test	ELISA	Thermo Fisher Scientific Oxoid Ltd	United Kingdom
Astrovirus RIDASCREEN® ELISA	ELISA	R-Biopharm AG	Germany
Astrovirus Ag (stool) ELISA	ELISA	Sierra Resources International, Inc.	USA

TABLE 9.3
Commercially Available Astrovirus Nucleic Acid Detection Kits

Product Name	Features	Producer	Country of Origin
Astrovirus@ceeramTools™ Astrovirus Test Kit Réf. KHASV for clinical samples (CE IVD)	Real-time RT-PCR	CEERAM S.A.S.	France
Astrovirus@ceeramTools™ Astrovirus Test Kit Réf. KASV for food and environmental samples	Real-time RT-PCR	CEERAM S.A.S.	France
Astrovirus Multiplex PCR—viral gastroenteritis symptom panel	Two tube multiplex real-time PCR kit	Fast-track Diagnostics Selenion Medical	The Benelux
PowerChek™ Astrovirus Real-time PCR kit	Real-time RT-PCR	BioBiZ	Korea

specificity and sensitivity when detecting HAstV in fecal samples compared with EM alone [79]. However, given that EM is relatively expensive, labor intensive, and insensitive, it is no longer widely used in routine laboratory diagnosis [6].

HAstV can be propagated in cell cultures in the presence of trypsin. Susceptible cell types include human embryonic kidney cells (HEK), rhesus monkey kidney cells ($LLCMK_2$), primary baboon kidney cells (PBK), and Caco-2 cells [6,80]. Given that not all HAstVs are readily culturable, and that cell culture takes several weeks, its usefulness in clinical virology and outbreak investigation is limited.

Immunofluorescence has been used for the diagnosis and serotyping of AstV infections. The specificity and sensitivity of this procedure have been shown to be comparable with EM and radioimmunoassays [81].

ANTIGEN DETECTION TESTS

In clinical laboratory settings, commercially available antigen detection methods have been the preferred choice for HAstV diagnosis (Table 9.2). For example, enzyme immunoassays (EIAs) and immunochromatography (IC) tests are used for detecting AstV antigens in human fecal specimens. These assays are based on the specific interaction between type-common antibodies and shared epitopes of surface structures of the HAstV particle. The specificity and sensitivity of EIAs and IC tests range from 98%–100% and 76%–100%, respectively [6,82,83]. Tests based on latex agglutination (LA) are also available, with sensitivity and specificity comparable with other rapid detection methods [84,85]. In addition, IC and LA tests have the added advantage that specialized equipment is not required, they are easy to perform, and results are available within a few minutes.

NUCLEIC ACID DETECTION METHODS

An evolution of nucleic acid detection methods for AstV has been witnessed during the past 20 years or so, from traditional hybridization methods, parallel PCR assays, and more recently, deep sequencing–based detection. During this evolution of diagnostic methods, various test principles were merged in an attempt to increase both specificity and sensitivity.

Hybridization Techniques

Northern blot assays were first used to detect HAstV genomic RNA at a time when EM was the only detection method available for HAstV [86]. Northern blotting of HAstV genomic RNA was among the first nucleic acid–based molecular methods to be used for serotyping HAstV [87]. However, liquid hybridization, a modified DNA-DNA hybridization method, proved to be a rapid, reliable, and less expensive method than Northern blotting, although it offered only moderate sensitivity [88]. DNA oligonucleotide microarray detection assays can distinguish between all eight HAstV serotypes. The assay utilizes RT-PCR products amplified with degenerate primers. The amplicon is hybridized to a large set of short oligonucleotide probes, representative of each serotype [89]. Jaaskelainen and Maunula have also developed a microarray assay, using RNA transcripts for detection [90].

Nucleic Acid Amplification Methods

The very first AstV RT-PCR assays were introduced in the early 1990s [6,56]. In these and subsequent studies, oligonucleotide primers were designed to detect different parts of the viral genome and were either type-common or type-specific. Type-common primers mainly target the nonstructural proteins and UTRs of the HAstV genome [52,91]. For genotyping, fragments of the capsid protein–encoding genomic region are amplified; these reactions are less sensitive, but provide information about the viral type without the need to sequence the PCR products [51,87]. Nested RT-PCR assays have shown a dramatic increase in sensitivity and specificity when compared with ELISAs, EM, or even single round PCR assays [6,52,92].

An improvement in the specificity of RT-PCR was achieved when cell culture was combined with RT-PCR (integrated cell culture-PCR or ICC-PCR). ICC-PCR was designed for the detection of AstV in environmental water and fecal samples [6]. This detection method allows HAstV replication in cell culture for a short period prior to PCR amplification. ICC-PCR has the advantage of detecting viable viruses and reduces the time needed for detection when using traditional virus culture methods [93]. ICC-PCR, when combined with nested PCR, minimizes the number of false negative tests and increases the specificity of the assay, as two sets of specific primers are utilized [94,95].

Quantitative real-time RT-PCR (RT-qPCR) was developed for the detection of all eight HAstV serotypes in feces using primers that bind to highly conserved regions of ORF1a in a

TaqMan-based assay [6]. Multiplex real-time PCR or RT-PCR can also be used for simultaneous detection of several viruses in a sample. Multiplex detection systems have similar sensitivity compared with uniplex RT-qPCR assays for HAstV and have been designed for parallel detection of HAstV, adenovirus, enterovirus, norovirus GI and GII, parechovirus, rotavirus (group A and C), and sapovirus. Extrinsic controls such as phocine herpesvirus and MS2 bacteriophage are usually spiked during the RNA/DNA extraction step to monitor the efficiency of extraction and amplification [6,96–98].

HAstV detection can also be performed using the TaqMan Array Card (TAC), which bypasses the limitations of some multiplex PCR systems. In this array, a 384-well platform is used with spatial multiplexing. In 2012, Liu developed this system to detect 19 enteropathogens, including HAstV and several other viruses, bacteria, and parasites. TAC is a sensitive, cost-effective, and quantitative method, which easily detects mixed infections [99].

Other Molecular Methods

Nucleic acid sequence-based amplification (NASBA) is an isothermal amplification assay adapted for HAstV detection by researchers from the Centers for Disease Control and Prevention (CDC) [100]. This test was compared with RT-PCR and no significant differences in detection rates were observed, thus establishing an alternative to traditional PCR-based methods [100]. Loop-mediated isothermal amplification (LAMP) or RT-LAMP is another isothermal amplification method. A real-time reverse-transcription LAMP (qRT-LAMP) assay for the detection of the HAstV-1 capsid gene has been reported [101]. In this assay, the turbidity of the mixture can be monitored with a spectrophotometer, providing an opportunity for quantitative measurement [101]. However, the major advantage of this method is that the results can be visualized by eye.

Next-generation sequencing (NGS) techniques may be the future of routine molecular diagnostics, provided that the individual reactions are cost-effective and that bioinformatics analysis becomes an easy, automated process. The current applications of NGS in AstV detection are mainly for pathogen discovery through viral metagenomics, where the amount of relevant data is highly sensitive to sample processing. Alternatively, unbiased random primed RT-PCR combined with high throughput sequencing has been used to describe novel AstVs in patients with diarrhea and CNS disease [64].

TREATMENT AND PREVENTION

HAstV gastroenteritis is typically a mild, self-limiting infection. More severe disease usually develops in patients with underlying diseases, malnutrition, immunodeficiency, or coinfection with other pathogens. Electrolyte and fluid loss in young children may require oral or intravenous rehydration therapy. Intravenous immunoglobulin can be a beneficial adjunct in patients with severe immunodeficiency who do not respond to supportive therapies.

AstVs are able to persist under extreme environmental conditions; they retain infectivity on inanimate surfaces and human hands, in dried fecal material, and in water. General hygienic procedures, wearing gloves during food handling process, and appropriate treatment of potable water are key factors in the prevention of HAstV outbreaks [78]. At present, vaccine development is not a priority for HAstV.

CONCLUSIONS

HAstVs are genetically diverse members of the *Mamastrovirus* genus. An etiology of gastroenteritis has been described for classical HAstVs, but disease association, if any, awaits further investigation for novel strains. HAstV is an important foodborne pathogen, often acquired through the consumption of contaminated shellfish. Monitoring shellfish using advanced diagnostic techniques is of interest for public health, and could be routinely performed to avoid foodborne diarrhea caused by HAstV.

REFERENCES

1. Madeley C, Cosgrove B. 1975. 28 nm particles in faeces in infantile gastroenteritis. *Lancet* 306:451–452.
2. Appleton H, Higgins PG. 1975. Viruses and gastroenteritis in infants. *Lancet* 305:1297.
3. Méndez E, Arias C. 2007. Astroviruses. *In* Knipe DM, Howley PM (eds), *Fields Virology*, pp. 981–1000, 6th edn, vol 1. Oxford Press, Philadelphia, PA.
4. International Committee on Taxonomy of Viruses (ICTV). International Committee on Taxonomy of Viruses (ICTV) online. http://ictvonline.org/index.asp?bhcp=1. Accessed November 2015.
5. Kumar S, Stecher G, Tamura K. 2016. MEGA7: Molecular Evolutionary Genetics Analysis version 7.0 for bigger datasets. *Mol Biol Evol* 33(7): 1870–1874.
6. Shinde SB, Fernandes CB, Patravale VB. 2012. Recent trends in *in-vitro* nanodiagnostics for detection of pathogens. *J Control Release* 159:164–180.
7. Herrmann JE, Hudson RW, Perron-Henry DM, Kurtz JB, Blacklow NR. 1988. Antigenic characterization of cell-cultivated astrovirus serotypes and development of astrovirus-specific monoclonal antibodies. *J Infect Dis* 158:182–185.
8. Méndez-Toss M, Romero-Guido P, Munguía ME, Méndez E, Arias CF. 2000. Molecular analysis of a serotype 8 human astrovirus genome. *J Gen Virol* 81:2891–2897.
9. Lee T, Kurtz J. 1994. Prevalence of human astrovirus serotypes in the Oxford region 1976–92, with evidence for two new serotypes. *Epidemiol Infect* 112:187–193.
10. Koopmans M, Bijen M, Monroe S, Vinje J. 1998. Age-stratified seroprevalence of neutralizing antibodies to astrovirus types 1 to 7 in humans in The Netherlands. *Clin Diagn Lab Immunol* 5:33–37.
11. Bányai K, Meleg E, Moschidou P, Martella V. 2010. Detection of newly described astrovirus MLB1 in stool samples from children. *Emerg Infect Diseases* 16:169–170.
12. Finkbeiner SR, Kirkwood CD, Wang D. 2008. Complete genome sequence of a highly divergent astrovirus isolated from a child with acute diarrhea. *Virol J* 5:117.
13. Finkbeiner SR, Holtz LR, Jiang Y, Rajendran P, Franz CJ, Zhao G, Kang G, Wang D. 2009. Human stool contains a previously unrecognized diversity of novel astroviruses. *Virol J* 6:161.
14. Kapoor A, Li L, Victoria J, Oderinde B, Mason C, Pandey P, Zaidi S, Delwart E. 2009. Multiple novel astrovirus species in human stool. *J Gen Virol* 90:2965–2972.
15. Holtz LR, Bauer IK, Rajendran P, Kang G, Wang D. 2011. Astrovirus MLB1 is not associated with diarrhea in a cohort of Indian children. *PLoS One* 6:e28647.
16. Burbelo PD, Ching KH, Esper F, Iadarola MJ, Delwart E, Lipkin WI, Kapoor A. 2011. Serological studies confirm the novel astrovirus HMOAstV-C as a highly prevalent human infectious agent. *PLoS One* 6:e22576.
17. Matsui SM, Kiang D, Ginzton N, Chew T, Geigenmiller- Gnirke, U. 2001. Molecular biology of astroviruses: selected highlights. In Chadwick, D. and Goode, J.A. (eds), *Novartis Foundation Symposium 238*, pp. 219–245. Wiley, Chichester; New York.
18. Risco C, Carrascosa JL, Pedregosa AM, Humphrey CD, Sánchez-Fauquier A. 1995. Ultrastructure of human astrovirus serotype 2. *J Gen Virol* 76:2075–2080.
19. Dryden KA, Tihova M, Nowotny N, Matsui SM, Mendez E, Yeager M. 2012. Immature and mature human astrovirus: Structure, conformational changes, and similarities to hepatitis E virus. *J Mol Biol* 422:650–658.
20. Wang QH, Kakizawa J, Wen LY, Shimizu M, Nishio O, Fang ZY, Ushijima H. 2001. Genetic analysis of the capsid region of astroviruses. *J Med Virol* 64:245–255.
21. Bass D, Qiu S. 2000. Proteolytic processing of the astrovirus capsid. *J Virol* 74:1810–1814.
22. Monroe SS, Stine S, Gorelkin L, Herrmann J, Blacklow N, Glass R. 1991. Temporal synthesis of proteins and RNAs during human astrovirus infection of cultured cells. *J Virol* 65:641–648.
23. Sánchez-Fauquier A, Carrascosa A, Carrascosa J, Otero A, Glass R, Lopez J, San Martin C, Melero J. 1994. Characterization of a human astrovirus serotype 2 structural protein (VP26) that contains an epitope involved in virus neutralization. *Virology* 201:312–320.
24. Bass DM, Upadhyayula U. 1997. Characterization of human serotype 1 astrovirus-neutralizing epitopes. *J Virol* 71:8666–8671.
25. Dong J, Dong L, Méndez E, Tao Y. 2011. Crystal structure of the human astrovirus capsid spike. *Proc Natl Acad Sci USA* 108:12681–12686.

26. Matsui SM, Kim J, Greenberg H, Young L, Smith L, Lewis T, Herrmann J, Blacklow N, Dupuis K, Reyes G. 1993. Cloning and characterization of human astrovirus immunoreactive epitopes. *J Virol* 67:1712–1715.
27. Willcocks MM, Carter MJ. 1993. Identification and sequence determination of the capsid protein gene of human astrovirus serotype 1. *FEMS Microbiol Lett* 114:1–7.
28. Bosch A, Pinto RM, Guix S. 2014. Human astroviruses. *Clin Microbiol Rev* 27:1048–1074.
29. SIB Swiss Institute of Bioinformatics: ViralZone. ViralZone online: Mamastrovirus *on* SIB Swiss Institute of Bioinformatics. http://viralzone.expasy.org/viralzone/all_by_species/281.html. Accessed 29 February 2016.
30. Koci MD, Seal BS, Schultz-Cherry S. 2000. Molecular characterization of an avian astrovirus. *J Virol* 74:6173–6177.
31. Jonassen CM, Jonassen TØ, Sveen TM, Grinde B. 2003. Complete genomic sequences of astroviruses from sheep and turkey: Comparison with related viruses. *Virus Res* 91:195–201.
32. Jiang B, Monroe SS, Koonin EV, Stine SE, Glass RI. 1993. RNA sequence of astrovirus: Distinctive genomic organization and a putative retrovirus-like ribosomal frameshifting signal that directs the viral replicase synthesis. *Proc Natl Acad Sci USA* 90:10539–10543.
33. Willcocks M, Brown T, Madeley C, Carter M. 1994. The complete sequence of a human astrovirus. *J Gen Virol* 75:1785–1788.
34. Lewis TL, Matsui SM. 1996. Astrovirus ribosomal frameshifting in an infection-transfection transient expression system. *J Virol* 70:2869–2875.
35. Lewis TL, Matsui SM. 1997. Studies of the astrovirus signal that induces (–1) ribosomal frameshifting. *Adv Exp Med Biol* 412:323–330.
36. Jonassen CM, Jonassen T, Grinde B. 1998. A common RNA motif in the 3' end of the genomes of astroviruses, avian infectious bronchitis virus and an equine rhinovirus. *J Gen Virol* 79:715–718.
37. Guix S, Caballero S, Villena C, Bartolome R, Latorre C, Rabella N, Simo M, Bosch A, Pinto RM. 2002. Molecular epidemiology of astrovirus infection in Barcelona, Spain. *J Clin Microbiol* 40:133–139.
38. Schultz-Cherry S, King DJ, Koci MD. 2001. Inactivation of an astrovirus associated with poult enteritis mortality syndrome. *Avian Dis* 45:76–82.
39. He X, Cheng L, Zhang D, Xie X, Wang D, Wang Z. 2011. One-year monthly survey of rotavirus, astrovirus and norovirus in three sewage treatment plants in Beijing, China and associated health risk assessment. *Water Sci Technol* 64:1202–1210.
40. Maunula L, Kalso S, Von Bonsdorff C-H, Pönkä A. 2004. Wading pool water contaminated with both noroviruses and astroviruses as the source of a gastroenteritis outbreak. *Epidemiol Infect* 132:737–743.
41. Meleg E, Jakab F, Kocsis B, Banyai K, Melegh B, Szűcs G. 2006. Human astroviruses in raw sewage samples in Hungary. *J Appl Microbiol* 101:1123–1129.
42. Pusch D, Oh DY, Wolf S, Dumke R, Schröter-Bobsin U, Höhne M, Röske I, Schreier E. 2005. Detection of enteric viruses and bacterial indicators in German environmental waters. *Arch Virol* 150:929–947.
43. Taylor MB, Walter J, Berke T, Cubitt WD, Mitchell DK, Matson DO. 2001. Characterisation of a South African human astrovirus as type 8 by antigenic and genetic analyses. *J Med Virol* 64:256–261.
44. Koci MD, Kelley LA, Larsen D, Schultz-Cherry S. 2004. Astrovirus-induced synthesis of nitric oxide contributes to virus control during infection. *J Virol* 78:1564–1574.
45. Nguyen TA, Hoang L, Pham LD, Hoang KT, Mizuguchi M, Okitsu S, Ushijima H. 2008. Identification of human astrovirus infections among children with acute gastroenteritis in the Southern Part of Vietnam during 2005–2006. *J Med Virol* 80:298–305.
46. De Grazia S, Platia MA, Rotolo V, Colomba C, Martella V, Giammanco G. 2011. Surveillance of human astrovirus circulation in Italy 2002–2005: Emergence of lineage 2c strains. *Clin Microbiol Infect* 17:97–101.
47. Tzipori S, Menzies J, Gray E. 1981. Detection of astrovirus in the faeces of red deer. *Vet Rec* 108:286–286.
48. Rivera R, Nollens HH, Venn-Watson S, Gulland FM, Wellehan JF. 2010. Characterization of phylogenetically diverse astroviruses of marine mammals. *J Gen Virol* 91:166–173.
49. Noel JS, Lee TW, Kurtz JB, Glass RI, Monroe SS. 1995. Typing of human astroviruses from clinical isolates by enzyme immunoassay and nucleotide sequencing. *J Clin Microbiol* 33:797–801.
50. Gabbay YB, Leite JPG, Oliveira DS, Nakamura LS, Nunes MR, Mascarenhas JDAP, Heinemann MB, Linhares AC. 2007. Molecular epidemiology of astrovirus type 1 in Belem, Brazil, as an agent of infantile gastroenteritis, over a period of 18 years (1982–2000): Identification of two possible new lineages. *Virus Res* 129:166–174.
51. Guix S, Bosch A, Pinto R. 2005. Human astrovirus diagnosis and typing: Current and future prospects. *Lett Appl Microbiol* 41:103–105.

52. Belliot G, Laveran H, Monroe S. 1997. Detection and genetic differentiation of human astroviruses: Phylogenetic grouping varies by coding region. *Arch Virol* 142:1323–1334.
53. De Grazia S, Medici MC, Pinto P, Moschidou P, Tummolo F, Calderaro A, Bonura F, Banyai K, Giammanco GM, Martella V. 2012. Genetic heterogeneity and recombination in human type 2 astroviruses. *J Clin Microbiol* 50:3760–3764.
54. Victoria M, Carvalho-Costa FA, Heinemann MB, Leite JPG, Miagostovich MP. 2007. Genotypes and molecular epidemiology of human astroviruses in hospitalized children with acute gastroenteritis in Rio de Janeiro, Brazil. *J Med Virol* 79:939–944.
55. Finkbeiner SR, Li Y, Ruone S, Conrardy C, Gregoricus N, Toney D, Virgin HW, Anderson LJ, Vinje J, Wang D. 2009. Identification of a novel astrovirus (astrovirus VA1) associated with an outbreak of acute gastroenteritis. *J Virol* 83:10836–10839.
56. Jonassen TØ, Monceyron C, Lee TW, Kurtz JB, Grinde B. 1995. Detection of all serotypes of human astrovirus by the polymerase chain reaction. *J Virol Methods* 52:327–334.
57. Naficy AB, Rao MR, Holmes JL, Abu-Elyazeed R, Savarino SJ, Wierzba TF, Frenck RW, Monroe SS, Glass RI, Clemens JD. 2000. Astrovirus diarrhea in Egyptian children. *J Infect Dis* 182:685–690.
58. De Benedictis P, Schultz-Cherry S, Burnham A, Cattoli G. 2011. Astrovirus infections in humans and animals–molecular biology, genetic diversity, and interspecies transmissions. *Infect Genet Evol* 11:1529–1544.
59. Glass R, Noel J, Mitchell D, Herrmann J, Blacklow N, Pickering L, Dennehy P, Ruiz-Palacios G, De Guerrero M, Monroe S. 1996. The changing epidemiology of astrovirus-associated gastroenteritis: A review. *In* Chiba S, Estes MK, Nakata S, Calisher CH (eds), *Viral Gastroenteritis*, pp. 287–300, vol 12. Arch Virol. Springer-Verlag GmBH Wien, Austria.
60. Walter JE, Mitchell DK. 2003. Astrovirus infection in children. *Curr Opin Infect Dis* 16:247–253.
61. Colomba C, De Grazia S, Giammanco G, Saporito L, Scarlata F, Titone L, Arista S. 2006. Viral gastroenteritis in children hospitalised in Sicily, Italy. *Eur J Clin Microbiol Infect Dis* 25:570–575.
62. Kurtz J, Lee T. 1987. Astroviruses: Human and animal. *In* Symposium CF (ed), *Novel Diarrhoea Viruses*, pp. 92–107. Wiley, Chichester, UK.
63. Englund L, Chriél M, Dietz H, Hedlund K. 2002. Astrovirus epidemiologically linked to pre-weaning diarrhoea in mink. *Vet Microbiol* 85:1–11.
64. Quan PL, Wagner TA, Briese T, Torgerson TR, Hornig M, Tashmukhamedova A, Firth C, Palacios GF, Baisre-de Leon A, Paddock CD. 2010. Astrovirus encephalitis in boy with X-linked agammaglobulinemia. *Emerg Infect Diseases* 17:918–925.
65. Imada T, Yamaguchi S, Mase M, Tsukamoto K, Kubo M, Morooka A. 2000. Avian nephritis virus (ANV) as a new member of the family Astroviridae and construction of infectious ANV cDNA. *J Virol* 74:8487–8493.
66. Todd D, Smyth VJ, Ball N, Donnelly BM, Wylie M, Knowles NJ, Adair B. 2009. Identification of chicken enterovirus-like viruses, duck hepatitis virus type 2 and duck hepatitis virus type 3 as astroviruses. *Avian Pathol* 38:21–29.
67. Snodgrass DR, Angus KW, Gray EW, Menzies JD, Paul G. 1979. Pathogenesis of diarrhoea caused by astrovirus infections in lambs. *Arch Virol* 60:217–226.
68. Jakab F, Peterfai J, Verebely T, Meleg E, Banyai K, Mitchell DK, Szûcs G. 2007. Human astrovirus infection associated with childhood intussusception. *Pediatr Int* 49:103–105.
69. Aminu M, Ameh E, Geyer A, Esona M, Taylor M, Steele A. 2009. Role of astrovirus in intussusception in Nigerian infants. *J Trop Pediatr* 55:192–194.
70. Wunderli W, Meerbach A, Guengoer T, Berger C, Greiner O, Caduff R, Trkola A, Bossart W, Gerlach D, Schibler M. 2011. Astrovirus infection in hospitalized infants with severe combined immunodeficiency after allogeneic hematopoietic stem cell transplantation. *PLoS One* 6:e27483.
71. Ruuska T, Vesikari T. 1990. Rotavirus disease in Finnish children: Use of numerical scores for clinical severity of diarrhoeal episodes. *Scand J Infect Dis* 22:259–267.
72. Woode GN, Pohlenz JF, Gourley N, Fagerland JA. 1984. Astrovirus and Breda virus infections of dome cell epithelium of bovine ileum. *J Clin Microbiol* 19:623–630.
73. Behling-Kelly E, Schultz-Cherry S, Koci M, Kelley L, Larsen D, Brown C. 2002. Localization of astrovirus in experimentally infected turkeys as determined by *in situ* hybridization. *Vet Pathol* 39:595–598.
74. Thouvenelle ML, Haynes JS, Reynolds DL. 1995. Astrovirus infection in hatchling turkeys: Histologic, morphometric, and ultrastructural findings. *Avian Dis* 39:328–336.
75. Sebire NJ, Malone M, Shah N, Anderson G, Gaspar HB, Cubitt WD. 2004. Pathology of astrovirus associated diarrhoea in a paediatric bone marrow transplant recipient. *J Clin Pathol* 57:1001–1003.

76. Moser LA, Carter M, Schultz-Cherry S. 2007. Astrovirus increases epithelial barrier permeability independently of viral replication. *J Virol* 81:11937–11945.
77. Nighot PK, Moeser A, Ali RA, Blikslager AT, Koci MD. 2010. Astrovirus infection induces sodium malabsorption and redistributes sodium hydrogen exchanger expression. *Virology* 401:146–154.
78. Meleg E, Banyai K. 2010. Astroviruses. In Liu D (ed), *Molecular Detection of Human Viral Pathogens*. pp. 89–99. CRC Press, Boca Raton, FL.
79. Noel J, Cubitt D. 1994. Identification of astrovirus serotypes from children treated at the Hospitals for Sick Children, London 1981–93. *Epidemiol Infect* 113:153–159.
80. Willcocks M, Carter M, Laidler F, Madeley C. 1990. Growth and characterisation of human faecal astrovirus in a continuous cell line. *Arch Virol* 113:73–81.
81. Kriston S, Willcocks M, Carter M, Cubitt W. 1996. Seroprevalence of astrovirus types 1 and 6 in London, determined using recombinant virus antigen. *Epidemiol Infect* 117:159–164.
82. Moe C, Allen J, Monroe S, Gary H, Humphrey C, Herrmann J, Blacklow N, Carcamo C, Koch M, Kim K. 1991. Detection of astrovirus in pediatric stool samples by immunoassay and RNA probe. *J Clin Microbiol* 29:2390–2395.
83. Tran A, Talmud D, Lejeune B, Jovenin N, Renois F, Payan C, Leveque N, Andreoletti L. 2010. Prevalence of rotavirus, adenovirus, norovirus, and astrovirus infections and coinfections among hospitalized children in northern France. *J Clin Microbiol* 48:1943–1946.
84. Araki K, Kobayashi S, Utagawa E, Kobayashi M, Shinozaki T, Abe T. 1998. Prevalence of human astrovirus serotypes in Shizuoka 1991–96. *Kansenshogaku Zasshi* 72:12–16.
85. Komoriya T, Kohno H, Kimura A, Ushijima H. 2002. The development of sensitive latex agglutination tests for detecting astroviruses (serotypes 1 and 3) from clinical stool specimen. *JARMAM* 13:103–114.
86. Willcocks M, Carter M, Silcock J, Madeley C. 1991. A dot-blot hybridization procedure for the detection of astrovirus in stool samples. *Epidemiol Infect* 107:405–410.
87. Palombo EA, Bishop RF. 1996. Annual incidence, serotype distribution, and genetic diversity of human astrovirus isolates from hospitalized children in Melbourne, Australia. *J Clin Microbiol* 34:1750–1753.
88. Belliot G, Fankhauser R, Monroe S. 2001. Characterization of "Norwalk-like viruses" and astroviruses by liquid hybridization assay. *J Virol Methods* 91:119–130.
89. Brown DW, Gunning KB, Henry DM, Awdeh ZL, Brinker JP, Tzipori S, Herrmann JE. 2008. A DNA oligonucleotide microarray for detecting human astrovirus serotypes. *J Virol Methods* 147:86–92.
90. Jääskeläinen AJ, Maunula L. 2006. Applicability of microarray technique for the detection of noro-and astroviruses. *J Virol Methods* 136:210–216.
91. Mitchell DK, Matson DO, Jiang X, Berke T, Monroe SS, Carter MJ, Willcocks MM, Pickering LK. 1999. Molecular epidemiology of childhood astrovirus infection in child care centers. *J Infect Dis* 180:514–517.
92. Shi M, Sikotra S, Lee T, Kurtz J, Getty B, Hart C, Myint S. 1994. Use of a nested PCR method for the detection of astrovirus serotype 1 in human faecal material. *Mol Cell Probes* 8:481–486.
93. Rodríguez RA, Pepper IL, Gerba CP. 2009. Application of PCR-based methods to assess the infectivity of enteric viruses in environmental samples. *Appl Environ Microbiol* 75:297–307.
94. Chapron CD, Ballester NA, Fontaine JH, Frades CN, Margolin AB. 2000. Detection of astroviruses, enteroviruses, and adenovirus types 40 and 41 in surface waters collected and evaluated by the information collection rule and an integrated cell culture-nested PCR procedure. *Appl Environ Microbiol* 66:2520–2525.
95. Chapron C, Ballester N, Margolin A. 2000. The detection of astrovirus in sludge biosolids using an integrated cell culture nested PCR technique. *J Appl Microbiol* 89:11–15.
96. Mori K, Hayashi Y, Akiba T, Nagano M, Tanaka T, Hosaka M, Nakama A, Kai A, Saito K, Shirasawa H. 2013. Multiplex real-time PCR assays for the detection of group C rotavirus, astrovirus, and Subgenus F adenovirus in stool specimens. *J Virol Methods* 191:141–147.
97. van Maarseveen NM, Wessels E, de Brouwer CS, Vossen AC, Claas EC. 2010. Diagnosis of viral gastroenteritis by simultaneous detection of Adenovirus group F, Astrovirus, Rotavirus group A, Norovirus genogroups I and II, and Sapovirus in two internally controlled multiplex real-time PCR assays. *J Clin Virol* 49:205–210.
98. Wolffs PF, Bruggeman CA, van Well GT, van Loo IH. 2011. Replacing traditional diagnostics of fecal viral pathogens by a comprehensive panel of real-time PCRs. *J Clin Microbiol* 49:1926–1931.
99. Liu J, Gratz J, Amour C, Kibiki G, Becker S, Janaki L, Verweij JJ, Taniuchi M, Sobuz SU, Haque R. 2013. A laboratory-developed TaqMan Array Card for simultaneous detection of 19 enteropathogens. *J Clin Microbiol* 51:472–480.

100. Tai JH, Ewert MS, Belliot G, Glass RI, Monroe SS. 2003. Development of a rapid method using nucleic acid sequence-based amplification for the detection of astrovirus. *J Virol Methods* 110:119–127.
101. Wei H, Zeng J, Deng C, Zheng C, Zhang X, Ma D, Yi Y. 2013. A novel method of real-time reverse-transcription loop-mediated isothermal amplification developed for rapid and quantitative detection of human astrovirus. *J Virol Methods* 188:126–131.

10 Rotavirus

Celeste Donato, Daniel Cowley, and Carl Kirkwood

CONTENTS

INTRODUCTION

Diarrheal disease is the second most common cause of death in children under 5 years of age worldwide. Diarrheal disease is also responsible for an estimated 1.5 million deaths/year, accounting for almost 20% of mortality within that age group [1–3]. There are more than 20 different bacterial, viral, and parasitic pathogens that cause diarrheal disease and the proportion and range of pathogens can vary depending on the country [4]. Rotavirus, calicivirus, enteropathogenic, and enterotoxigenic *Escherichia coli* cause more than half of all diarrheal deaths in children under 5 years of age globally [5].

The *Rotavirus* genus within the *Reoviridae* family encompasses a large and diverse group of viruses capable of causing diarrheal disease in infants, children, adults, and the young of a wide variety of animal species [6]. Human group A rotavirus strains were identified in 1973 by Ruth Bishop and colleagues in Australia and rapidly gained recognition as the predominant cause of severe acute gastroenteritis in young children worldwide [7,8]. This chapter will outline the importance of group A rotavirus strains as human pathogens and provide a description of their classification, epidemiology, clinical presentation, diagnosis, treatment, pathogenesis, immunity, transmission, genome, replication, diversity, and vaccines.

CLASSIFICATION AND HOST RANGE

The *Reoviridae* family includes the genera *Rotavirus*, *Aquareovirus*, *Coltivirus*, *Cypovirus*, *Fijivirus*, *Idnoreovirus*, *Mycoreovirus*, *Orbivirus*, *Orthoreovirus*, *Oryzavirus*, *Phytoreovirus*, and *Seadornavirus* [9]. Within the *Rotavirus* genus, there are seven known rotavirus species (groups A–H) that are defined by distinct antigenic and genetic differences in the viral protein 6 (VP6) gene [10]. Group A rotaviruses are the predominant cause of disease in humans, as well as in a wide range of mammalian and some avian species [11]. The numerous animal species include pigs, cattle (including buffalo), goats, cats, dogs, horses, and birds (including chickens, pigeons, turkeys, pheasants, guinea fowl, and partridges) [12,13]. There is a considerable burden of disease and economic loss associated with group A rotaviruses in calves, weaning and postweaning piglets, and foals. Species-specific vaccines are currently available and are administered to female animals in the late stages of pregnancy [14–16]. Group A rotavirus strains have been sporadically detected in numerous other species including monkeys, mice, guanacos, bats, sugar gliders, pandas, giraffes, Japanese raccoon dogs, masked palm civets, Galapagos sea lions, raccoons, skunks, emus, common opossums, ostriches, reindeer, camels, red squirrels, and striped bass [6,17–28]. The sporadic or limited detection in some species may reflect limited surveillance.

Group B strains have primarily been detected in cattle, sheep, pigs, goats, and rats [29], and have also been detected in humans during outbreaks of severe gastroenteritis in adults in China, Bangladesh, India, Myanmar, and Nepal [11,30,31]. Group C strains have been detected in cattle, dogs, pigs, and ferrets [29], and cause sporadic outbreaks, predominantly in children, in numerous countries worldwide including Australia, Brazil, Argentina, and the United States [32–34]. Group D strains (chickens, pheasants, and turkeys), group E strains (pigs), and group F and G strains (chickens and turkeys) have limited host ranges [29,35]. Only seven group H strains (previously referred to as new adult diarrhea rotavirus, or ADRV) have been isolated worldwide, three human strains from China and Bangladesh, a porcine strain from Japan, and three porcine strains from Brazil [10,36].

Group A rotavirus strains can be classified into subgroups (SG I, II, I + II, non-I, or non-II) based on the antigenic specificity of the VP6 protein. Genetic analysis allows classification into genogroups I (previously subgroup I) and II (previously subgroups II, I + II or non-I, non-II) [37]. A binary classification system for rotaviruses was established in 1989 and initially used the antigenic reactivity of the two outer capsid proteins to denote G serotypes for VP7 (glycoprotein) and P serotypes for VP4 (protease sensitive) [38]. This antigenic classification has been replaced by classification based on genetic sequence, resulting in the designation of G and P genotypes rather than serotypes. To date,

TABLE 10.1
Nucleotide Percentage Identity Cut-Off Values Defining Genotypes

Gene Product	Percentage Identity	Number of Genotypes	Genotype Denotation	Genotype Name
VP7	80	27	G	Glycosylated
VP4	80	37	P	Protease sensitive
VP6	85	17	I	Inner capsid
VP1	83	9	R	RNA-dependent RNA polymerase
VP2	84	9	C	Core protein
VP3	81	8	M	Methyltransferase
NSP1	79	18	A	Interferon agonist
NSP2	85	10	N	NTPase
NSP3	85	12	T	Translation enhancer
NSP4	85	15	E	Enterotoxin
NSP5	91	11	H	PHosphoprotein

Source: From Guo, D. et al., *Infect Genet Evol,* 12, 1567–1576, 2012; Jere, K.C. et al., *Infect Genet Evol,* 21C, 58–66, 2013; Matthijnssens, J. et al., *Arch Virol,* 156, 1397–1413, 2011; Papp, H. et al., *Acta Microbiol Immunol Hung,* 59, 411–421, 2012; Trojnar, E. et al., *J Gen Virol,* 94, 136–142, 2013.

27 G genotypes and 37 P genotypes have been identified [6,39]. In 2008, a genotyping classification system based on the open reading frame (ORF) of each gene was adopted. The nomenclature for the complete genome assignment, VP7-VP4-VP6-VP1-VP2-VP3-NSP1-NSP2-NSP3-NSP4-NSP5/6, is Gx-P[x]-Ix-Rx-Cx-Mx-Ax-Nx-Tx-Ex-Hx, respectively [38]. To date, 27 G (VP7), 37 P (VP4), 17 I (VP6), 9 R (VP1), 9 C (VP2), 8 M (VP3), 18 A (NSP1), 10 N (NSP2), 12 T (NSP3), 15 E (NSP4), and 11 H (NSP5) genotypes have been described (Table 10.1) [6,26,39–41]. There are three major genotype constellations: Wa (G1-P[8]-I1-R1-C1-M1-A1-N1-T1-E1-H1), DS-1 (G2-P[4]-I2-R2-C2-M2-A2-N2-T2-E2-H2), and AU-1 (G3-P[9]-I3-R3-C3-M3-A3-N3-T3-E3-H3) [6].

A nomenclature for strain names has also been proposed: RV group/species of origin/country of identification/common name/year of identification/G- and P-type [6]. A web-based tool, RotaC, has been developed to facilitate the fast and accurate classification of rotavirus genes [42].

The 27 G genotypes and 37 P genotypes described to date in humans and various animal species are genetically diverse and unevenly distributed among species, suggesting host species barriers and host range restrictions [13]. Some genotypes have wider host ranges than others; G3 has the widest host range, and other genotypes such as G13–G27 and P16–37 appear to have a degree of host restriction, with limited detection observed in predominantly avian species, cattle, and pigs (Tables 10.2 and 10.3) [6,18].

Some species appear to be more permissive to infection, with a broad range of rotavirus strains identified, and the greatest number of diverse G and P genotypes has been identified in cattle and pigs (Tables 10.2 and 10.3). However, this may also reflect increased surveillance in some species. In humans, 12 G and 16 P genotypes have been identified and detected as at least 70 G–P genotype combinations. Some genotypes are endemic in the human population, circulating worldwide at high frequencies, while other genotypes are sporadically detected and may represent zoonotic transmissions [43]. A review of 100,000 strains from 100 countries between 1996 and 2005 revealed that G1P[8], G2P[4], G3P[8], G4P[8], and G9P[8] strains accounted for 74.7% of all strains [43]. Strains with these five genotype combinations cause over 90% of rotavirus disease in North America, Europe, and Australia. However, in South America and Africa, G1P[8], G2P[4], G3P[8], G4P[8], and G9P[8] strains represent 83% and 55% of isolates, respectively, and unusual genotypes are more frequently identified [44]. Novel genotypes have been observed in developing and developed countries,

TABLE 10.2
Distribution of Group A Rotavirus VP7 G Genotypes Detected in Humans and Various Animal Species

Species	1	2	3	4	5	6	7	8	9	10	11	12	13	14	15	16	17	18	19	20	21	22	23	24	25	26	27
Human	●	●	●	●	○	○		○	●	○	○	●								○							
Pig	○	●	●	●	●	○		○	●	○	●	○														○	
Monkey			●					○																			
Cow	○		○		○	●	○	●		●				○	○			○		○	○			○			
Goat			○			●																					
Sheep								●		○																	
Horse			●	○	●			○		○			●	●													
Dog			●																								
Cat			●																								
Mouse																○											
Rabbit			●																								
Birds*							○										●	●	●			○	●				
Panda	○																										
Giraffe								○																			
Guanaco								○																			
Bat			○																						○		
Sugar Glider																											○

Source: Table adapted from Estes, M.K. and Kapikian, A.Z. *Fields Virology*, Lippincott Williams & Wilkins, Philadelphia, PA, 2007; Kobayashi, N. et al., *Communicating Current Research and Educational Topics and Trends in Applied Microbiology*, Vol. 1, Formatex, Badajoz, 2007 [46]; Martella, V. et al., *Vet Microbiol,* 140, 246–255, 2010; Matthijnssens, J. et al., *Arch Virol,* 156, 1397–1413, 2011.

* Birds: Chicken, turkey, pheasant, and pigeon.

● Endemic genotype routinely identified.

○ Sporadic detection.

TABLE 10.3
Distribution of Group A Rotavirus VP4 P Genotypes Detected in Humans and Various Animal Species

Species	1	2	3	4	5	6	7	8	9	10	11	12	13	14	15	16	17	18	19	20	21	22	23	24	25	26	27	28	29	30	31	32	33	34	35	36	37
Human	○	○	○	●	○	●	○	●	○	○	○		○	○					○						○			○									
Pig	○				○	●	●	○			○		●	○					●				●			●	○	○				○		○			
Monkey	○	●	○		●																			○													
Cow	●		○		●		○				●	○		●	●		○				○								○				○				
Goat	○														○																						
Sheep														○	○																						
Horse	○						○				○	●						○																			
Dog			●																																		
Cat			●						○																												
Mouse																●				●																	
Rabbit											○			●								○															
Birds*																	●													○	○				○		○
Panda							○																														
Giraffe	○																																				
Guanaco	○													○																							
Bat			○			○				○																											
Sugar Glider																																				○	

Source: Table adapted from Estes, M.K. and Kapikian, A.Z. *Fields Virology*, Lippincott Williams & Wilkins, Philadelphia, PA, 2007; Kobayashi, N. et al., *Communicating Current Research and Educational Topics and Trends in Applied Microbiology*, Vol. 1, Formatex, Badajoz, 2007; Martella, V. et al., *Vet Microbiol,* 140, 246–255, 2010; Matthijnssens, J. et al., *Arch Virol,* 156, 1397–1413, 2011.

* Birds: Chicken, turkey, pheasant, and pigeon.

● Endemic genotype routinely identified.

○ Sporadic detection.

highlighting the diversity of rotavirus strains in circulation worldwide. Globally, rare genotypes can be detected at high frequencies in some countries, such as G8P[6] and G8P[4] in Malawi, G5P[8] in Brazil, G6P[8] in Hungary, and G10P[11] in India [44,45].

EPIDEMIOLOGY OF HUMAN ROTAVIRUS DISEASE

Rotavirus is a ubiquitous pathogen, and almost every child experienced a rotavirus infection in the first 2 or 3 years of life before the widespread introduction of a vaccine in 2006 [47]. Rotavirus is the most common cause of severe gastroenteritis in children worldwide [8] and has been estimated to cause 114 million episodes of diarrhea annually, resulting in 24 million clinic visits and 2.4 million hospitalizations in children under 5 years of age [8]. The incidence of rotavirus disease is similar in developing and developed countries, although the mortality rates are unevenly distributed [48,49]. Of the estimated 453,000 annual deaths, the overwhelming majority occur in developing nations in Asia and sub-Saharan Africa [49]. The higher mortality rates in developing countries may be due to a greater prevalence of malnutrition, limited access to treatment, higher exposure to the virus, and coinfections with other enteric pathogens [50]. Children from developed countries predominantly experience infection between 3 and 35 months of age. The overwhelming majority of disease in developing countries affects younger children, with 60%–80% of severe disease occurring by 12–15 months of age [51].

Prior to vaccine introduction in children under 59 months of age in the United States, an estimated 3 million cases of rotavirus infection occurred each year, resulting in 2.3 million cases that required home care, 387,351 general practitioner (GP) visits, 213,946 emergency department (ED) visits, 67,033 hospitalizations, and 30 deaths [52]. This resulted in direct medical costs of $319 million and nonmedical costs to society of $893 million each year [52].

SEASONALITY

Rotavirus disease exhibits a seasonal association in numerous countries with temperate climates, whereby the majority of rotavirus disease occurs in the winter months [53,54]. However, in regions with a tropical climate, rotavirus infection occurs year-round, with higher rates of detection observed in cooler, drier months [55,56].

CLINICAL PRESENTATION

Rotavirus infection in infants and young children typically leads to more severe symptoms than with other enteric pathogens [57]. The incubation period can be 1–4 days and is typically less than 48 h, compared with 4.5 days for astrovirus, 1.2 days for norovirus genotypes I and II, and 1.7 days for sapovirus infections [58]. Infection typically results in vomiting, followed by acute, profuse, nonbloody, watery diarrhea that can last 3–7 days, involving as many as 10–20 bowel movements a day, which may be associated with abdominal pain and low-grade fever [4,59]. In approximately 10% of primary infections, a child will have symptoms severe enough to require medical attention and be administered treatment with oral rehydration fluids [4,60,61]. In approximately 1%–3% of infections, the symptoms are so severe that intravenous rehydration is required [4]. Illness is typically more severe in children residing in developing countries due to numerous factors including malnutrition and concurrent infections that lead to severe dehydration and shock, which can prove fatal [4]. Dehydration and electrolyte abnormalities resulting in cardiovascular failure are the most common cause of rotavirus mortality. Death may also occur from seizures and the aspiration of vomitus [62].

Rotavirus antigenemia and viremia are commonly detected in children hospitalized with rotavirus gastroenteritis and may be associated with increased severity of fever and vomiting [63]. Small quantities of rotavirus antigens, RNA, and infectious virus have been detected in serum, cerebrospinal fluid, and in several organs (liver, heart, lungs, and kidneys) [64]. Viremia occurs during symptomatic and asymptomatic infections and has unknown clinical significance in immunocompetent children [65]. Chronic rotavirus infection is commonly observed in immunocompromised children with various conditions, including severe combined immunodeficiency (SCID), X-linked agammaglobulinemia, cartilage hair hypoplasia, acquired immunodeficiency syndrome (AIDS), and DiGeorge syndrome [66–69]. Immunocompromised children routinely shed virus in their stool for protracted periods of time (from 6 weeks up to a year) and the strains excreted have often undergone genomic rearrangement [67,69]. Severe and protracted rotavirus infections have also been observed in adult patients with immunosuppression related to bone marrow or renal transplantation and AIDS [70–72].

In addition to gastroenteritis, rotavirus infection has been associated with a number of other conditions including respiratory infections, necrotizing enterocolitis, pneumatosis intestinalis, hepatic abscess, myocarditis, seizures, and meningoencephalitis [73–78]. However, due to the pervasive nature of rotavirus infection, an etiological association has not been confirmed and the association may be coincidental rather than causative. Although some rotavirus strains have been documented to cause biliary atresia (inflammation and blockage of the bile duct) in mouse models, and group C rotavirus strains have been detected in liver biopsies from biliary atresia patients, a true association is yet to be demonstrated [79–81].

DIAGNOSIS

It is difficult to differentiate rotavirus gastroenteritis from other causes of gastroenteritis, particularly viral causes, based on clinical symptoms [82]. Fecal samples obtained from children with rotavirus infection do not contain blood or leukocytes [57,82], which can be associated with bacterial infections. Initially, rotavirus was diagnosed using electron microscopy (EM), although this method was laborious, insensitive, time consuming, required a high level of expertise, and could not differentiate between the different rotavirus genotypes. Improvements in diagnostic assays have facilitated increased detection and characterization, resulting in the advent of epidemiological studies. Antigen-based immunoassay methods (e.g., enzyme-linked immunosorbent assay [ELISA]) are reliable, convenient, and inexpensive [59]. RNA–RNA hybridization, G and P genotyping using reverse transcription polymerase chain reaction (RT-PCR; approximately 1000 times more sensitive than immunoassays), real-time RT-PCR, and Sanger sequencing are routinely used in the identification and characterization of rotavirus strains. With the use of next-generation sequencing, a new era in rotavirus characterization is foreseen [83,84]. DNA oligonucleotide microarrays are being developed to overcome the issue of genetic drift affecting the efficacy of genotyping primer binding, and these techniques have an increased ability to distinguish mixed infections [85,86].

PATHOGENESIS

The pathogenesis of rotavirus gastroenteritis is poorly understood and likely due to multiple complex processes [87–91]. An examination of duodenal biopsy specimens from children with rotavirus infection revealed short and blunted villi, crypt hypertrophy, mitochondrial swelling, distension of the endoplasmic reticulum (ER), and mononuclear cell infiltration of the lamina propria [7,92,93]. There is a strong correlation between the severity of diarrhea and the degree of mucosal damage, suggesting that malabsorption due to the loss of absorptive cells may contribute to diarrhea [92]. However, it is likely that factors other than damage to the epithelium play a role in the onset of disease. Animal models have revealed that rotavirus infection results in the net secretion of fluid, sodium, and chloride from the intestine and the inhibition of glucose cotransport of electrolytes [88,94]. In addition, the rotavirus nonstructural protein 4 (NSP4) protein acts as an enterotoxin to produce a secretory state in some animal models [95]. When applied extracellularly, the enterotoxin peptide (aa 112–175) inhibits glucose-coupled sodium transport and stimulates elevated intracellular calcium levels, which activates an anion channel resulting in a secretory state [96–98]. The role of an intracellular enterotoxin in human rotavirus gastroenteritis is not yet established [87]. In *ex vivo* studies, NSP4 stimulates the secretion of serotonin by enterochromaffin cells, resulting in the activation of vagal afferent nerves connected to brainstem structures associated with nausea and vomiting [91]. The enteric nervous system (ENS) also plays a role in the pathogenesis of rotavirus gastroenteritis by stimulating the intestinal secretion of fluids and electrolytes as well as intestinal motility, resulting in deceased intestinal transit time [88,90].

TREATMENT

Rotavirus gastroenteritis is usually self-limiting, and the primary treatment is to rectify dehydration and electrolyte imbalance. Oral rehydration solutions are effective in treating dehydration and are preferred to intravenous rehydration [99,100]. Intravenous hydration is recommended when oral rehydration cannot be maintained due to excessive vomiting, reduced consciousness, intestinal ileus, and when dehydration becomes severe [99]. Zinc supplementation for 10–14 days has been shown to significantly decrease the prevalence and duration of diarrheal disease in developing country settings [101,102]. Although not routine, the oral administration of immunoglobulin has been shown to be beneficial in treating chronic rotavirus diarrhea [103]. Probiotic administration appears to reduce the duration of rotavirus gastroenteritis via an unknown mechanism of action [104,105]. Antiviral therapies such as nitazoxanide and racecadotril have been shown to decease the duration of diarrhea and stool output [106].

TRANSMISSION TO CONTACTS

Modes of Transmission

Rotavirus is transmitted via the fecal–oral route, with as few as nine infectious particles demonstrated to initiate an infection in adults [107]. Although rotavirus has been detected in urine and upper respiratory samples, these bodily fluids are not typically associated with transmission [73,108], although airborne transmission has been proposed for the epizootic diarrhea of infant mice (EDIM) strain in mice [109]. Foodborne transmission occurs less frequently than other enteric viruses, but is still an important transmission route. Rotavirus has been detected in shellfish, and is likely due to fecal contamination [110–113]. Rotavirus is a minor cause of traveler's diarrhea (predominantly in Mexico, Jamaica, and Central America) and contamination of water supplies, particularly with group B strains [59,114–116].

Transmission of the virus primarily occurs through close personal contact and contact with contaminated surfaces and objects (fomites). Rotavirus virions are robust and resistant to physical inactivation by treatment with fluorocarbons, ether, and concentrations of chlorine at levels used in the treatment of sewage effluent and drinking water [117]. However, rotavirus virions are susceptible to calcium chelators and antiseptic preparations that contain high concentrations of alcohols (>40%), free chlorine (>20,000 parts per million [ppm]), or iodophores (>10,000 ppm iodine) [118]. Rotavirus virions can survive at ambient temperatures for long periods of time, but survival is diminished at relatively high humidity [11,117,119].

SHEDDING OF ROTAVIRUS IN STOOL FACILITATING TRANSMISSION

Rotavirus is routinely shed in the stool for 7–10 days, and approximately one-third of infants have been observed to continue to shed the virus for up to 21 days [120]. The amount of virus shed in the stool can be as high as 10^{10} virions per gram of stool [119]. The duration of shedding can be longer in children with symptomatic infection compared with asymptomatic infection, with higher levels of virus shed in symptomatic infections [121,122].

ROTAVIRUS DISEASE IN ADULTS

Rotavirus infection may be transmitted to family members. Studies have reported that up to 75% of children and 33%–55% of adults develop rotavirus infection following contact with an infected child within the family [123,124]. The incidence of sporadic rotavirus diarrhea in adults has been reported to vary from 2% to 22% in numerous countries worldwide [59,125–127]. Higher rates have been reported in some settings, including 42% in Indonesia and 63% in Mexico [128,129]. In the United States, 81,000 ED visits and 18,000 hospitalizations annually are due to rotavirus infection, resulting in inpatient costs of approximately $152 million per year [130–132]. Following a 2–6 day incubation period, symptoms including malaise, headache, abdominal cramping, diarrhea, and fever are commonly reported and persist for 1–4 days [59]. Infection is frequently asymptomatic, as healthy adults generally have protective immunity acquired from previous exposure [133]. Disease is more commonly detected in the elderly and the immunosuppressed, and the seasonal pattern of disease seen in the pediatric population is also observed in adults. The prevalence of rotavirus in adults has been reported to be similar to the prevalence of salmonella, shigella, and campylobacter [134]. The same rotavirus genotypes have been reported to circulate in similar proportions in adults and children [135,136].

PROTECTION AGAINST ROTAVIRUS DISEASE

Immunity to rotavirus infection varies over a person's life. Neonatal infections are typically asymptomatic, likely due to the transfer of maternal antibodies that confer protection during the first months of life [137]. The emergence of symptomatic infections often correlates with the decline of these antibodies. Breastfeeding may provide a level of passive protection in infants [137].

The first rotavirus infection is often associated with the most severe disease symptoms, and does not protect against subsequent rotavirus reinfection but does lessen the severity of subsequent disease [138]. A seminal study in Mexican infants showed that, by the age of 2, 96% of infants had been infected with rotavirus at least once, 69% had experienced at least two infections, and 13% had experienced five infections. Following primary infection, 40% of infants were protected against any subsequent reinfection, 75% were protected against diarrhea caused by subsequent reinfections, and 88% were protected against severe rotavirus disease [47]. Subsequent rotavirus infections were usually due to a different rotavirus strain with a different G and/or P genotype [47]. Primary infection typically generates homotypic protection (genotype specific), while subsequent infection results in the broadening of immune responses and the production of a heterotypic immune response (genotype cross-reactive) [139]. Symptomatic reinfections can occur in developing country settings; a study in India revealed that >70% of children developed several symptomatic rotavirus infections during the first 3 years of life [140].

IMMUNE RESPONSE TO ROTAVIRUS INFECTION

Immunity to rotavirus infection involves both cellular and humoral immune responses. Rotavirus antigens are transported to Peyer's patches and are processed by B cells, macrophages, or dendritic cells, a process that results in the presentation to T cells. This results in the stimulation of rotavirus-specific B cells and cytotoxic T-lymphocyte–precursor expansion [141]. The most frequent B cell response is directed at VP6, and these antibodies are able to bind double-layered virus particles and inhibit viral transcription [142].

The innate immune system response to viral pathogens is often characterized by the induction of interferon (IFN) and other pro-inflammatory cytokines [143]. However, the role of the innate immune response to rotavirus infection is largely unknown. Several cytokines, including interleukin-8 (IL-8), IL-2, and gamma IFN (IFN-γ), have been detected in the plasma and stools of children with acute rotavirus infection [144–146]. *In vitro* studies reveal that rotavirus particles induce an early IFN response, as do ssRNA transcripts, due to the ineffective 5' capping of viral transcripts by VP3 [143]. However, the virus is able to rapidly suppress IFN signaling via the action of NSP1 [147–149].

ROTAVIRUS STRUCTURE

Rotavirus has a distinct wheel-like appearance when viewed under EM (Figure 10.1) and the mature, infectious rotavirus virion (virus particle) is approximately 75 nm in diameter [93,150]. Rotavirus virions possess a nonenveloped capsid, exhibit icosahedral symmetry, and are composed of three concentric protein layers that encase the genome [151].

The innermost protein layer (core) encasing the genome is predominately composed of 120 copies of VP2 arranged into 60 dimers, forming an icosahedral lattice (triangulation number (T) = 1) [152]. VP2 acts as the skeletal structure of the viral core in conjunction with VP1 and VP3, which are anchored within the VP2 shell [153].

The middle protein layer (inner capsid) is composed of 780 copies of VP6 (arranged into 260 trimers) that form pillars in the icosahedral lattice (T = 13; levo form) [153,154]. The VP6 layer acts to stabilize the otherwise fragile core by binding to VP2 and as an adaptor for the outer capsid proteins VP7 and VP4 [155].

The outer capsid is similarly composed of 780 copies of the glycoprotein VP7 (arranged into 260 trimers), forming an icosahedral lattice (T = 13; levo form) into which 160 copies of the protease-sensitive protein VP4 (arranged as 60 trimers) are embedded [155]. The outer and middle capsids are perforated by 132 aqueous channels, 12 type I channels, 60 type II channels, and 60 type III

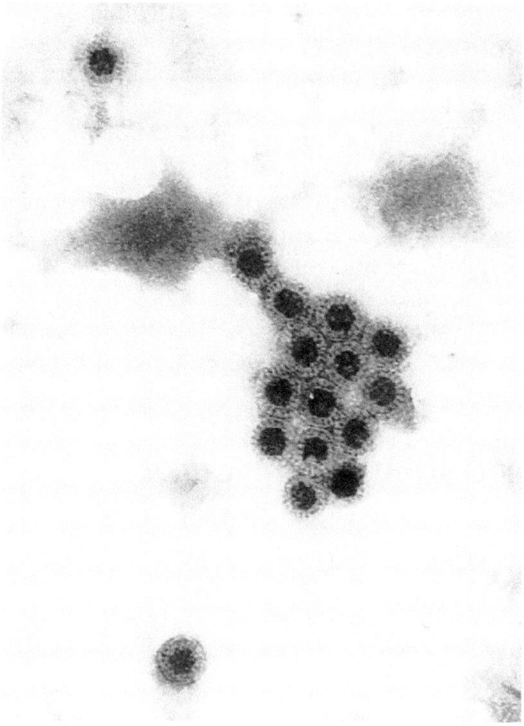

FIGURE 10.1 Electron micrograph of rotavirus particles. The rotavirus particle comprises three concentric protein layers encasing the 11-segment dsRNA genome. The inner core is composed of VP1, VP2, and VP3; the middle layer is composed of VP6; and the outer capsid is composed of VP7 and VP4. Image sourced from RF Bishop.

channels [156]. There are three morphologic appearances of rotavirus particles visible by EM. The mature, infectious particle is referred to as a triple-layered particle (TLP), which can lose the outer capsid and become a noninfectious double-layered particle (DLP). The loss of the inner capsid results in single-layered particles (SLPs) [11].

ROTAVIRUS GENOME

The rotavirus genome comprises 11 segments of double-stranded RNA (dsRNA). Ten genes are monocistronic and encode a single protein (VP1–VP4, VP6, VP7, and NSP1–NSP4), with genome segment 11 encoding two proteins (NSP5 and NSP6) [11]. The rotavirus genome is approximately 18,555 base pairs (bp) in length, smaller than other mammalian reoviruses including reovirus type 3 (23,549 bp) and the ruminant bluetongue virus (19,218 bp) [157]. Like other members of the *Reoviridae* virus family, each gene has short 5' and 3' terminal noncoding sequences. These are highly conserved sequences, with a 5' GGC sequence found in all strains and a UGUGACC sequence 3' found in most strains [157].

ROTAVIRUS REPLICATION CYCLE

Rotavirus replication is a complex process occurring within the mature epithelial cells positioned at the tip of villi [158]. The replication cycle can be separated into nine stages: (1) infection of the host cell, (2) uncoating of the viral particle, (3) mRNA production, (4) protein production, (5) viroplasm formation, (6) genome packaging, (7) ER interaction with DLPs, (8) coating of DLPs, and finally, (9) release of nascent viral particles.

Rotavirus cell entry is not completely understood. To facilitate membrane penetration, trypsin-like proteases present in the host gastrointestinal tract proteolytically cleave the VP4 protein into two subunits (VP8* and VP5*, where * denotes a cleavage product), resulting in a sevenfold increase in viral infectivity [159]. VP8* appears to mediate host cell attachment for many strains. Although sialic acid has been proposed as a cell receptor for attachment for some animal strains, numerous human and animal strains demonstrate sialic acid-independent receptor mechanisms [160–163]. Alternative cellular surface molecules, including glycans with internal sialic acid, or interactions between VP5* and specific integrins may function in receptor binding [155,164]. The VP8* protein of some P genotypes has been shown to interact with the secretor histo-blood group antigen (HBGA). Various strains of rotavirus with genotype P[4], P[6], P[8], P[11], P[14], and P[25] have shown specific recognition of the neutral oligosaccharide of HBGAs, and most bind either to the H type 1, Lewis b, or a antigens [165–168]. It has been reported that "nonsecretor" individuals may be resistant to infection by P[8] rotavirus strains [169].

Viral entry into epithelial cells is pH independent, and there are two hypotheses describing the mechanisms of entry. The first hypothesis involves the direct fusion and subsequent penetration of the viral particle through the plasma membrane [170]. The second and favored hypothesis involves internalization via Ca^{2+}-dependent receptor-mediated endocytosis, trafficking to, and subsequent penetration of the early endosomal membrane. Once calcium equilibrium is reached between the endosome and the cytoplasm, the outer capsid layer dissociates from the virus particle, resulting in the formation of DLPs [171]. As a consequence, VP5* embedded in the VP7 layer undergoes a conformational change, facilitating membrane destabilization and lysis of the vesicle membrane to release the virus into the cytoplasm [151,172,173].

The released DLPs become transcriptionally active on reaching the cytoplasm. Structural evidence suggests that the disassociation of VP7 results in dilation of the channels in the DLP, thus facilitating the influx of ions and nucleotides and the efflux of transcripts [155]. Within the core, the RNA-dependent RNA polymerase (RdRp) uses the negative-sense RNA strand of the dsRNA genome as a template for the synthesis of nascent positive-sense RNA transcripts of each of the 11 rotavirus genes [174]. The nascent positive-sense RNA transcripts are directed toward VP3 to acquire a 3' cap via an RNA exit tunnel, while the genomic negative-sense RNA is directed back into the core to reassociate with the genomic positive-sense RNA [155,175].

Type I aqueous channels present in the core and inner capsids facilitate the release of positive-sense RNA into the cytoplasm [156]. The positive-sense RNA has a dual role in the replication cycle, acting as the mRNA template for protein synthesis as well as the template for genome replication [155]. Viroplasms form large, semiregular networks designed to sequester viral RNAs and capsid proteins for assembly into nascent virions. The transcription products released from DLPs that escape the viroplasm accumulate in the cytosol and are translated by free ribosomes, with the exception of the VP7 and NSP4 proteins, which are translated by ribosomes associated with the rough ER [176,177]. The positive-sense RNA within the viroplasms is selectively packaged into assembling VP2 cores and replicated by the RdRp (negative-sense RNA strand synthesis) into the dsRNA genome [155]. The DLPs are released from the viroplasm, and NSP4 may recruit DLPs from the viroplasm and VP4 to the ER membrane, resulting in deformation of the ER membrane and budding of the DLP/VP4/NSP4 complex into the ER. The transient membrane is lost (via an unknown mechanism) and the VP7 and VP4 outer capsid assembles to form TLPs [155].

In vitro studies suggest that TLPs egress from infected cells via multiple mechanisms. While direct lysis has been observed in cell culture, it may not reflect the principal mechanism [155,178]. The TLPs may be released via trafficking and secretion, possibly utilizing a novel secretion pathway that bypasses the Golgi apparatus and lysosomes or a preexisting secretory pathway [155].

GENOMIC DIVERSITY OF ROTAVIRUSES

Rotavirus strains are able to evolve rapidly, employing a myriad of mechanisms to generate diversity in the wild-type strain population [157].

Genetic Drift

The generation of spontaneous sequential point mutations (genetic drift) occurs due to the error-prone nature of the RdRp of RNA viruses, including rotaviruses [157]. The mutation rates of RNA viruses have been found to vary considerably between viruses of different hosts, from as low as 1×10^{-6} to as high as 1×10^{-3} substitutions/site/year [179]. *In vitro* studies suggest that a progeny rotavirus genome differs from its parental strain by at least one mutation [180]. The overall mutation rate for all genes from a porcine strain through multiple passages of cell culture was calculated to be 5×10^{-5} substitutions/site/year [180]. The rate of mutation has been calculated for VP7 genes from different G genotypes; 1.8×10^{-3} (global G9), 1.66×10^{-3} (global G12), 0.9×10^{-3} (Vietnam, G1, 1998–2008), 7.25×10^{-4} (Japan, G1, 1987–2000), 1.31×10^{-3} (global G1), and 5.8×10^{-4} substitutions/site/year for the VP4 gene of human rotaviruses [179,181–183]. Similar rates have been calculated for the VP7 genes of group B rotavirus strains in China (7.9×10^{-4}) and Bangladesh (1.75×10^{-3}) [184,185]. Global NSP2 N1 genotype strains have a calculated mutation rate of 8.7×10^{-4} [186]. Although replicated by the same RdRp, each gene may be subjected to different immune pressures, host-selection pressures, and constraints, resulting in different rates of mutation. Affected by host-neutralizing antibody responses, the VP7 and VP4 proteins are thought to evolve more rapidly than the internal structural and nonstructural proteins. The latter are thought to be more stable because of structural and functional constraints and diminished exposure to the immune system [157].

Rearrangement

Genomic rearrangement involves alterations within a single genome segment in the form of deletions, duplications, or insertions in the sequence [187]. Rearrangement is most commonly observed in rotavirus strains excreted by chronically infected immunocompromised children and animals [187,188]. Rearrangement has also been observed *in vitro* after the serial passage of strains at high multiplicity in cell culture systems [189]. Rearrangement is frequently detected in NSP1, NSP3, and NSP5 genes and reported sporadically in VP6, NSP2, and NSP4 genes [190,191]. Most rearranged genes are not defective and replace nonrearranged RNA segments structurally and functionally, as most instances involve head-to-tail duplication occurring downstream of the ORF [190]. Infrequently, rearrangements occur within the ORF [192]. The molecular mechanism leading to rearrangement is unknown. One hypothesis suggests that, during either transcription or replication, the viral RdRp could interrupt RNA synthesis, revert to its template, and reinitiate RNA synthesis [187]. Another proposed mechanism, the "loop model," proposes that during transcription the 3' end of the negative RNA strand is inserted into the polymerase catalytic core before the complete negative strand is transcribed, creating a loop [193].

Recombination

Recombination is the formation of new, covalently linked combinations of genetic material from two different parental genomes or between different sites of the same genome [194]. A small number of recombination events have been reported for group A rotavirus strains. Intragenic recombination has been identified in G4 and G1 VP7 genes, where the parental strains belonged to different sublineages of the same genotype [195,196]. Intergenotype recombinant VP7 genes have been identified, derived from G9 and G1, G1 and G3, and G3 and G2 parental strains [197–200]. Intergenogroup recombinant VP6, NSP2, and NSP4 genes have identified between G9P[8] or G1P[8] and G2P[4] parental strains [201,202]. An intergenic/intersegmental recombination event between NSP2 and NSP5 genes created a rearranged NSP5 gene, derived from the serial passage of the porcine rotavirus strain SB1A [203]. Recombination is dependent on numerous intracellular and immunological factors, including the successful coinfection of a cell by two parental strains, the efficient replication of parental genomes with template switching, and the adaptation to different environments to facilitate transmission [204].

Reassortment

Genomic reassortment (genetic shift) occurs when single cells are infected with multiple different yet compatible rotavirus strains. The segmented nature of the rotavirus genome facilitates the exchange of genes between strains to produce new gene combinations, often leading to the generation of novel viral strains [205]. Reassortment occurs more frequently and provides greater diversity than genetic drift, and is most efficient when the strains are closely related [157]. Reassortment between strains belonging to different rotavirus species (group A–H) has not been detected; however, the NSP1 gene of avian group A strains clusters closer to group D strains than mammalian group A strains, suggesting that reassortment may have occurred between group A and D strains [157,206]. It has been shown that all 11 genes can undergo reassortment and, in theory, the coinfection of a single cell by two strains could give rise to 2^{11} (2046) distinct viruses [207]. However, some genome constellations are more favorable than others, and more commonly isolated. The VP4, VP6, VP7, NSP4, and NSP5 genes are often inherited together [208,209]. The VP1, VP2, VP3, and VP6 genes also show a degree of genetic linkage, likely due to the critical interactions between these proteins [210]. Intergenogroup and intragenotypic reassortment of single and multiple genes also provides a mechanism to generate significant diversity in rotavirus strain populations [202,211–229].

INTERSPECIES AND ZOONOTIC TRANSMISSION

Rotavirus strains exhibit a degree of host species restriction, with species-specific strains (homologous) more commonly identified than interspecies strains (heterologous) [13]. Numerous studies have shown that heterologous strains can infect and/or induce diarrhea under experimental conditions in animal models, including human group A rotavirus strains in a gnobiotic piglet model, RVA/Simian-tc/USA/RRV/1975/G3P[3] in rabbits, RVA/Pigeon-tc/JPN/PO-13/1983/G18P[17], and RVA/Bat-tc/CHN/MSLH14/2011/G3P[3] in mice [230–233]. The natural attenuation of some animal strains in humans (the *Jennerian* approach) has been utilized to generate potential vaccine candidates [234–236].

The introduction of animal rotavirus strains into the human population (zoonotic transmission) also increases the genetic diversity within strains causing human infection. Interspecies transmissions of complete animal strains are uncommon [13]. The formation of reassortant viruses that possess genes of human and animal origin could create viruses more capable of replication and transmission among humans than the original animal parental strain [13]. Numerous strains have been reported as reassortant strains derived from human and animal parental strains, and may include the exchange of single or multiple genes [6,237,238]. Initially, RNA–RNA hybridization studies were seminal in the characterization of reassortant human–animal strains, but the advent of genomic sequencing has resulted in an exponential increase in the detection of zoonotic strains [239]. Reassortant rotavirus strains are not solely identified as sporadic cases. Porcine–human G9P[19] and G5P[8] reassortant strains have been identified as associated with high disease incidence in India and Brazil, Paraguay, and Argentina, respectively [240–246]. Bovine–human G8P[8] and G10P[11] strains have been identified in Africa and India, respectively [247,248]. Some globally circulating rotavirus genotypes, including G9 and G12, originated as zoonotic transmissions [229,249–251]. Zoonotic transmission tends to occur more often in countries where contact with domestic animals is higher, or where animals reside in close contact with humans, such as cats and dogs or domestic farm animals such as cattle and pigs [252,253].

ROTAVIRUS OUTBREAKS IN INSTITUTIONAL SETTINGS

The burden of rotavirus disease in older adults is not fully understood, as rotavirus testing is not routine. Rotavirus outbreaks in care homes have been attributed to G1P[8], G4P[8], and G9P[8] strains in Ireland; G1P[8] strains in Japan, the United Kingdom, and Hungary; and G9P[8] strains in Spain [254–259]. Numerous outbreaks have also been reported without determining the genotype of the strain responsible, including outbreaks in nursing homes in Slovenia, Japan, Norway, the United States, Canada, Germany, and a long-term care facility in France [260–266].

Multiple outbreaks due to G2P[4] rotavirus strains have been reported in mental health facilities, restaurants, school trips, schools, a rehabilitation institution in Japan, a prison in Brazil, and a tsunami relief camp in India [256,266–272]. G2P[4] strains are the most common genotype identified during outbreaks in adults, suggesting heterotypic immunity following natural infection may be less protective against G2P[4] strains compared with other genotypes. Alternatively this may be due to G2P[4] strains being more virulent in adults than other genotypes [267]. In addition, rotavirus outbreaks have been reported in hospitals in numerous countries and attributed to nosocomial transmissions in pediatric and adult settings, particularly in cardiology, oncology, and geriatric wards where they may result in fatalities and ward closures [273–282]. Transmission in the hospital setting in winter is largely facilitated by staff, with pediatric nurses, medical students, and other staff reporting symptoms of gastroenteritis during the winter peaks of rotavirus disease [283].

COMMUNITY SETTINGS

Community outbreaks of rotavirus gastroenteritis in children and adults have been reported, but are less common than outbreaks in institutional settings. Outbreaks reported in schools and colleges in the United States have been attributed to G2P[4] strains, and a G1P[8] strain was reported in an elementary school in Japan [267,284,285]. A large outbreak was caused by a G9P[8] rotavirus in 2005 in Rio Branco City, Brazil, with 816 cases of acute gastroenteritis reported over 8 days, resulting in eight child deaths. The outbreak was likely caused by a contaminated water source [286].

Waterborne and shellfish-borne outbreaks, usually due to sewage contamination, have been reported in the United States, France, Turkey, Greece, and Finland [110,287–290]. Another large outbreak in 1964 in an isolated area of Micronesia involved almost 3500 people [291]. An outbreak involving 64,000 adults and children in Nicaragua in 2005 was attributed to a G4P[8] virus that possessed several unusual amino acid mutations [292]. Other outbreaks have occurred among closed communities, including a Finnish military base, an Israeli kibbutz, and an isolated indigenous South American community [293–295].

FOODBORNE TRANSMISSION

The foodborne and waterborne transmission of pathogenic viruses is well recognized, with outbreaks of norovirus and hepatitis A virus extensively reported [296–300]. Rotavirus infection occurs mainly via the fecal–oral route, and food handlers lacking sufficient hygiene can contaminate food during production, preparation, and serving. Furthermore, fomites act as a mechanism for viral transmission and can result in rotavirus contamination of food and water used in irrigation or food processing. Rotavirus has been detected in several food types including strawberries, vegetables, lettuce, leafy greens, packaged beef, and shellfish [301–306].

In several of these reports, the number of samples that tested positive for rotavirus was low when compared with other enteric pathogens. As detection methods for rotavirus within food matrices can prove ineffective due to low sensitivity, most studies also rely on epidemiological parameters to explore outbreaks [113,307]. One study, published in 2001, found that the foodborne route of rotavirus transmission accounted for just 1% of all reported cases in the United States [115]. Another study in 2003 estimated that the population-attributed risk of rotavirus-contaminated food entering a dwelling accounted for 4% of all rotavirus-associated gastroenteritis cases [308]. Although rotavirus may not be the most important or most prevalent viral agent in terms of foodborne viral disease, it is still capable of causing widespread outbreaks through this transmission route.

TRANSMISSION FROM FRESH PRODUCE

Rotavirus contamination of water sources used to irrigate crops has been identified and documented [302,309]. As rotavirus is nonenveloped, it has shown resilience in the environment, surviving in water for weeks [117]. The virus poses a risk to fresh produce for human consumption when contaminated water is used for the irrigation of fresh produce. A study in Canada sampled field strawberries in Quebec at various time points after irrigation from a local river source. Using conventional RT-PCR and nested RT-PCR assays, rotavirus was detected on 3.33% (2/60) of strawberry samples at 1 and 6 h postirrigation with river water [301]. The river samples tested had no detectable rotavirus, indicating low levels of contamination in the water [301].

In another study, following high incidences of rotavirus-associated diarrhea in Costa Rica, a dot ELISA was used to detect rotavirus in lettuce samples from local markets [303]. In ten pooled samples, of five lettuce heads per pool, 30% (3/10) of pooled samples tested positive for rotavirus. It was concluded that the lettuce crops were likely contaminated with sewerage, highlighting the risk of rotavirus transmission into the food chain by applying contaminated water onto food crops meant for human consumption [303].

TRANSMISSION FROM PREPARED FOOD

Reports of rotavirus outbreaks associated with contaminated foods are also documented, with cases associated with sandwiches, potato stew, salad, and raw cabbage [113,267,310,311]. Preparation of food by infected handlers is a known risk for rotavirus infection and poses a potential route for contaminants to enter the food chain [308].

Following a rotavirus outbreak in 2007 within a long-term care facility in Germany, several food samples were tested for human pathogens [113]. Using nested RT-PCR and sequencing, human group A rotavirus was detected in potato stew samples [113]. A stool sample from one affected child was confirmed positive by rotavirus ELISA. Nested RT-PCR analysis showed that the rotavirus detected in the stool sample was 100% identical to the rotavirus detected in the stew, confirming the cause of the outbreak [113]. As rotavirus could not have remained viable during cooking, it was concluded that the contamination of the stew occurred afterwards, but the exact cause was not established [113]. This case highlights the importance of food handler hygiene during food preparation and while serving food to prevent contamination with rotavirus.

In another rotavirus outbreak at a university in the United States in 1999, 1.6% (85/5453) of students became ill with diarrhea and/or vomiting after eating sandwiches from the same dining hall. Six stool specimens from dining hall staff and affected students were tested using a combination of EM, RT-PCR, and enzyme immunoassays. All six specimens tested positive for group A rotavirus. An asymptomatic university employee was implicated as the source during preparation of delicatessen sandwiches [267].

Frequently, in an outbreak setting, it is difficult to conclude that a particular food was the source of infection, since other pathogens are often detected at the same time and rotavirus is not always detected in the implicated food source [113,267,310,311].

ROTAVIRUS TRANSMISSION FROM SEAFOOD

Rotavirus has been detected in water sediment and shellfish in several studies [110–112,312]. Le Guyader et al. [313] carried out a 3-year study aimed at detecting enteric pathogens in shellfish in southern France between 1995 and 1998. RT-PCR and dot blot hybridization were used to detect viral pathogens, including hepatitis A, rotavirus, and norovirus, in mussels and oysters from five collection sites. Rotavirus was detected in 27% of oyster samples and 52% of mussels tested [313]. Interestingly, despite causing seasonal peaks of human infections in winter, rotavirus was detected all year round at one of the sites [313].

Despite its detection in shellfish intended for human consumption, rotavirus has not always been directly linked to outbreaks in humans. Several explanations for the lack of causal data have been proposed [252], including the fact that rotavirus usually only causes symptomatic infections in children, who generally are less likely to consume shellfish [117]; possible loss of infectivity upon shellfish uptake [314]; low viral survival rates; and poor diagnostics [315].

Together, these reports show that both contaminated food and water represent a potential source for rotavirus infection, although they may not be the primary transmission routes for this virus. This nevertheless emphasizes the importance of maintaining stringent hygiene while preparing and handling food, as well as raising general public awareness about disease transmission to prevent disease outbreaks.

ROTAVIRUS VACCINES

The ubiquitous nature of rotavirus, coupled with the low infectious dose and robustness of the virion, means that improved sanitation and hygiene have had little impact on decreasing the incidence of rotavirus infection [64]. The development of rotavirus vaccines has been a World Health Organization (WHO) priority, with an aim to achieve the United Nations Millennium Goal 4 of reducing childhood mortality [1,316]. Estimates suggest that vaccination would prevent 2.46 million rotavirus deaths between 2007 and 2025 in countries eligible for vaccine introduction through the Global Alliance for Vaccines and Immunization (GAVI) [317].

Early vaccine development was based on the classical Jennerian approach, considered advantageous because animal strains are often naturally attenuated in humans or are easily attenuated through cell culture passage [13]. This approach was applied to several candidate strains including

RIT4237 (derived from the NCDV bovine strain), the bovine strain Wister calf-3 (WC3), and MMU18006 (derived from the RRV monkey strain) [318]. These candidate vaccines were well tolerated and provided a level of cross protection [319,320]. The commercial development of the RIT4237 strain was ultimately abandoned and the RRV and WC3 strains were further developed as vaccine candidates using a modified Jennerian approach; creating human–animal reassortant strains using cell culture [321].

CURRENTLY LICENSED VACCINES

Two second-generation live-oral vaccines, Rotarix® (GlaxoSmithKline Biologicals, Rixensart, Belgium) and RotaTeq® (Merck & Co. Pennsylvania), are licensed and available for use globally. These vaccines have been introduced into the childhood immunization programs of 58 countries worldwide.

ROTATEQ®

RotaTeq® is a live-attenuated, pentavalent oral vaccine that contains five genetically distinct human–bovine reassortant virus strains. The exact molecular basis of attenuation is unknown, but is thought to be due to host range restriction [158]. Each reassortant strain contains a gene encoding one of the outer capsid proteins of the human strains VP7 (G1 [WI79-9], G2 [SC2-9], G3 [WI78-8], G4 [BrB-9], or P[8] [WI79-4]) within a bovine rotavirus backbone ([WC3], G6P[5]) [322]. RotaTeq® is administered as a three-dose schedule oral vaccine, recommended at 2, 4, and 6 months of age, with the first dose to be administered before 15 weeks of age [323]. Large phase III clinical trials have demonstrated the efficacy of three doses of RotaTeq® to be 74% against rotavirus gastroenteritis of any severity and 98% against severe rotavirus gastroenteritis [324]. Sustained efficacy has been demonstrated for 3 years after the final vaccine dose, and significant protection against hospitalizations and ED visits associated with rotavirus genotypes G1 (95.5%), G2 (81.9%), G3 (89.0%), G4 (83.4%), and G9 (94.2%) has also been demonstrated [325].

ROTARIX®

Rotarix® is a monovalent oral vaccine containing a single strain (RIX4414, derived from the human strain 89-12 isolated in Cincinnati) of the most prevalent human genotype worldwide (G1P[8]). The strain was attenuated by serial passage in cell culture and several genetic alterations were observed compared with the parental strain, which have been suggested to be associated with attenuation [158,326,327]. Rotarix® is administered orally in a two-dose schedule at 2 and 4 months of age, with the first dose administered before 15 weeks of age [323]. The safety and efficacy of Rotarix® was demonstrated in phase II and III trials in Latin America and Europe [326,328–332]. Integrated analysis of these clinical trials reported vaccine efficacy against rotavirus gastroenteritis of any severity, and severe rotavirus gastroenteritis was 82.57% and 87.43% for G1P[8] strains, 87.66% and 90.19% for G3P[8] strains, 84.86% and 93.37% for G4[8] strains, 60.64% and 83.76% for G9P[8] strains, and 81.04% and 71.42% against fully heterotypic G2P[4] strains in phase II and phase III trials, respectively [333]. The liquid formulation of Rotarix® has been shown to be immunogenic, reactogenic, and safe at levels similar to the lyophilized formulation [334]. Rotarix® has also been shown to be safe, reactogenic, and immunogenic in European preterm infants [335].

A trial of Rotarix® was conducted in South Africa and Malawi, countries that experience intermediate and high mortality rates in children under 5 years, respectively. The overall efficacy observed in South Africa was similar to that observed in developed countries. However, a lower vaccine efficacy and a lower rate of seroconversion measured by serum IgA was observed in Malawian children [336]. Importantly, sustained vaccine efficacy was observed into the second year of life in South African children [337]. The liquid formulation of Rotarix® was shown to be safe and immunogenic when coadministered with the diphtheria–tetanus–whole-cell pertussis, hepatitis B, and oral polio virus vaccines in infants in Vietnam and the Philippines [338]. Similar levels of vaccine efficacy were observed between well-nourished and malnourished infants in Brazil, Mexico, and Venezuela [339].

Clinical trials of RotaTeq® were conducted in Bangladesh, Vietnam, Ghana, Kenya, and Mali. The efficacy of RotaTeq® against severe rotavirus gastroenteritis in Bangladesh and Vietnam was 71.0% and 68.1%, respectively, in the first year of life. Efficacy decreased in the second year of life to 39.3% in Bangladesh and 64.6% in Vietnam [340]. Between trials in Ghana, Kenya, and Mali, the immunogenicity of the RotaTeq® vaccine was similar, although the efficacy against severe rotavirus gastroenteritis in the first year of life was 65.0% in Ghana, 83.4% in Kenya, and 1.0% in Mali. The efficacy of the vaccine decreased in the second year of life in Ghana and Kenya and increased in Mali [341]. The efficacy against severe gastroenteritis varied for the individual genotypes: G1 (32.3%), G2 (27.1%), G3 (62.3%), G8 (87.5%), G9 (49.7%), P[8] (36.1%), P[4] (18.2%), and P[6] (47.7%) [341].

The primary protection afforded by RotaTeq® is based on homotypic protection specific to each of the component strains, providing protection against the most common genotypes in circulation. The protection afforded by Rotarix® is based on the concept that heterotypic protection can be generated as a result of cross-reactive antigens between different genotypes [45]. The mechanism by which heterotypic immunity develops following homotypic immunization has not been determined. Possible mechanisms include the induction of protective antibody responses to heterotypic epitopes on VP4 or VP7, protective effects of antibodies to common antigens on VP6 or NSP1, or cross-reactive protective T cell responses [342]. Interestingly, the efficacy of Rotarix® did not increase with a three-dose immunization schedule compared with a two-dose schedule, for reasons that remain unclear [336].

The observed decrease in efficacy of rotavirus vaccines in developing countries is not well understood and may be attributed to numerous factors. The effectiveness of the vaccine may be influenced by environmental factors, including an increased diversity of circulating strains, an increased occurrence of mixed infections, and seasonality, as equatorial regions experience year-round rotavirus seasons. Biological factors include differences in age at primary infection, coinfections of the gut, and comorbidities such as malnutrition, HIV, malaria, and tuberculosis. In addition, high levels of humorally transferred maternal antibodies against rotavirus may also play a role. Higher doses of the vaccines or additional doses may be required to provide an adequate level of protection in developing countries [343]. Numerous studies have revealed that concomitant oral polio vaccine administration has no effect on rotavirus vaccine efficacy [344–347].

OTHER CANDIDATE VACCINES

There are several third-generation vaccines in various stages of development and licensure. In China, the Lanzhou lamb vaccine (Lanzhou Institute of Biological Products, Lanzhou, China) consists of a G10P[12] strain. The vaccine was licensed in 2000 and over 30 million doses have been administered to children under 5 years of age [348]. One dose is administered at 2–36 months, followed by yearly boosters until the age of 3 years [349]. No clinical trial results or vaccine efficacy data have been reported in prelicensure studies. Limited postlicensure studies have identified varying degrees of vaccine effectiveness, with one study reporting 73.3% protection against rotavirus hospitalization, and higher effectiveness was found in older children, possibly due to the yearly booster [348–351]. Rotavin-M1 (Institute of Vaccines and Biologicals, Vietnam) is a live-attenuated vaccine derived from a Vietnamese G1P[8] strain (KH0118-2003). Phase I and II clinical trials revealed that the vaccine was well tolerated, safe, and immunogenic in adults and infants [352]. The Indian rotavirus vaccine 116E (Bharat Biotech, India) is a live-attenuated vaccine derived from a single G9P[11] reassortant strain isolated from an Indian child. Clinical trials revealed that the vaccine is well tolerated, safe, and immunogenic in adults and infants [353,354]. Phase III clinical trials have shown that the vaccine is efficacious in infants, which has subsequently led to its licensure in 2014 [355].

An asymptomatic neonatal rotavirus strain isolated in Australia (G3P[6], RV3) is currently undergoing phase I and phase IIa/b clinical trials in Australia, New Zealand and Indonesia [353,354]. The NIH BRV-TV vaccine is a tetravalent human-bovine reassortant vaccine comprising a G6P[7] backbone strain with G1 (strain D), G2 (DS-1), G3 (P), and G4 (ST-3) VP7 genes and expanded to hexavalent with the inclusion of G8 (1290) and G9 (AU32) components [235,356,357]. Other vaccine candidates such as synthetic viral proteins, inactivated DLPs, empty virus-like particles, synthetic peptides, transgenic rice-expressed rotavirus antigens, VP6-specific llama-derived single domain nanoantibodies, and DNA vaccines are also in the early stages of development [358–363].

IMPACT OF VACCINE INTRODUCTION

Burden of Disease

The introduction of rotavirus vaccines into routine vaccination programs has resulted in a significant reduction in rotavirus-associated hospitalizations, ED visits, and episodes of gastroenteritis in numerous countries worldwide. Several studies in Belgium report a significant decrease in rotavirus hospitalizations and other indicators of disease burden following the introduction of Rotarix® and RotaTeq® [364–367]. A 58%–77% decrease in rotavirus hospitalizations in children under 5 years and a 33% decrease in all-cause acute gastroenteritis hospitalizations in children under 2 years have been reported in Belgium [367]. Significant vaccine effectiveness has been demonstrated in numerous European countries following vaccine introduction, including France, Austria, Greece, Finland, Spain, and Germany, resulting in decreased hospitalizations and ED presentations due to rotavirus gastroenteritis [368–374].

Over 20 vaccine effectiveness studies from various regions of the United States have reported significant decreases in the burden of rotavirus disease following vaccine introduction [375–398]. Vaccine effectiveness against rotavirus gastroenteritis hospitalization and ED visits was 88% (three doses), 81% (two doses), and 69% (one dose) [376]. The substantial decrease in rotavirus hospitalizations and ED visits in the United States has reduced annual rotavirus-associated medical costs by $187 million each year [399].

In Central and South American countries, vaccine effectiveness has been lower than in European countries and the United States. However, significant decreases in the burden of rotavirus disease have been observed in Nicaragua, Brazil, Mexico, Panama, and El Salvador [400–418]. In some developing countries, a decrease in protection in the second year of life has been reported [340,341,418]. A comprehensive review of studies from Mexico, Brazil, El Salvador, and Panama revealed that vaccine introduction has resulted in a 22%–41% reduction in diarrhea-associated mortality, a 17%–51% decrease in hospitalizations for diarrhea, and a 59%–81% decrease in rotavirus-associated gastroenteritis hospitalizations in children less than 5 years of age [419]. Following vaccine introduction in Bolivia, Honduras, Venezuela, and Argentina, there has been a marked decrease in rotavirus-related deaths [420].

In Australia, there was a 71% and 38% decline, respectively, in rotavirus-coded and nonrotavirus-coded acute gastroenteritis hospitalizations in children under 5 years of age following vaccine introduction. This represented the avoidance of more than 7700 hospitalizations in 2009–2010. Reductions were also observed in children aged 5–19 years, suggesting that the transmission of virus was reduced at a population level. Decreases in hospitalizations of indigenous children were smaller than those in the general population, and fluctuated by location and year [421].

Age-Specific Incidence

Epidemiological shifts in age distribution have been observed following measles, varicella, and pertussis vaccine introduction [422–424]. In Philadelphia, a shift in age has been observed following rotavirus vaccine introduction. Compared with the previous 13 rotavirus seasons, the median age of patients increased from 11 months to 23 months in the 2008–2009 season, and almost 50% of cases were in infants too old to be vaccinated [385]. Similar data have been reported for other regions of the United States, South Africa, and Belgium [365,390,425].

Herd Immunity

Concurrent with the reported decrease in the burden of rotavirus disease in the vaccine-eligible population, there has been a marked reduction in disease burden in older, nonimmunized populations and in infants too young to be vaccinated. Herd immunity occurs as a result of decreased transmission of wild-type rotavirus in the community, and provides indirect protection to unvaccinated individuals [426]. Herd immunity has been observed after vaccine introduction in El Salvador, Panama, Mexico, the United States, Finland, Germany, and Austria [130,369,374,383,384,387,388,390–392,403,414,416,417,427,428]. In the United States, the prevalence of rotavirus disease in adults has declined by 48.8% since vaccine introduction [429]. This phenomenon is not limited to rotavirus, as the implementation of pediatric vaccination against *Bordetella pertussis*, *Neisseria meningitidis*, *Haemophilus influenzae type B*, and *Streptococcus pneumonia* has also resulted in a significant decline in adult disease [430–433]. Herd immunity is important in developing country settings where vaccine efficacy and coverage tend to be lower [426].

Seasonality

Changes in the rotavirus season have been detected in some countries following vaccine introduction. In the United States, there has been a shift in the onset of the season by 1–2 months, the season is considerably shorter and reduced in magnitude, and the geographic pattern of spread is no longer identified [391,434,435]. Diminished and delayed peaks in disease have been observed in Belgium, southeastern Brazil, Austria, South Africa, and Taiwan following vaccine introduction [366,370,408,425,436].

Intussusception in the Vaccine-Era

Intussusception (IS) is a potentially life-threatening bowel obstruction caused by the invagination of a section of bowel into the adjacent proximal bowel. While intussusception can occur at any age, most cases occur in infants aged 4–9 months [437]. Cases of idiopathic IS are uncommon in infancy and rates vary worldwide for infants during the first year of life, from 9 per 100,000 in Bangladesh to as high as 328 per 100,000 in South Korea. Globally, the mean incidence of IS is 74 per 100,000, with peak incidence among infants 5–7 months of age [438]. Following the administration of more than one million doses, RotaShield® was withdrawn in late 1999 due to an association with IS, where an estimated one excess case per 10,000–15,000 infants vaccinated was identified. The greatest risk was in children who received the first dose when older than 3 months of age [8,439].

Clinical trial and early postlicensure studies did not report an elevated risk of IS following vaccination with RotaTeq® or Rotarix® [440]. Postmarketing IS surveillance from Australia suggested that there was no overall increase in IS following vaccine introduction, although there was some evidence of an elevated risk following the first dose of both Rotarix® and RotaTeq® [441]. Recent data reveal that the relative incidence of IS in the 1–7-day period after the first vaccine dose is 6.8 per 100,000 for Rotarix® and 9.9 per 100,000 for RotaTeq®. There is a smaller increased risk 1–7 days after the second dose of each vaccine. This equates to an estimated excess of 14 IS cases per year [442]. Data from the United States supported these findings, where the first dose of RotaTeq® was associated with approximately 0.75–1.5 excess cases of IS per 100,000 vaccine recipients in the 21-day window following the first dose. No significant increased risk was observed after the second or third dose [443,444]. An underpowered study indicated a potential risk associated with Rotarix® [444]. A study conducted in Mexico and Brazil found an excess risk of 1 in 51,000 after the first dose of Rotarix® in Mexican infants, but no increased risk in Brazilian infants [445]. Importantly, in all these studies, the benefits of rotavirus vaccination for reducing rotavirus disease far outweighed the very minor risk of IS.

REFERENCES

1. Wardlaw T, Salama P, Brocklehurst C, Chopra M, Mason E. 2010. Diarrhoea: Why children are still dying and what can be done. *Lancet* 375:870–872.
2. Liu L, Johnson HL, Cousens S, Perin J, Scott S, Lawn JE, Rudan I, et al. 2012. Global, regional, and national causes of child mortality: An updated systematic analysis for 2010 with time trends since 2000. *Lancet* 379:2151–2161.
3. Walker CL, Aryee MJ, Boschi-Pinto C, Black RE. 2012. Estimating diarrhea mortality among young children in low and middle income countries. *PLoS One* 7:e29151.
4. Glass RI, Parashar U, Patel M, Gentsch J, Jiang B. 2013. Rotavirus vaccines: Successes and challenges. *J Infect* 68:S9–S18. doi:10.1016/j.jinf.2013.09.010.
5. Lanata CF, Fischer-Walker CL, Olascoaga AC, Torres CX, Aryee MJ, Black RE. 2013. Global causes of diarrheal disease mortality in children <5 years of age: A systematic review. *PLoS One* 8:e72788.
6. Matthijnssens J, Ciarlet M, McDonald SM, Attoui H, Banyai K, Brister JR, Buesa J, et al. 2011. Uniformity of rotavirus strain nomenclature proposed by the Rotavirus Classification Working Group (RCWG). *Arch Virol* 156:1397–1413.
7. Bishop RF, Davidson GP, Holmes IH, Ruck BJ. 1973. Virus particles in epithelial cells of duodenal mucosa from children with acute nonbacterial gastroenteritis. *Lancet* 7841:1281–1283.
8. Glass RI, Parashar UD, Bresee JS, Turcios R, Fischer TK, Widdowson M-A, Jiang B, Gentsch JR. 2006. Rotavirus vaccines: Current prospects and future challenges. *Lancet* 368:323–332.
9. Mertens P. 2004. The dsRNA viruses. *Virus Res* 101:3–13.
10. Matthijnssens J, Otto PH, Ciarlet M, Desselberger U, Van Ranst M, Johne R. 2012. VP6-sequence-based cutoff values as a criterion for rotavirus species demarcation. *Arch Virol* 157:1177–1182.
11. Estes MK, Kapikian AZ. 2007. *Fields Virology*, 5th edn., vol 2. Lippincott Williams & Wilkins, Philadelphia, PA.
12. Legrottaglie R, Rizzi V, Agrimi P. 1997. Isolation and identification of avian rotavirus from pheasant chicks with signs of clinical enteritis. *Comp Immunol Microbiol Infect Dis* 20:205–210.
13. Martella V, Banyai K, Matthijnssens J, Buonavoglia C, Ciarlet M. 2010. Zoonotic aspects of rotaviruses. *Vet Microbiol* 140:246–255.
14. Lecce JG, King MW. 1978. Role of rotavirus (reo-like) in weanling diarrhea of pigs. *J Clin Microbiol* 8:454–458.
15. Conner ME, Darlington RW. 1980. Rotavirus infection in foals. *Am J Vet Res* 41:1699–1703.
16. Papp H, Laszlo B, Jakab F, Ganesh B, De Grazia S, Matthijnssens J, Ciarlet M, Martella V, Banyai K. 2013. Review of group A rotavirus strains reported in swine and cattle. *Vet Microbiol* 165:190–199.
17. Mulherin E, Bryan J, Beltman M, O'Grady L, Pidgeon E, Garon L, Lloyd A, et al. 2008. Molecular characterisation of a bovine-like rotavirus detected from a giraffe. *BMC Vet Res* 4:46.
18. Abe M, Yamasaki A, Ito N, Mizoguchi T, Asano M, Okano T, Sugiyama M. 2010. Molecular characterization of rotaviruses in a Japanese raccoon dog (*Nyctereutes procyonoides*) and a masked palm civet (*Paguma larvata*) in Japan. *Vet Microbiol* 146:253–259.
19. Coria-Galindo E, Rangel-Huerta E, Verdugo-Rodriguez A, Brousset D, Salazar S, Padilla-Noriega L. 2009. Rotavirus infections in Galapagos sea lions. *J Wildl Dis* 45:722–728.
20. Guo L, Yan Q, Yang S, Wang C, Chen S, Yang X, Hou R, Quan Z, Hao Z. 2013. Full genome sequence of giant panda rotavirus strain CH-1. *Genome Announc* 1: pii: e00241–12.
21. Evans RH. 1984. Rotavirus-associated diarrhea in young raccoons (*Procyon lotor*), striped skunks (*Mephitis mephitis*) and red foxes (*Vulpes vulpes*). *J Wildl Dis* 20:79–85.
22. Hines ME, II, Styer EL, Baldwin CA, Cole JR, Jr. 1995. Combined adenovirus and rotavirus enteritis with *Escherichia coli* septicemia in an emu chick (*Dromaius novaehollandiae*). *Avian Dis* 39:646–651.
23. Silva LC, Sanches AA, Gregori F, Brandao PE, Alfieri AA, Headley SA, Jerez JA. 2012. First description of group A rotavirus from fecal samples of ostriches (*Struthio camelus*). *Res Vet Sci* 93:1066–1069.
24. Shibahara T, Wada Y, Tsunemitsu H, Kubo M, Ishikawa Y, Kadota K. 2001. Gastroenteritis associated with *Helicobacter*-like organisms and rotavirus in a reindeer (*Rangifer tarandus*). *Aust Vet J* 79:133–135.

25. Samal SK, Dopazo CP, McPhillips TH, Baya A, Mohanty SB, Hetrick FM. 1990. Molecular characterization of a rotaviruslike virus isolated from striped bass (*Morone saxatilis*). *J Virol* 64:5235–5240.
26. Jere KC, Esona MD, Ali YH, Peenze I, Roy S, Bowen MD, Saeed IK, et al. 2013. Novel NSP1 genotype characterised in an African camel G8P[11] rotavirus strain. *Infect Genet Evol* 21C:58–66.
27. Everest DJ, Duff JP, Higgins RJ. 2011. Rotavirus in a wild English red squirrel (*Sciurus vulgaris*) identified by electron microscopy. *Vet Rec* 169:160.
28. Linhares AC, Pereira JD, Nakauth CM, Gabbay YB. 1986. Rotavirus infection in wild marsupials (*Didelphis marsupialis*) of the Amazon region. *Trans R Soc Trop Med Hyg* 80:20–24.
29. Kapikian AZ, Chanock RM. 1996. Rotaviruses*In* Fields BN, Knipe DM, Howley PM (eds), *Fields Virology*, 3rd edn, pp. 1657–1708. Lippincott-Raven, Philadelphia, PA.
30. Sanekata T, Ahmed MU, Kader A, Taniguchi K, Kobayashi N. 2003. Human group B rotavirus infections cause severe diarrhea in children and adults in Bangladesh. *J Clin Microbiol* 41:2187–2190.
31. Alam MM, Pun SB, Gauchan P, Yokoo M, Doan YH, Tran TN, Nakagomi T, Nakagomi O, Pandey BD. 2013. The first identification of rotavirus B from children and adults with acute diarrhoea in Kathmandu, Nepal. *Trop Med Health* 41:129–134.
32. Bridger JC, Pedley S, McCrae MA. 1986. Group C rotaviruses in humans. *J Clin Microbiol* 23:760–763.
33. Castello AA, Arguelles MH, Rota RP, Humphrey CD, Olthoff A, Gentsch JR, Glass RI, Glikmann G, Jiang B. 2009. Detection and characterization of group C rotavirus in Buenos Aires, Argentina, 1997–2003. *J Med Virol* 81:1109–1116.
34. Jiang B, Dennehy PH, Spangenberger S, Gentsch JR, Glass RI. 1995. First detection of group C rotavirus in fecal specimens of children with diarrhea in the United States. *J Infect Dis* 172:45–50.
35. Kindler E, Trojnar E, Heckel G, Otto PH, Johne R. 2012. Analysis of rotavirus species diversity and evolution including the newly determined full-length genome sequences of rotavirus F and G. *Infect Genet Evol* 14C:58–67.
36. Molinari BL, Lorenzetti E, Otonel RA, Alfieri AF, Alfieri AA. 2014. Species H rotavirus detected in piglets with diarrhea, Brazil, 2012. *Emerg Infect Dis* 20:1019–1022.
37. Iturriza Gómara M, Wong C, Blome S, Desselberger U, Gray J. 2002. Molecular characterization of VP6 genes of human rotavirus isolates: Correlation of genogroups with subgroups and evidence of independent segregation. *J Virol* 76:6596–6601.
38. Matthijnssens J, Ciarlet M, Rahman M, Attoui H, Banyai K, Estes MK, Gentsch JR, et al. 2008. Recommendations for the classification of group A rotaviruses using all 11 genomic RNA segments. *Arch Virol* 153:1621–1629.
39. Trojnar E, Sachsenroder J, Twardziok S, Reetz J, Otto PH, Johne R. 2013. Identification of an avian group A rotavirus containing a novel VP4 gene with a close relationship to those of mammalian rotaviruses. *J Gen Virol* 94:136–142.
40. Guo D, Liu J, Lu Y, Sun Y, Yuan D, Jiang Q, Lin H, Li C, Si C, Qu L. 2012. Full genomic analysis of rabbit rotavirus G3P[14] strain N5 in China: Identification of a novel VP6 genotype. *Infect Genet Evol* 12:1567–1576.
41. Papp H, Al-Mutairi LZ, Chehadeh W, Farkas SL, Lengyel G, Jakab F, Martella V, Szucs G, Banyai K. 2012. Novel NSP4 genotype in a camel G10P[15] rotavirus strain. *Acta Microbiol Immunol Hung* 59:411–421.
42. Maes P, Matthijnssens J, Rahman M, Van Ranst M. 2009. RotaC: A web-based tool for the complete genome classification of group A rotaviruses. *BMC Microbiol* 9:238.
43. Bányai K, László B, Duque J, Steele AD, Nelson EAS, Gentsch JR, Parashar UD. 2012. Systematic review of regional and temporal trends in global rotavirus strain diversity in the pre rotavirus vaccine era: Insights for understanding the impact of rotavirus vaccination programs. *Vaccine* 30 Suppl 1:A122–A130.
44. Santos N, Hoshino Y. 2005. Global distribution of rotavirus serotypes/genotypes and its implication for the development and implementation of an effective rotavirus vaccine. *Rev Med Virol* 15:29–56.
45. Gentsch JR, Laird AR, Bielfelt B, Griffin DD, Banyai K, Ramachandran M, Jain V, et al. 2005. Serotype diversity and reassortment between human and animal rotavirus strains: Implications for rotavirus vaccine programs. *J Infect Dis* 192 Suppl 1:S146–159.
46. Kobayashi N, Ishino M, Wang HY, Chawla-Sarkar M, Krishnan T, Naik TN. 2007. Diversity of G-type and P-type of human and animal rotaviruses and its genetic background. *In* Mendez-Vilas A (ed), *Communicating Current Research and Educational Topics and Trends in Applied Microbiology,* Vol 1. Formatex, Badajoz.
47. Velazquez FR, Matson DO, Calva JJ, Guerrero L, Morrow AL, Carter-Campbell S, Glass RI, Estes MK, Pickering LK, Ruiz-Palacios GM. 1996. Rotavirus infections in infants as protection against subsequent infections. *N Engl J Med* 335:1022–1028.

48. Parashar UD, Gibson CJ, Bresee JS, Glass RI. 2006. Rotavirus and severe childhood diarrhea. *Emerg Infect Dis* 12:304–306.
49. Tate JE, Burton AH, Boschi-Pinto C, Steele AD, Duque J, Parashar UD. 2012. 2008 estimate of worldwide rotavirus-associated mortality in children younger than 5 years before the introduction of universal rotavirus vaccination programmes: A systematic review and meta-analysis. *Lancet Infect Dis* 12:136–141.
50. Dagan R, Bar-David Y, Sarov B, Katz M, Kassis I, Greenberg D, Glass RI, Margolis CZ, Sarov I. 1990. Rotavirus diarrhea in Jewish and Bedouin children in the Negev region of Israel: Epidemiology, clinical aspects and possible role of malnutrition in severity of illness. *Pediatr Infect Dis J* 9:314–321.
51. Dennehy PH. 2012. Effects of vaccine on rotavirus disease in the pediatric population. *Curr Opin Pediatr* 24:76–84.
52. Widdowson MA, Meltzer MI, Zhang X, Bresee JS, Parashar UD, Glass RI. 2007. Cost-effectiveness and potential impact of rotavirus vaccination in the United States. *Pediatrics* 119:684–697.
53. Atchison CJ, Tam CC, Hajat S, van Pelt W, Cowden JM, Lopman BA. 2010. Temperature-dependent transmission of rotavirus in Great Britain and The Netherlands. *Proc Biol Sci* 277:933–942.
54. Cook SM, Glass RI, LeBaron CW, Ho MS. 1990. Global seasonality of rotavirus infections. *Bull WHO* 68:171–177.
55. Levy K, Hubbard AE, Eisenberg JN. 2009. Seasonality of rotavirus disease in the tropics: A systematic review and meta-analysis. *Int J Epidemiol* 38:1487–1496.
56. Jagai JS, Sarkar R, Castronovo D, Kattula D, McEntee J, Ward H, Kang G, Naumova EN. 2012. Seasonality of rotavirus in South Asia: A meta-analysis approach assessing associations with temperature, precipitation, and vegetation index. *PLoS One* 7:e38168.
57. Kovacs A, Chan L, Hotrakitya C, Overturf G, Portnoy B. 1987. Rotavirus gastroenteritis: Clinical and laboratory features and use of the Rotazyme test. *Am J Dis Child* 141:161–166.
58. Lee RM, Lessler J, Lee RA, Rudolph KE, Reich NG, Perl TM, Cummings DA. 2013. Incubation periods of viral gastroenteritis: A systematic review. *BMC Infect Dis* 13:446.
59. Anderson EJ, Weber SG. 2004. Rotavirus infection in adults. *Lancet Infect Dis* 4:91–99.
60. Rodriguez WJ, Kim HW, Brandt CD, Schwartz RH, Gardner MK, Jeffries B, Parrott RH, Kaslow RA, Smith JI, Kapikian AZ. 1987. Longitudinal study of rotavirus infection and gastroenteritis in families served by a pediatric medical practice: Clinical and epidemiologic observations. *Pediatr Infect Dis J* 6:170–176.
61. Raul Velazquez F, Calva JJ, Lourdes Guerrero M, Mass D, Glass RI, Pickering LK, Ruiz-Palacios GM. 1993. Cohort study of rotavirus serotype patterns in symptomatic and asymptomatic infections in Mexican children. *Pediatr Infect Dis J* 12:54–61.
62. Carlson JA, Middleton PJ, Szymanski MT, Huber J, Petric M. 1978. Fatal rotavirus gastroenteritis: An analysis of 21 cases. *Am J Dis Child* 132:477–479.
63. Hemming M, Huhti L, Rasanen S, Salminen M, Vesikari T. 2014. Rotavirus antigenemia in children is associated with more severe clinical manifestations of acute gastroenteritis. *Pediatr Infect Dis J* 33:366–371.
64. Grimwood K, Lambert SB. 2009. Rotavirus vaccines: Opportunities and challenges. *Hum Vaccin* 5:57–69.
65. Blutt SE, Matson DO, Crawford SE, Staat MA, Azimi P, Bennett BL, Piedra PA, Conner ME. 2007. Rotavirus antigenemia in children is associated with viremia. *PLoS Med* 4:e121.
66. Gilger MA, Matson DO, Conner ME, Rosenblatt HM, Finegold MJ, Estes MK. 1992. Extraintestinal rotavirus infections in children with immunodeficiency. *J Pediatr* 120:912–917.
67. Wood DJ, David TJ, Chrystie IL, Totterdell B. 1988. Chronic enteric virus infection in two T-cell immunodeficient children. *J Med Virol* 24:435–444.
68. Saulsbury FT, Winkelstein JA, Yolken RH. 1980. Chronic rotavirus infection in immunodeficiency. *J Pediatr* 97:61–65.
69. Oishi I, Kimura T, Murakami T, Haruki K, Yamazaki K, Seto Y, Minekawa Y, Funamoto H. 1991. Serial observations of chronic rotavirus infection in an immunodeficient child. *Microbiol Immunol* 35:953–961.
70. Yolken RH, Bishop CA, Townsend TR, Bolyard EA, Bartlett J, Santos GW, Saral R. 1982. Infectious gastroenteritis in bone-marrow-transplant recipients. *N Engl J Med* 306:1010–1012.
71. Peigue-Lafeuille H, Henquell C, Chambon M, Gazuy N, De Champs C, Cluzel R. 1991. Nosocomial rotavirus infections in adult renal transplant recipients. *J Hosp Infect* 18:67–70.
72. Thomas PD, Pollok RC, Gazzard BG. 1999. Enteric viral infections as a cause of diarrhoea in the acquired immunodeficiency syndrome. *HIV Med* 1:19–24.

73. Zheng BJ, Chang RX, Ma GZ, Xie JM, Liu Q, Liang XR, Ng MH. 1991. Rotavirus infection of the oropharynx and respiratory tract in young children. *J Med Virol* 34:29–37.
74. Rotbart HA, Nelson WL, Glode MP, Triffon TC, Kogut SJ, Yolken RH, Hernandez JA, Levin MJ. 1988. Neonatal rotavirus-associated necrotizing enterocolitis: Case control study and prospective surveillance during an outbreak. *J Pediatr* 112:87–93.
75. Capitanio MA, Greenberg SB. 1991. Pneumatosis intestinalis in two infants with rotavirus gastroenteritis. *Pediatr Radiol* 21:361–362.
76. Grunow JE, Dunton SF, Waner JL. 1985. Human rotavirus-like particles in a hepatic abscess. *J Pediatr* 106:73–76.
77. Cioc AM, Nuovo GJ. 2002. Histologic and *in situ* viral findings in the myocardium in cases of sudden, unexpected death. *Mod Pathol* 15:914–922.
78. Lynch M, Lee B, Azimi P, Gentsch J, Glaser C, Gilliam S, Chang HG, Ward R, Glass RI. 2001. Rotavirus and central nervous system symptoms: Cause or contaminant? Case reports and review. *Clin Infect Dis* 33:932–938.
79. Riepenhoff-Talty M, Gouvea V, Evans MJ, Svensson L, Hoffenberg E, Sokol RJ, Uhnoo I, et al. 1996. Detection of group C rotavirus in infants with extrahepatic biliary atresia. *J Infect Dis* 174:8–15.
80. Allen SR, Jafri M, Donnelly B, McNeal M, Witte D, Bezerra J, Ward R, Tiao GM. 2007. Effect of rotavirus strain on the murine model of biliary atresia. *J Virol* 81:1671–1679.
81. Rauschenfels S, Krassmann M, Al-Masri AN, Verhagen W, Leonhardt J, Kuebler JF, Petersen C. 2009. Incidence of hepatotropic viruses in biliary atresia. *Eur J Pediatr* 168:469–476.
82. Poulton J, Tarlow MJ. 1987. Diagnosis of rotavirus gastroenteritis by smell. *Arch Dis Child* 62:851–852.
83. Wilde J, Yolken R, Willoughby R, Eiden J. 1991. Improved detection of rotavirus shedding by polymerase chain reaction. *Lancet* 337:323–326.
84. Pang XL, Lee B, Boroumand N, Leblanc B, Preiksaitis JK, Yu Ip CC. 2004. Increased detection of rotavirus using a real time reverse transcription-polymerase chain reaction (RT-PCR) assay in stool specimens from children with diarrhea. *J Med Virol* 72:496–501.
85. Honma S, Chizhikov V, Santos N, Tatsumi M, Timenetsky Mdo C, Linhares AC, Mascarenhas JD, et al. 2007. Development and validation of DNA microarray for genotyping group A rotavirus VP4 (P[4], P[6], P[8], P[9], and P[14]) and VP7 (G1 to G6, G8 to G10, and G12) genes. *J Clin Microbiol* 45:2641–2648.
86. Chizhikov V, Wagner M, Ivshina A, Hoshino Y, Kapikian AZ, Chumakov K. 2002. Detection and genotyping of human group A rotaviruses by oligonucleotide microarray hybridization. *J Clin Microbiol* 40:2398–2407.
87. Ball JM, Tian P, Zeng CQ, Morris AP, Estes MK. 1996. Age-dependent diarrhea induced by a rotaviral nonstructural glycoprotein. *Science* 272:101–104.
88. Lundgren O, Svensson L. 2001. Pathogenesis of rotavirus diarrhea. *Microbes Infect* 3:1145–1156.
89. Ramig RF. 2004. Pathogenesis of intestinal and systemic rotavirus infection. *J Virol* 78:10213–10220.
90. Lundgren O, Peregrin AT, Persson K, Kordasti S, Uhnoo I, Svensson L. 2000. Role of the enteric nervous system in the fluid and electrolyte secretion of rotavirus diarrhea. *Science* 287:491–495.
91. Hågbom M, Istrate C, Engblom D, Karlsson T, Rodriguez-Diaz J, Buesa J, Taylor JA, et al. 2011. Rotavirus stimulates release of serotonin (5-HT) from human enterochromaffin cells and activates brain structures involved in nausea and vomiting. *PLoS Pathog* 7:e1002115.
92. Davidson GP, Barnes GL. 1979. Structural and functional abnormalities of the small intestine in infants and young children with rotavirus enteritis. *Acta Paediatr Scand* 68:181–186.
93. Holmes IH, Ruck BJ, Bishop RF, Davidson GP. 1975. Infantile enteritis viruses: Morphogenesis and morphology. *J Virol* 16:937–943.
94. Davidson GP, Gall DG, Petric M, Butler DG, Hamilton JR. 1977. Human rotavirus enteritis induced in conventional piglets: Intestinal structure and transport. *J Clin Invest* 60:1402–1409.
95. Zhang M, Zeng CQ, Morris AP, Estes MK. 2000. A functional NSP4 enterotoxin peptide secreted from rotavirus-infected cells. *J Virol* 74:11663–11670.
96. Halaihel N, Lievin V, Ball JM, Estes MK, Alvarado F, Vasseur M. 2000. Direct inhibitory effect of rotavirus NSP4(114-135) peptide on the Na(+)-D-glucose symporter of rabbit intestinal brush border membrane. *J Virol* 74:9464–9470.
97. Dong Y, Zeng CQ, Ball JM, Estes MK, Morris AP. 1997. The rotavirus enterotoxin NSP4 mobilizes intracellular calcium in human intestinal cells by stimulating phospholipase C-mediated inositol 1,4,5-trisphosphate production. *Proc Natl Acad Sci USA* 94:3960–3965.

98. Morris AP, Scott JK, Ball JM, Zeng CQ, O'Neal WK, Estes MK. 1999. NSP4 elicits age-dependent diarrhea and Ca(2+)mediated I(−) influx into intestinal crypts of CF mice. *Am J Physiol* 277:G431–444.
99. King CK, Glass R, Bresee JS, Duggan C. 2003. Managing acute gastroenteritis among children: Oral rehydration, maintenance, and nutritional therapy. *MMWR* 52:1–16.
100. Santosham M, Daum RS, Dillman L, Rodriguez JL, Luque S, Russell R, Kourany M, et al. 1982. Oral rehydration therapy of infantile diarrhea: A controlled study of well-nourished children hospitalized in the United States and Panama. *N Engl J Med* 306:1070–1076.
101. Bhutta ZA, Bird SM, Black RE, Brown KH, Gardner JM, Hidayat A, Khatun F, et al. 2000. Therapeutic effects of oral zinc in acute and persistent diarrhea in children in developing countries: Pooled analysis of randomized controlled trials. *Am J Clin Nutr* 72:1516–1522.
102. Bhutta ZA, Black RE, Brown KH, Gardner JM, Gore S, Hidayat A, Khatun F, et al. 1999. Prevention of diarrhea and pneumonia by zinc supplementation in children in developing countries: Pooled analysis of randomized controlled trials—Zinc Investigators' Collaborative Group. *J Pediatr* 135:689–697.
103. Guarino A, Guandalini S, Albano F, Mascia A, De Ritis G, Rubino A. 1991. Enteral immunoglobulins for treatment of protracted rotaviral diarrhea. *Pediatr Infect Dis J* 10:612–614.
104. Van Niel CW, Feudtner C, Garrison MM, Christakis DA. 2002. Lactobacillus therapy for acute infectious diarrhea in children: A meta-analysis. *Pediatrics* 109:678–684.
105. Sindhu KN, Sowmyanarayanan TV, Paul A, Babji S, Ajjampur SS, Priyadarshini S, Sarkar R, et al. 2014. Immune response and intestinal permeability in children with acute gastroenteritis treated with Lactobacillus rhamnosus GG: A randomized, double-blind, placebo-controlled trial. *Clin Infect Dis* 58(8):1107–1115. doi:10.1093/cid/ciu065.
106. Kang G. 2013. New-generation treatment? Targeted antiviral therapy for rotavirus. *Gastroenterology* 145:711–714.
107. Ward RL, Bernstein DI, Young EC, Sherwood JR, Knowlton DR, Schiff GM. 1986. Human rotavirus studies in volunteers: Determination of infectious dose and serological response to infection. *J Infect Dis* 154:871–880.
108. Vollet JJ, III, DuPont HL, Pickering LK. 1981. Nonenteric sources of rotavirus in acute diarrhea. *J Infect Dis* 144:495.
109. Adams WR, Kraft LM. 1963. Epizootic diarrhea of infant mice: Indentification of the etiologic agent. *Science* 141:359–360.
110. Le Guyader FS, Le Saux JC, Ambert-Balay K, Krol J, Serais O, Parnaudeau S, et al. 2008. Aichi virus, norovirus, astrovirus, enterovirus, and rotavirus involved in clinical cases from a French oyster-related gastroenteritis outbreak. *J Clin Microbiol* 46:4011–4017.
111. Le Guyader F, Dubois E, Menard D, Pommepuy M. 1994. Detection of hepatitis A virus, rotavirus, and enterovirus in naturally contaminated shellfish and sediment by reverse transcription-seminested PCR. *Appl Environ Microbiol* 60:3665–3671.
112. Kittigul L, Panjangampatthana A, Rupprom K, Pombubpa K. 2014. Genetic diversity of rotavirus strains circulating in environmental water and bivalve shellfish in Thailand. *Int J Environ Res Public Health* 11:1299–1311.
113. Mayr C, Strohe G, Contzen M. 2009. Detection of rotavirus in food associated with a gastroenteritis outbreak in a mother and child sanatorium. *Int J Food Microbiol* 135:179–182.
114. Hung T, Chen GM, Wang CG, Yao HL, Fang ZY, Chao TX, Chou ZY, et al. 1984. Waterborne outbreak of rotavirus diarrhoea in adults in China caused by a novel rotavirus. *Lancet* 1:1139–1142.
115. Mead PS, Slutsker L, Dietz V, McCaig LF, Bresee JS, Shapiro C, Griffin PM, Tauxe RV. 1999. Food-related illness and death in the United States. *Emerg Infect Dis* 5:607–625.
116. Black RE. 1990. Epidemiology of travelers' diarrhea and relative importance of various pathogens. *Rev Infect Dis* 12 Suppl 1:S73–S79.
117. Ansari SA, Springthorpe VS, Sattar SA. 1991. Survival and vehicular spread of human rotaviruses: Possible relation to seasonality of outbreaks. *Rev Infect Dis* 13:448–461.
118. Lloyd-Evans N, Springthorpe VS, Sattar SA. 1986. Chemical disinfection of human rotavirus-contaminated inanimate surfaces. *J Hyg (Lond)* 97:163–173.
119. Abad FX, Pinto RM, Bosch A. 1994. Survival of enteric viruses on environmental fomites. *Appl Environ Microbiol* 60:3704–3710.
120. Richardson S, Grimwood K, Gorrell R, Palombo E, Barnes G, Bishop R. 1998. Extended excretion of rotavirus after severe diarrhoea in young children. *Lancet* 351:1844–1848.
121. Mukhopadhya I, Sarkar R, Menon VK, Babji S, Paul A, Rajendran P, Sowmyanarayanan TV, et al. 2013. Rotavirus shedding in symptomatic and asymptomatic children using reverse transcription-quantitative PCR. *J Med Virol* 85:1661–1668.

122. Kang G, Iturriza-Gomara M, Wheeler JG, Crystal P, Monica B, Ramani S, Primrose B, et al. 2004. Quantitation of group A rotavirus by real-time reverse-transcription-polymerase chain reaction: Correlation with clinical severity in children in South India. *J Med Virol* 73:118–122.
123. Grimwood K, Abbott GD, Fergusson DM, Jennings LC, Allan JM. 1983. Spread of rotavirus within families: A community based study. *BMJ* 287:575–577.
124. Kim HW, Brandt CD, Kapikian AZ, Wyatt RG, Arrobio JO, Rodriguez WJ, Chanock RM, Parrott RH. 1977. Human reovirus-like agent infection: Occurrence in adult contacts of pediatric patients with gastroenteritis. *JAMA* 238:404–407.
125. Svenungsson B, Lagergren A, Ekwall E, Evengard B, Hedlund KO, Karnell A, Lofdahl S, Svensson L, Weintraub A. 2000. Enteropathogens in adult patients with diarrhea and healthy control subjects: A 1-year prospective study in a Swedish clinic for infectious diseases. *Clin Infect Dis* 30:770–778.
126. Nakajima H, Nakagomi T, Kamisawa T, Sakaki N, Muramoto K, Mikami T, Nara H, Nakagomi O. 2001. Winter seasonality and rotavirus diarrhoea in adults. *Lancet* 357:1950.
127. Friesema IH, De Boer RF, Duizer E, Kortbeek LM, Notermans DW, Smeulders A, Bogerman J, et al. 2012. Aetiology of acute gastroenteritis in adults requiring hospitalization in The Netherlands. *Epidemiol Infect* 140:1780–1786.
128. del Refugio Gonzalez-Losa M, Polanco-Marin GG, Manzano-Cabrera L, Puerto-Solis M. 2001. Acute gastroenteritis associated with rotavirus in adults. *Arch Med Res* 32:164–167.
129. Oyofo BA, Subekti D, Tjaniadi P, Machpud N, Komalarini S, Setiawan B, Simanjuntak C, et al. 2002. Enteropathogens associated with acute diarrhea in community and hospital patients in Jakarta, Indonesia. *FEMS Immunol Med Microbiol* 34:139–146.
130. Lopman BA, Curns AT, Yen C, Parashar UD. 2011. Infant rotavirus vaccination may provide indirect protection to older children and adults in the United States. *J Infect Dis* 204:980–986.
131. Gastanaduy PA, Hall AJ, Curns AT, Parashar UD, Lopman BA. 2013. Burden of norovirus gastroenteritis in the ambulatory setting—United States, 2001–2009. *J Infect Dis* 207:1058–1065.
132. Lopman BA, Hall AJ, Curns AT, Parashar UD. 2011. Increasing rates of gastroenteritis hospital discharges in US adults and the contribution of norovirus, 1996–2007. *Clin Infect Dis* 52:466–474.
133. Awachat PS, Kelkar SD. 2006. Dual infection due to simian G3: Human reassortant and human G9 strains of rotavirus in a child and subsequent spread of serotype G9, leading to diarrhea among grandparents. *J Med Virol* 78:134–138.
134. Anderson EJ, Katz BZ, Polin JA, Reddy S, Weinrobe MH, Noskin GA. 2012. Rotavirus in adults requiring hospitalization. *J Infect* 64:89–95.
135. Wang Y-H, Kobayashi N, Zhou X, Nagashima S, Zhu Z-R, Peng J-S, Liu M-Q, et al. 2009. Phylogenetic analysis of rotaviruses with predominant G3 and emerging G9 genotypes from adults and children in Wuhan, China. *J Med Virol* 81:382–389.
136. Ianiro G, Delogu R, Bonomo P, Fiore L, Ruggeri FM. 2014. Molecular analysis of group A rotaviruses detected in adults and adolescents with severe acute gastroenteritis in Italy in 2012. *J Med Virol* 86(6):1073–1082. doi:10.1002/jmv.23871.
137. Desselberger U, Huppertz HI. 2011. Immune responses to rotavirus infection and vaccination and associated correlates of protection. *J Infect Dis* 203:188–195.
138. Bishop RF, Barnes GL, Cipriani E, Lund JS. 1983. Clinical immunity after neonatal rotavirus infection: A prospective longitudinal study in young children. *N Engl J Med* 309:72–76.
139. Heaton PM, Goveia MG, Miller JM. 2005. Development of a pentavalent rotavirus vaccine against prevalent serotypes of rotavirus gastroenteritis. *J Infect Dis* 192:S17–S21.
140. Gladstone BP, Ramani S, Mukhopadhya I, Muliyil J, Sarkar R, Rehman AM, Jaffar S, et al. 2011. Protective effect of natural rotavirus infection in an Indian birth cohort. *N Engl J Med* 365:337–346.
141. Offit PA. 1996. Host factors associated with protection against rotavirus disease: The skies are clearing. *J Infect Dis* 174 Suppl 1:S59–S64.
142. Aiyegbo MS, Sapparapu G, Spiller BW, Eli IM, Williams DR, Kim R, Lee DE, et al. 2013. Human rotavirus VP6-specific antibodies mediate intracellular neutralization by binding to a quaternary structure in the transcriptional pore. *PLoS One* 8:e61101.
143. Uzri D, Greenberg HB. 2013. Characterization of rotavirus RNAs that activate innate immune signaling through the RIG-I-like receptors. *PLoS One* 8:e69825.
144. De Boissieu D, Lebon P, Badoual J, Bompard Y, Dupont C. 1993. Rotavirus induces alpha-interferon release in children with gastroenteritis. *J Pediatr Gastroenterol Nutr* 16:29–32.
145. Kutukculer N, Caglayan S. 1997. Tumor necrosis factor-alpha and interleukin-6 in stools of children with bacterial and viral gastroenteritis. *J Pediatr Gastroenterol Nutr* 25:556–557.

146. Yasukawa M, Nakagomi O, Kobayashi Y. 1990. Rotavirus induces proliferative response and augments non-specific cytotoxic activity of lymphocytes in humans. *Clin Exp Immunol* 80:49–55.
147. Arnold MM, Patton JT. 2011. Diversity of interferon antagonist activities mediated by NSP1 proteins of different rotavirus strains. *J Virol* 85:1970–1979.
148. Barro M, Patton JT. 2007. Rotavirus NSP1 inhibits expression of type I interferon by antagonizing the function of interferon regulatory factors IRF3, IRF5, and IRF7. *J Virol* 81:4473–4481.
149. Barro M, Patton JT. 2005. Rotavirus nonstructural protein 1 subverts innate immune response by inducing degradation of IFN regulatory factor 3. *Proc Natl Acad Sci USA* 102:4114–4119.
150. Flewett TH, Bryden AS, Davies H, Woode GN, Bridger JC, Derrick JM. 1974. Relation between viruses from acute gastroenteritis of children and newborn calves. *Lancet* 2:61–63.
151. Settembre EC, Chen JZ, Dormitzer PR, Grigorieff N, Harrison SC. 2011. Atomic model of an infectious rotavirus particle. *EMBO J* 30:408–416.
152. Li Z, Baker ML, Jiang W, Estes MK, Prasad BV. 2009. Rotavirus architecture at subnanometer resolution. *J Virol* 83:1754–1766.
153. McClain B, Settembre E, Temple BR, Bellamy AR, Harrison SC. 2010. X-ray crystal structure of the rotavirus inner capsid particle at 3.8 A resolution. *J Mol Biol* 397:587–599.
154. Zhang X, Settembre E, Xu C, Dormitzer PR, Bellamy R, Harrison SC, Grigorieff N. 2008. Near-atomic resolution using electron cryomicroscopy and single-particle reconstruction. *Proc Natl Acad Sci USA* 105:1867–1872.
155. Trask SD, McDonald SM, Patton JT. 2012. Structural insights into the coupling of virion assembly and rotavirus replication. *Nat Rev Microbiol* 10:165–177.
156. Prasad BV, Rothnagel R, Zeng CQ, Jakana J, Lawton JA, Chiu W, Estes MK. 1996. Visualization of ordered genomic RNA and localization of transcriptional complexes in rotavirus. *Nature* 382:471–473.
157. Taniguchi K, Urasawa S. 1995. Diversity in rotavirus genomes. *Semin Virol* 6:123–131.
158. Greenberg HB, Estes MK. 2009. Rotaviruses: From pathogenesis to vaccination. *Gastroenterology* 136:1939–1951.
159. Estes MK, Graham DY, Mason BB. 1981. Proteolytic enhancement of rotavirus infectivity: Molecular mechanisms. *J Virol* 39:879–888.
160. Ciarlet M, Ludert JE, Iturriza-Gomara M, Liprandi F, Gray JJ, Desselberger U, Estes MK. 2002. Initial interaction of rotavirus strains with N-acetylneuraminic (sialic) acid residues on the cell surface correlates with VP4 genotype, not species of origin. *J Virol* 76:4087–4095.
161. Dormitzer PR, Sun ZY, Blixt O, Paulson JC, Wagner G, Harrison SC. 2002. Specificity and affinity of sialic acid binding by the rhesus rotavirus VP8* core. *J Virol* 76:10512–10517.
162. Ciarlet M, Estes MK. 1999. Human and most animal rotavirus strains do not require the presence of sialic acid on the cell surface for efficient infectivity. *J Gen Virol* 80 (Pt 4):943–948.
163. Mendez E, Arias CF, Lopez S. 1993. Binding to sialic acids is not an essential step for the entry of animal rotaviruses to epithelial cells in culture. *J Virol* 67:5253–5259.
164. Zarate S, Espinosa R, Romero P, Mendez E, Arias CF, Lopez S. 2000. The VP5 domain of VP4 can mediate attachment of rotaviruses to cells. *J Virol* 74:593–599.
165. Huang P, Xia M, Tan M, Zhong W, Wei C, Wang L, Morrow A, Jiang X. 2012. Spike protein VP8* of human rotavirus recognizes histo-blood group antigens in a type-specific manner. *J Virol* 86:4833–4843.
166. Hu L, Crawford SE, Czako R, Cortes-Penfield NW, Smith DF, Le Pendu J, Estes MK, Prasad BVV. 2012. Cell attachment protein VP8* of a human rotavirus specifically interacts with A-type histo-blood group antigen. *Nature* 485:256–259.
167. Liu Y, Huang P, Tan M, Biesiada J, Meller J, Castello AA, Jiang B, Jiang X. 2012. Rotavirus VP8*: Phylogeny, host range, and interaction with histo-blood group antigens. *J Virol* 86:9899–9910.
168. Ramani S, Cortes-Penfield NW, Hu L, Crawford SE, Czako R, Smith DF, Kang G, et al. 2013. The VP8* domain of neonatal rotavirus strain G10P[11] binds to type II precursor glycans. *J Virol* 87:7255–7264.
169. Imbert-Marcille BM, Barbe L, Dupe M, Le Moullac-Vaidye B, Besse B, Peltier C, Ruvoen-Clouet N, Le Pendu J. 2014. A FUT2 gene common polymorphism determines resistance to rotavirus A of the P[8] genotype. *J Infect Dis* 209:1227–1230.
170. Estes MK, Greenberg H. 2013. *Rotaviruses*, 6th edn., vol 2. Lippincott Williams & Wilkins, Philadelphia, PA.
171. Patton JT, Silvestri LS, Tortorici MA, Vasquez-Del Carpio R, Taraporewala ZF. 2006. Rotavirus genome replication and morphogenesis: Role of the viroplasm. *Curr Top Microbiol Immunol* 309:169–187.
172. Kim IS, Trask SD, Babyonyshev M, Dormitzer PR, Harrison SC. 2010. Effect of mutations in VP5 hydrophobic loops on rotavirus cell entry. *J Virol* 84:6200–6207.

173. Tihova M, Dryden KA, Bellamy AR, Greenberg HB, Yeager M. 2001. Localization of membrane permeabilization and receptor binding sites on the VP4 hemagglutinin of rotavirus: Implications for cell entry. *J Mol Biol* 314:985–992.
174. Lawton JA, Estes MK, Prasad BV. 2000. Mechanism of genome transcription in segmented dsRNA viruses. *Adv Virus Res* 55:185–229.
175. Lu X, McDonald SM, Tortorici MA, Tao YJ, Vasquez-Del Carpio R, Nibert ML, Patton JT, Harrison SC. 2008. Mechanism for coordinated RNA packaging and genome replication by rotavirus polymerase VP1. *Structure* 16:1678–1688.
176. Au KS, Mattion NM, Estes MK. 1993. A subviral particle binding domain on the rotavirus nonstructural glycoprotein NS28. *Virology* 194:665–673.
177. Stirzaker SC, Poncet D, Both GW. 1990. Sequences in rotavirus glycoprotein VP7 that mediate delayed translocation and retention of the protein in the endoplasmic reticulum. *J Cell Biol* 111:1343–1350.
178. Musalem C, Espejo RT. 1985. Release of progeny virus from cells infected with simian rotavirus SA11. *J Gen Virol* 66 (Pt 12):2715–2724.
179. Jenkins GM, Rambaut A, Pybus OG, Holmes EC. 2002. Rates of molecular evolution in RNA viruses: A quantitative phylogenetic analysis. *J Mol Evol* 54:156–165.
180. Blackhall J, Fuentes A, Magnusson G. 1996. Genetic stability of a porcine rotavirus RNA segment during repeated plaque isolation. *Virology* 225:181–190.
181. Matthijnssens J, Heylen E, Zeller M, Rahman M, Lemey P, Van Ranst M. 2010. Phylodynamic analyses of rotavirus genotypes G9 and G12 underscore their potential for swift global spread. *Mol Biol Evol* 27:2431–2436.
182. Trang NV, Yamashiro T, Anh le TK, Hau VT, Luan le T, Anh DD. 2012. Genetic variation in the VP7 gene of rotavirus G1P[8] strains isolated in Vietnam, 1998–2009. *Virus Res* 165:190–196.
183. Nagaoka Y, Tatsumi M, Tsugawa T, Yoto Y, Tsutsumi H. 2012. Phylogenetic and computational structural analysis of VP7 gene of group a human rotavirus G1P[8] strains obtained in Sapporo, Japan from 1987 to 2000. *J Med Virol* 84:832–838.
184. Yang JH, Kobayashi N, Wang YH, Zhou X, Li Y, Zhou DJ, Hu ZH, et al. 2004. Phylogenetic analysis of a human group B rotavirus WH-1 detected in China in 2002. *J Med Virol* 74:662–667.
185. Rahman M, Hassan ZM, Zafrul H, Saiada F, Banik S, Faruque AS, Delbeke T, Matthijnssens J, Van Ranst M, Azim T. 2007. Sequence analysis and evolution of group B rotaviruses. *Virus Res* 125:219–225.
186. Donker NC, Kirkwood CD. 2012. Selection and evolutionary analysis in the nonstructural protein NSP2 of rotavirus A. *Infect Genet Evol* 12:1355–1361.
187. Desselberger U. 1996. Genome rearrangements of rotaviruses. *Arch Virol Suppl* 12:37–51.
188. Pedley S, Hundley F, Chrystie I, McCrae MA, Desselberger U. 1984. The genomes of rotaviruses isolated from chronically infected immunodeficient children. *J Gen Virol* 65:1141–1150.
189. Hundley F, Biryahwaho B, Gow M, Desselberger U. 1985. Genome rearrangements of bovine rotavirus after serial passage at high multiplicity of infection. *Virology* 143:88–103.
190. Patton JT, Vasquez-Del Carpio R, Tortorici MA, Taraporewala ZF. 2007. Coupling of rotavirus genome replication and capsid assembly. *Adv Virus Res* 69:167–201.
191. Zeller M, Heylen E, De Coster S, Van Ranst M, Matthijnssens J. 2012. Full genome characterization of a porcine-like human G9P[6] rotavirus strain isolated from an infant in Belgium. *Infect Genet Evol* 12:1492–1500.
192. Taniguchi K, Kojima K, Urasawa S. 1996. Nondefective rotavirus mutants with an NSP1 gene which has a deletion of 500 nucleotides, including a cysteine-rich zinc finger motif-encoding region (nucleotides 156 to 248), or which has a nonsense codon at nucleotides 153–155. *J Virol* 70:4125–4130.
193. Matthijnssens J, Rahman M, Van Ranst M. 2006. Loop model: Mechanism to explain partial gene duplications in segmented dsRNA viruses. *Biochem Biophys Res Commun* 340:140–144.
194. Holmes EC. 2013. *Virus Evolution*, 6th edn., vol 1. Lippincott Williams & Wilkins, Philadelphia, PA.
195. Parra GI, Bok K, Martinez M, Gomez JA. 2004. Evidence of rotavirus intragenic recombination between two sublineages of the same genotype. *J Gen Virol* 85:1713–1716.
196. Phan TG, Okitsu S, Maneekarn N, Ushijima H. 2007. Evidence of intragenic recombination in G1 rotavirus VP7 genes. *J Virol* 81:10188–10194.
197. Martinez-Laso J, Roman A, Head J, Cervera I, Rodriguez M, Rodriguez-Avial I, Picazo JJ. 2009. Phylogeny of G9 rotavirus genotype: A possible explanation of its origin and evolution. *J Clin Virol* 44:52–57.
198. Phan TG, Okitsu S, Maneekarn N, Ushijima H. 2007. Genetic heterogeneity, evolution and recombination in emerging G9 rotaviruses. *Infect Genet Evol* 7:656–663.

199. Suzuki Y, Gojobori T, Nakagomi O. 1998. Intragenic recombinations in rotaviruses. *FEBS letters* 427:183–187.
200. Martinez-Laso J, Roman A, Rodriguez M, Cervera I, Head J, Rodriguez-Avial I, Picazo JJ. 2009. Diversity of the G3 genes of human rotaviruses in isolates from Spain from 2004 to 2006: Cross-species transmission and inter-genotype recombination generates alleles. *J Gen Virol* 90:935–943.
201. Jere KC, Mlera L, Page NA, van Dijk AA, O'Neill HG. 2011. Whole genome analysis of multiple rotavirus strains from a single stool specimen using sequence-independent amplification and 454(R) pyrosequencing reveals evidence of intergenotype genome segment recombination. *Infect Genet Evol* 11:2072–2082.
202. Donker NC, Boniface K, Kirkwood CD. 2011. Phylogenetic analysis of rotavirus A NSP2 gene sequences and evidence of intragenic recombination. *Infect Genet Evol* 11:1602–1607.
203. Cao D, Barro M, Hoshino Y. 2008. Porcine rotavirus bearing an aberrant gene stemming from an intergenic recombination of the NSP2 and NSP5 genes is defective and interfering. *J Virol* 82:6073–6077.
204. Worobey M, Holmes EC. 1999. Evolutionary aspects of recombination in RNA viruses. *J Gen Virol* 80 (Pt 10):2535–2543.
205. Ramig RF, Ward RL. 1991. Genomic segment reassortment in rotaviruses and other reoviridae. *Adv Virus Res* 39:163–207.
206. Trojnar E, Otto P, Roth B, Reetz J, Johne R. 2010. The genome segments of a group D rotavirus possess group A-like conserved termini but encode group-specific proteins. *J Virol* 84:10254–10265.
207. Cunliffe NA, Bresee JS, Gentsch JR, Glass RI, Hart CA. 2002. The expanding diversity of rotaviruses. *Lancet* 359:640–641.
208. Khamrin P, Maneekarn N, Malasao R, Nguyen TA, Ishida S, Okitsu S, Ushijima H. 2010. Genotypic linkages of VP4, VP6, VP7, NSP4, NSP5 genes of rotaviruses circulating among children with acute gastroenteritis in Thailand. *Infect Genet Evol* 10:467–472.
209. Chaimongkol N, Khamrin P, Malasao R, Thongprachum A, Ushijima H, Maneekarn N. 2012. Genotypic linkages of gene segments of rotaviruses circulating in pediatric patients with acute gastroenteritis in Thailand. *Infect Genet Evol* 12:1381–1391.
210. Heiman EM, McDonald SM, Barro M, Taraporewala ZF, Bar-Magen T, Patton JT. 2008. Group A human rotavirus genomics: Evidence that gene constellations are influenced by viral protein interactions. *J Virol* 82:11106–11116.
211. Maunula L, von Bonsdorff C-H. 2002. Frequent reassortments may explain the genetic heterogeneity of rotaviruses: Analysis of Finnish rotavirus strains. *J Virol* 76:11793–11800.
212. McDonald SM, Davis K, McAllen JK, Spiro DJ, Patton JT. 2011. Intra-genotypic diversity of archival G4P[8] human rotaviruses from Washington, DC. *Infect Genet Evol* 11:1586–1594.
213. McDonald SM, Matthijnssens J, McAllen JK, Hine E, Overton L, Wang S, Lemey P, et al. 2009. Evolutionary dynamics of human rotaviruses: Balancing reassortment with preferred genome constellations. *PLoS Pathog* 5:e1000634.
214. Iturriza-Gomara M, Green J, Brown DW, Ramsay M, Desselberger U, Gray JJ. 2000. Molecular epidemiology of human group A rotavirus infections in the United Kingdom between 1995 and 1998. *J Clin Microbiol* 38:4394–4401.
215. Watanabe M, Nakagomi T, Koshimura Y, Nakagomi O. 2001. Direct evidence for genome segment reassortment between concurrently-circulating human rotavirus strains. *Arch Virol* 146:557–570.
216. Kaga E, Nakagomi O. 1994. Recurrent circulation of single nonstructural gene substitution reassortants among human rotaviruses with a short RNA pattern. *Arch Virol* 136:63–71.
217. Doan YH, Nakagomi T, Nakagomi O. 2012. Repeated circulation over 6 years of intergenogroup mono-reassortant G2P[4] rotavirus strains with genotype N1 of the NSP2 gene. *Infect Genet Evol* 12:1202–1212.
218. Medici MC, Abelli LA, Martella V, Martinelli M, Lorusso E, Buonavoglia C, Dettori G, Chezzi C. 2007. Characterization of inter-genogroup reassortant rotavirus strains detected in hospitalized children in Italy. *J Med Virol* 79:1406–1412.
219. Nakagomi O, Nakagomi T. 2002. Genomic relationships among rotaviruses recovered from various animal species as revealed by RNA-RNA hybridization assays. *Res Vet Sci* 73:207–214.
220. Ward RL, Nakagomi O, Knowlton DR, McNeal MM, Nakagomi T, Clemens JD, Sack DA, Schiff GM. 1990. Evidence for natural reassortants of human rotaviruses belonging to different genogroups. *J Virol* 64:3219–3225.
221. Kim HJ, Park JG, Alfajaro MM, Kim DS, Hosmillo M, Son KY, Lee JH, et al. 2012. Pathogenicity characterization of a bovine triple reassortant rotavirus in calves and piglets. *Vet Microbiol* 159:11–22.

222. Ahmed K, Nakagomi T, Nakagomi O. 2005. Isolation and molecular characterization of a naturally occurring non-structural protein 5 (NSP5) gene reassortant of group A rotavirus of serotype G2P[4] with a long RNA pattern. *J Med Virol* 77:323–330.
223. Ghosh S, Kobayashi N. 2011. Whole-genomic analysis of rotavirus strains: Current status and future prospects. *Future Microbiol* 6:1049–1065.
224. Esona MD, Geyer A, Page N, Trabelsi A, Fodha I, Aminu M, Agbaya VA, et al. 2009. Genomic characterization of human rotavirus G8 strains from the African rotavirus network: Relationship to animal rotaviruses. *J Med Virol* 81:937–951.
225. Ghosh S, Paul SK, Hossain MA, Alam MM, Ahmed MU, Kobayashi N. 2011. Full genomic analyses of two human G2P[4] rotavirus strains detected in 2005: Identification of a caprine-like VP3 gene. *J Gen Virol* 92:1222–1227.
226. Jere KC, Mlera L, O'Neill HG, Potgieter AC, Page NA, Seheri ML, van Dijk AA. 2011. Whole genome analyses of African G2, G8, G9, and G12 rotavirus strains using sequence-independent amplification and 454® pyrosequencing. *J Med Virol* 83:2018–2042.
227. Esona MD, Banyai K, Foytich K, Freeman M, Mijatovic-Rustempasic S, Hull J, Kerin T, et al. 2011. Genomic characterization of human rotavirus G10 strains from the African Rotavirus Network: Relationship to animal rotaviruses. *Infect Genet Evol* 11:237–241.
228. Kuzuya M, Fujii R, Hamano M, Kida K, Mizoguchi Y, Kanadani T, Nishimura K, Kishimoto T. 2014. Prevalence and molecular characterization of G1P[8] human rotaviruses possessing DS-1-like VP6, NSP4, and NSP5/6 in Japan. *J Med Virol* 86:1056–1064.
229. Rahman M, Matthijnssens J, Yang X, Delbeke T, Arijs I, Taniguchi K, Itturiza-Gomara M, Iftekharuddin N, Azim T, Van Ranst M. 2007. Evolutionary history and global spread of the emerging g12 human rotaviruses. *J Virol* 81:2382–2390.
230. Azevedo MS, Yuan L, Jeong KI, Gonzalez A, Nguyen TV, Pouly S, Gochnauer M, Zhang W, Azevedo A, Saif LJ. 2005. Viremia and nasal and rectal shedding of rotavirus in gnotobiotic pigs inoculated with Wa human rotavirus. *J Virol* 79:5428–5436.
231. He B, Yang F, Yang W, Zhang Y, Feng Y, Zhou J, Xie J, et al. 2013. Characterization of a novel G3P[3] rotavirus isolated from a lesser horseshoe bat: A distant relative of feline/canine rotaviruses. *J Virol* 87:12357–12366.
232. Mori Y, Sugiyama M, Takayama M, Atoji Y, Masegi T, Minamoto N. 2001. Avian-to-mammal transmission of an avian rotavirus: Analysis of its pathogenicity in a heterologous mouse model. *Virology* 288:63–70.
233. Ciarlet M, Estes MK, Conner ME. 2000. Simian rhesus rotavirus is a unique heterologous (non-lapine) rotavirus strain capable of productive replication and horizontal transmission in rabbits. *J Gen Virol* 81:1237–1249.
234. Clark HF, Offit PA, Ellis RW, Eiden JJ, Krah D, Shaw AR, Pichichero M, et al. 1996. The development of multivalent bovine rotavirus (strain WC3) reassortant vaccine for infants. *J Infect Dis* 174 Suppl 1:S73–S80.
235. Clements-Mann ML, Dudas R, Hoshino Y, Nehring P, Sperber E, Wagner M, Stephens I, Karron R, Deforest A, Kapikian AZ. 2001. Safety and immunogenicity of live attenuated quadrivalent human-bovine (UK) reassortant rotavirus vaccine administered with childhood vaccines to infants. *Vaccine* 19:4676–4684.
236. Vesikari T. 1996. Trials of oral bovine and rhesus rotavirus vaccines in Finland: A historical account and present status. *Arch Virol Suppl* 12:177–186.
237. Afrad MH, Matthijnssens J, Moni S, Kabir F, Ashrafi A, Rahman MZ, Faruque AS, Azim T, Rahman M. 2013. Genetic characterization of a rare bovine-like human VP4 mono-reassortant G6P[8] rotavirus strain detected from an infant in Bangladesh. *Infect Genet Evol* 19:120–126.
238. Dong HJ, Qian Y, Huang T, Zhu RN, Zhao LQ, Zhang Y, Li RC, Li YP. 2013. Identification of circulating porcine-human reassortant G4P[6] rotavirus from children with acute diarrhea in China by whole genome analyses. *Infect Genet Evol* 20:155–162.
239. Nakagomi O, Kaga E, Gerna G, Sarasini A, Nakagomi T. 1992. Subgroup I serotype 3 human rotavirus strains with long RNA pattern as a result of naturally occurring reassortment between members of the bovine and AU-1 genogroups. *Arch Virol* 126:337–342.
240. Bok K, Castagnaro N, Borsa A, Nates S, Espul C, Fay O, Fabri A, et al. 2001. Surveillance for rotavirus in Argentina. *J Med Virol* 65:190–198.
241. Carmona RC, Timenetsky Mdo C, da Silva FF, Granato CF. 2004. Characterization of rotavirus strains from hospitalized and outpatient children with acute diarrhoea in Sao Paulo, Brazil. *J Med Virol* 74:166–172.

242. Coluchi N, Munford V, Manzur J, Vazquez C, Escobar M, Weber E, Marmol P, Racz ML. 2002. Detection, subgroup specificity, and genotype diversity of rotavirus strains in children with acute diarrhea in Paraguay. *J Clin Microbiol* 40:1709–1714.
243. Gouvea V, de Castro L, Timenetsky MC, Greenberg H, Santos N. 1994. Rotavirus serotype G5 associated with diarrhea in Brazilian children. *J Clin Microbiol* 32:1408–1409.
244. Ghosh SK, Naik TN. 1989. Evidence for a new rotavirus subgroup in India. *Epidemiol Infect* 102:523–530.
245. Varghese V, Das S, Singh NB, Kojima K, Bhattacharya SK, Krishnan T, Kobayashi N, Naik TN. 2004. Molecular characterization of a human rotavirus reveals porcine characteristics in most of the genes including VP6 and NSP4. *Arch Virol* 149:155–172.
246. Okada J, Urasawa T, Kobayashi N, Taniguchi K, Hasegawa A, Mise K, Urasawa S. 2000. New P serotype of group A human rotavirus closely related to that of a porcine rotavirus. *J Med Virol* 60:63–69.
247. Kelkar SD, Ayachit VL. 2000. Circulation of group A rotavirus subgroups and serotypes in Pune, India, 1990–1997. *J Health Popul Nutr* 18:163–170.
248. Matthijnssens J, Rahman M, Yang X, Delbeke T, Arijs I, Kabue JP, Muyembe JJ, Van Ranst M. 2006. G8 rotavirus strains isolated in the Democratic Republic of Congo belong to the DS-1-like genogroup. *J Clin Microbiol* 44:1801–1809.
249. Ghosh S, Varghese V, Samajdar S, Bhattacharya SK, Kobayashi N, Naik TN. 2006. Molecular characterization of a porcine Group A rotavirus strain with G12 genotype specificity. *Arch Virol* 151:1329–1344.
250. Hoshino Y, Honma S, Jones RW, Ross J, Santos N, Gentsch JR, Kapikian AZ, Hesse RA. 2005. A porcine G9 rotavirus strain shares neutralization and VP7 phylogenetic sequence lineage 3 characteristics with contemporary human G9 rotavirus strains. *Virology* 332:177–188.
251. Matthijnssens J, Ciarlet M, Heiman E, Arijs I, Delbeke T, McDonald SM, Palombo EA, et al. 2008. Full genome-based classification of rotaviruses reveals a common origin between human Wa-Like and porcine rotavirus strains and human DS-1-like and bovine rotavirus strains. *J Virol* 82:3204–3219.
252. Cook N, Bridger J, Kendall K, Gomara MI, El-Attar L, Gray J. 2004. The zoonotic potential of rotavirus. *J Infect* 48:289–302.
253. Esona MD, Page NA, Akran VA, Armah GE, Steele DA. 2010. Characterization of 2 human genotype G10 rotavirus strains, 3008CM and 1784/CI/1999, isolated in Cameroon and Cote d'Ivoire during the 1999–2000 rotavirus season. *J Infect Dis* 202:S212–S219.
254. Gunn L, Feeney SA, Cashman O, Collins PJ, Coyle PV, O'Shea H. 2012. Molecular characterization of group A rotavirus found in elderly patients in Ireland; predominance of G1P[8], continued presence of G9P[8], and emergence of G2P[4]. *J Med Virol* 84:2008–2017.
255. Feeney SA, Mitchell SJ, Mitchell F, Wyatt DE, Fairley D, McCaughey C, Coyle PV, O'Neill HJ. 2006. Association of the G4 rotavirus genotype with gastroenteritis in adults. *J Med Virol* 78:1119–1123.
256. Yamagami T, Yokoyama H. 2007. Two cases of group A rotavirus outbreak among adults. *Kansenshogaku Zasshi* 81:256–260.
257. Sanz JC, Barbas JF, Lasheras MD, Jimenez M, Ramos B, Sanchez-Fauquier A. 2009. Detection of a rotavirus G9P[8] outbreak causing gastroenteritis in a geriatric nursing home. *Enferm Infecc Microbiol Clin* 27:219–221.
258. Banyai K, Angyal M, Kormendi E, Lakatos F, Uj M, Szucs G. 2002. Outbreak of human rotavirus infection in an adult community. *Orvosi Hetilap* 143:1347–1352.
259. Steel HM, Garnham S, Beards GM, Brown DW. 1992. Investigation of an outbreak of rotavirus infection in geriatric patients by serotyping and polyacrylamide gel electrophoresis (PAGE). *J Med Virol* 37:132–136.
260. Trop Skaza A, Beskovnik L, Zohar Cretnik T. 2011. Outbreak of rotavirus gastroenteritis in a nursing home, Slovenia, December 2010. *Euro Surveill* 16:pii: 19837.
261. Chimura Y, Annaka M, Shibazaki S, Adachi K, Shinkai T, Sadamasu K, Noguchi Y, Masuda Y, Inamatsu T. 2002. An epidemic of rotavirus infection in a nursing home for the elderly in Japan. *Kansenshogaku Zasshi* 76:450–454.
262. Edmonson LM, Ebbert JO, Evans JM. 2000. Report of a rotavirus outbreak in an adult nursing home population. *J Am Med Dir Assoc* 1:175–179.
263. Halvorsrud J, Orstavik I. 1980. An epidemic of rotavirus-associated gastroenteritis in a nursing home for the elderly. *Scand J Infect Dis* 12:161–164.
264. Lederer P, Muller R. 1995. Rotavirus gastroenteritis in homes for the aged: A neglected problem? *Gesundheitswesen* 57:326–328.
265. Piednoir E, Borderan GC, Borgey F, Thibon P, Lesellier P, Leservoisier R, Verger P, Le Coutour X. 2010. Direct costs associated with a hospital-acquired outbreak of rotaviral gastroenteritis infection in a long term care institution. *J Hosp Infect* 75:295–298.

266. Cardemil CV, Cortese MM, Medina-Marino A, Jasuja S, Desai R, Leung J, Rodriguez-Hart C, et al. 2012. Two rotavirus outbreaks caused by genotype G2P[4] at large retirement communities: Cohort studies. *Ann Intern Med* 157:621–631.
267. Griffin DD, Fletcher M, Levy ME, Ching-Lee M, Nogami R, Edwards L, Peters H, Montague L, Gentsch JR, Glass RI. 2002. Outbreaks of adult gastroenteritis traced to a single genotype of rotavirus. *J Infect Dis* 185:1502–1505.
268. Yan H, Abe T, Phan TG, Nguyen TA, Iso T, Ikezawa Y, Ishii K, Okitsu S, Ushijima H. 2005. Outbreak of acute gastroenteritis associated with group A rotavirus and genogroup I sapovirus among adults in a mental health care facility in Japan. *J Med Virol* 75:475–481.
269. Luchs A, Morillo SG, Ribeiro CD, Vilanova BC, Calux SJ, Carmona Rde C, Timenetsky Mdo C. 2011. Gastroenteritis outbreak due to G2P[4] rotavirus and GII norovirus at two correctional facilities in Brazil, 2010. *J Clin Virol* 51:213–214.
270. Iijima Y, Iwamoto T, Nukuzuma S, Ohishi H, Hayashi K, Kobayashi N. 2006. An outbreak of rotavirus infection among adults in an institution for rehabilitation: Long-term residence in a closed community as a risk factor for rotavirus illnessl. *Scand J Infect Dis* 38:490–496.
271. Mikami T, Nakagomi T, Tsutsui R, Ishikawa K, Onodera Y, Arisawa K, Nakagomi O. 2004. An outbreak of gastroenteritis during school trip caused by serotype G2 group A rotavirus. *J Med Virol* 73:460–464.
272. Samajdar S, Ghosh S, Naik TN, Roy S, Sugunan AP. 2008. The post-tsunami outbreak of diarrhoeal diseases in Car Nicobar Island, India, was caused by human group A rotavirus G2 strains. *J Infect* 57:357–359.
273. Ratner AJ, Neu N, Jakob K, Grumet S, Adachi N, Della-Latta P, Marvel E, Saiman L. 2001. Nosocomial rotavirus in a pediatric hospital. *Infect Control Hosp Epidemiol* 22:299–301.
274. Rogers M, Weinstock DM, Eagan J, Kiehn T, Armstrong D, Sepkowitz KA. 2000. Rotavirus outbreak on a pediatric oncology floor: Possible association with toys. *Am J Infect Control* 28:378–380.
275. Marrie TJ, Lee SH, Faulkner RS, Ethier J, Young CH. 1982. Rotavirus infection in a geriatric population. *Arch Intern Med* 142:313–316.
276. Lewis DC, Lightfoot NF, Cubitt WD, Wilson SA. 1989. Outbreaks of astrovirus type 1 and rotavirus gastroenteritis in a geriatric in-patient population. *J Hosp Infect* 14:9–14.
277. Abbas AM, Denton MD. 1987. An outbreak of rotavirus infection in a geriatric hospital. *J Hosp Infect* 9:76–80.
278. Wandless I, Narang HK, Evans JG. 1981. Rotavirus infection in elderly patients. *Age Ageing* 10:36–39.
279. Cubitt WD, Holzel H. 1980. An outbreak of rotavirus infection in a long-stay ward of a geriatric hospital. *J Clin Pathol* 33:306–308.
280. Holzel H, Cubitt DW, McSwiggan DA, Sanderson PJ, Church J. 1980. An outbreak of rotavirus infection among adults in a cardiology ward. *J Infect* 2:33–37.
281. Widdowson MA, van Doornum GJ, van der Poel WH, de Boer AS, van de Heide R, Mahdi U, Haanen P, Kool JL, Koopmans M. 2002. An outbreak of diarrhea in a neonatal medium care unit caused by a novel strain of rotavirus: Investigation using both epidemiologic and microbiological methods. *Infect Control Hosp Epidemiol* 23:665–670.
282. Rochi GU, Vella S, Resta S, Cochi S, Donelli G, Tangucci F, Menichella D, Varveri A, Inglese R. 1981. Outbreak of rotavirus gastroenteritis among premature infants. *BMJ* 283:886.
283. von Bonsdorff CH, Hovi T, Makela P, Tevalvoto-Aarnio M. 1976. Rotavirus associated with acute gastroenteritis in adults. *Lancet* 2:423.
284. JAMA. 2001. From the Centers for Disease Control and Prevention. Foodborne outbreak of group A rotavirus gastroenteritis among college students: District of Columbia, March–April 2000. *JAMA* 285:405–406.
285. Inoue Y, Imanishi Y, Kitahori Y. 2008. An outbreak of group A rotavirus G1P[8] in an elementary school, Nara prefecture. *Jpn J Infect Dis* 61:426.
286. Siqueira AA, Santelli AC, Alencar LR, Jr., Dantas MP, Dimech CP, Carmo GM, Santos DA, et al. 2010. Outbreak of acute gastroenteritis in young children with death due to rotavirus genotype G9 in Rio Branco, Brazilian Amazon region, 2005. *Int J Infect Dis* 14:e898–903.
287. Rasanen S, Lappalainen S, Kaikkonen S, Hamalainen M, Salminen M, Vesikari T. 2010. Mixed viral infections causing acute gastroenteritis in children in a waterborne outbreak. *Epidemiol Infect* 138:1227–1234.
288. Koroglu M, Yakupogullari Y, Otlu B, Ozturk S, Ozden M, Ozer A, Sener K, Durmaz R. 2011. A waterborne outbreak of epidemic diarrhea due to group A rotavirus in Malatya, Turkey. *New Microbiol* 34:17–24.

289. Mellou K, Katsioulis A, Potamiti-Komi M, Pournaras S, Kyritsi M, Katsiaflaka A, Kallimani A, et al. 2014. A large waterborne gastroenteritis outbreak in central Greece, March 2012: Challenges for the investigation and management. *Epidemiol Infect* 142:40–50.
290. Hopkins RS, Gaspard GB, Williams FP, Jr., Karlin RJ, Cukor G, Blacklow NR. 1984. A community waterborne gastroenteritis outbreak: Evidence for rotavirus as the agent. *Am J Public Health* 74:263–265.
291. Foster SO, Palmer EL, Gary GW, Jr., Martin ML, Herrmann KL, Beasley P, Sampson J. 1980. Gastroenteritis due to rotavirus in an isolated Pacific island group: An epidemic of 3,439 cases. *J Infect Dis* 141:32–39.
292. Bucardo F, Karlsson B, Nordgren J, Paniagua M, Gonzalez A, Amador JJ, Espinoza F, Svensson L. 2007. Mutated G4P[8] rotavirus associated with a nationwide outbreak of gastroenteritis in Nicaragua in 2005. *J Clin Microbiol* 45:990–997.
293. Meurman OH, Laine MJ. 1977. Rotavirus epidemic in adults. *N Engl J Med* 296:1298–1299.
294. Linhares AC, Pinheiro FP, Freitas RB, Gabbay YB, Shirley JA, Beards GM. 1981. An outbreak of rotavirus diarrhea among a nonimmune, isolated South American Indian community. *Am J Epidemiol* 113:703–710.
295. Friedman MG, Galil A, Sarov B, Margalith M, Katzir G, Midthun K, Taniguchi K, et al. 1988. Two sequential outbreaks of rotavirus gastroenteritis: Evidence for symptomatic and asymptomatic reinfections. *J Infect Dis* 158:814–822.
296. Hall AJ, Wikswo ME, Pringle K, Gould LH, Parashar UD. 2014. Vital signs: Foodborne norovirus outbreaks—United States, 2009–2012. *MMWR* 63:491–495.
297. Rakesh P, Sherin D, Sankar H, Shaji M, Subhagan S, Salila S. 2014. Investigating a community-wide outbreak of hepatitis A in India. *J Glob Infect Dis* 6:59–64.
298. Swinkels HM, Kuo M, Embree G, Andonov A, Henry B, Buxton JA. 2014. Hepatitis A outbreak in British Columbia, Canada: The roles of established surveillance, consumer loyalty cards and collaboration, February to May 2012. *Euro Surveill* 19(20792):2.
299. Guzman-Herrador B, Jensvoll L, Einoder-Moreno M, Lange H, Myking S, Nygard K, Stene-Johansen K, Vold L. 2014. Ongoing hepatitis A outbreak in Europe 2013 to 2014: Imported berry mix cake suspected to be the source of infection in Norway. *Euro Surveill* 19(15):20775.
300. Franck KT, Lisby M, Fonager J, Schultz AC, Bottiger B, Villif A, Absalonsen H, Ethelberg S. 2014. Sources of calicivirus contamination in foodborne outbreaks, Denmark, 2005 2011: —The role of the asymptomatic food handler. *J Infect Dis* 211(4):563–570. doi:10.1093/infdis/jiu479.
301. Brassard J, Gagne MJ, Genereux M, Cote C. 2012. Detection of human food-borne and zoonotic viruses on irrigated, field-grown strawberries. *Appl Environ Microbiol* 78:3763–3766.
302. Van Zyl WB, Page NA, Grabow WOK, Steele AD, Taylor MB. 2006. Molecular epidemiology of group A rotaviruses in water sources and selected raw vegetables in southern Africa. *Appl Environ Microbiol* 72:4554–4560.
303. Hernandez F, Monge R, Jimenez C, Taylor L. 1997. Rotavirus and hepatitis A virus in market lettuce (Latuca sativa) in Costa Rica. *Int J Food Microbiol* 37:221–223.
304. Mattison K, Harlow J, Morton V, Cook A, Pollari F, Bidawid S, Farber JM. 2010. Enteric viruses in ready-to-eat packaged leafy greens. *Emerg Infect Dis* 16:1815–1817; discussion 1817.
305. Jones TH, Nattress FM, Dilts B, Olsen D, Muehlhauser V. 2014. Numbers of coliforms, *Escherichia coli*, F-RNA phage, rotavirus, bovine enteric calicivirus and presence of non-O157 STEC on commercial vacuum packaged beef. *Food Microbiol* 42:225–231.
306. Gabrieli R, Macaluso A, Lanni L, Saccares S, Di Giamberardino F, Cencioni B, Petrinca AR, Divizia M. 2007. Enteric viruses in molluscan shellfish. *New Microbiol* 30:471–475.
307. Carter M. 2005. Enterically infecting viruses: Pathogenicity, transmission and significance for food and waterborne infection. *J Appl Microbiol* 98:1354–1380.
308. de Wit MA, Koopmans MP, van Duynhoven YT. 2003. Risk factors for norovirus, Sapporo-like virus, and group A rotavirus gastroenteritis. *Emerg Infect Dis* 9:1563.
309. Di Bartolo I, Monini M, Losio MN, Pavoni E, Lavazza A, Ruggeri FM. 2011. Molecular characterization of noroviruses and rotaviruses involved in a large outbreak of gastroenteritis in Northern Italy. *Appl Environ Microbiol* 77:5545–5548.
310. Gallimore CI, Pipkin C, Shrimpton H, Green AD, Pickford Y, McCartney C, Sutherland G, Brown DW, Gray JJ. 2005. Detection of multiple enteric virus strains within a foodborne outbreak of gastroenteritis: An indication of the source of contamination. *Epidemiol Infect* 133:41–47.
311. Mizukoshi F, Kuroda M, Tsukagoshi H, Sekizuka T, Funatogawa K, Morita Y, Noda M, Katayama K, Kimura H. 2014. A food-borne outbreak of gastroenteritis due to genotype G1P[8] rotavirus among adolescents in Japan. *Microbiol Immunol* 58(9):536–539. doi:10.1111/1348-0421.12176.

312. Keller R, Justino JF, Cassini ST. 2013. Assessment of water and seafood microbiology quality in a mangrove region in Vitoria, Brazil. *J Water Health* 11:573–580.
313. Le Guyader F, Haugarreau L, Miossec L, Dubois E, Pommepuy M. 2000. Three-year study to assess human enteric viruses in shellfish. *Appl Environ Microbiol* 66:3241–3248.
314. Metcalf TG, Melnick JL, Estes MK. 1995. Environmental virology: From detection of virus in sewage and water by isolation to identification by molecular biology—A trip of over 50 years. *Annu Rev Microbiol* 49:461–487.
315. Lipp EK, Rose JB. 1997. The role of seafood in foodborne diseases in the United States of America. *Rev Sci Tech* 16:620–640.
316. Anonymous. 1999. Rotavirus vaccines. *Wkly Epidemiol Rec* 74:33–38.
317. Atherly DE, Lewis KD, Tate J, Parashar UD, Rheingans RD. 2012. Projected health and economic impact of rotavirus vaccination in GAVI-eligible countries: 2011–2030. *Vaccine* 30 Suppl 1:A7–A14.
318. Ward RL, McNeal MM, Steele AD. 2008. Why does the world need another rotavirus vaccine? *Ther Clin Risk Manag* 4:49–63.
319. Ruuska T, Vesikari T, Delem A, Andre FE, Beards GM, Flewett TH. 1990. Evaluation of RIT 4237 bovine rotavirus vaccine in newborn infants: Correlation of vaccine efficacy to season of birth in relation to rotavirus epidemic period. *Scand J Infect Dis* 22:269–278.
320. Vesikari T, Isolauri E, D'Hondt E, Delem A, Andre FE, Zissis G. 1984. Protection of infants against rotavirus diarrhoea by RIT 4237 attenuated bovine rotavirus strain vaccine. *Lancet* 1:977–981.
321. Heaton PM, Ciarlet M. 2007. Vaccines: The pentavalent rotavirus vaccine–Discovery to licensure and beyond. *Clin Infect Dis* 45:1618–1624.
322. Matthijnssens J, Joelsson DB, Warakomski DJ, Zhou T, Mathis PK, van Maanen MH, Ranheim TS, Ciarlet M. 2010. Molecular and biological characterization of the 5 human-bovine rotavirus (WC3)-based reassortant strains of the pentavalent rotavirus vaccine, RotaTeq. *Virology* 403:111–127.
323. Cortese MM, Parashar UD. 2009. Prevention of rotavirus gastroenteritis among infants and children: Recommendations of the Advisory Committee on Immunization Practices (ACIP). *MMWR* 58:1–25.
324. Vesikari T, Matson DO, Dennehy P, Van Damme P, Santosham M, Rodriguez Z, Dallas MJ, et al. 2006. Safety and efficacy of a pentavalent human-bovine (WC3) reassortant rotavirus vaccine. *N Engl J Med* 354:23–33.
325. Vesikari T, Karvonen A, Ferrante SA, Kuter BJ, Ciarlet M. 2010. Sustained efficacy of the pentavalent rotavirus vaccine, RV5, up to 3.1 years following the last dose of vaccine. *Pediatr Infect Dis J* 29:957–963.
326. Ruiz-Palacios GM, Perez-Schael I, Velazquez FR, Abate H, Breuer T, Clemens SC, Cheuvart B, et al. 2006. Safety and efficacy of an attenuated vaccine against severe rotavirus gastroenteritis. *N Engl J Med* 354:11–22.
327. McCormack PL, Keam SJ. 2009. Rotavirus vaccine RIX4414 (Rotarix): A review of its use in the prevention of rotavirus gastroenteritis. *Paediatr Drugs* 11:75–88.
328. Vesikari T, Karvonen A, Puustinen L, Zeng SQ, Szakal ED, Delem A, De Vos B. 2004. Efficacy of RIX4414 live attenuated human rotavirus vaccine in Finnish infants. *Pediatr Infect Dis J* 23:937–943.
329. Salinas B, Perez Schael I, Linhares AC, Ruiz Palacios GM, Guerrero ML, Yarzabal JP, Cervantes Y, et al. 2005. Evaluation of safety, immunogenicity and efficacy of an attenuated rotavirus vaccine, RIX4414: A randomized, placebo-controlled trial in Latin American infants. *Pediatr Infect Dis J* 24:807–816.
330. Phua KB, Emmanuel SC, Goh P, Quak SH, Lee BW, Han HH, Ward RL, Bernstein DI, De Vos B, Bock HL. 2006. A rotavirus vaccine for infants: The Asian experience. *Ann Acad Med* 35:38–44.
331. Vesikari T, Karvonen A, Prymula R, Schuster V, Tejedor JC, Cohen R, Meurice F, Han HH, Damaso S, Bouckenooghe A. 2007. Efficacy of human rotavirus vaccine against rotavirus gastroenteritis during the first 2 years of life in European infants: Randomised, double-blind controlled study. *Lancet* 370:1757–1763.
332. Linhares AC, Velázquez FR, Pérez-Schael I, Sáez-Llorens X, Abate H, Espinoza F, López P, et al. 2008. Efficacy and safety of an oral live attenuated human rotavirus vaccine against rotavirus gastroenteritis during the first 2 years of life in Latin American infants: A randomised, double-blind, placebo-controlled phase III study. *Lancet* 371:1181–1189.
333. De Vos B, Han HH, Bouckenooghe A, Debrus S, Gillard P, Ward R, Cheuvart B. 2009. Live attenuated human rotavirus vaccine, RIX4414, provides clinical protection in infants against rotavirus strains with and without shared G and P genotypes: Integrated analysis of randomized controlled trials. *Pediatr Infect Dis J* 28:261–266.

334. Vesikari T, Karvonen A, Bouckenooghe A, Suryakiran PV, Smolenov I, Han HH. 2011. Immunogenicity, reactogenicity and safety of the human rotavirus vaccine RIX4414 oral suspension (liquid formulation) in Finnish infants. *Vaccine* 29:2079–2084.
335. Omenaca F, Sarlangue J, Szenborn L, Nogueira M, Suryakiran PV, Smolenov IV, Han HH. 2012. Safety, reactogenicity and immunogenicity of the human rotavirus vaccine in preterm European Infants: A randomized phase IIIb study. *Pediatr Infect Dis J* 31:487–493.
336. Madhi SA, Cunliffe NA, Steele D, Witte D, Kirsten M, Louw C, Ngwira B, et al. 2010. Effect of human rotavirus vaccine on severe diarrhea in African infants. *N Engl J Med* 362:289–298.
337. Madhi SA, Kirsten M, Louw C, Bos P, Aspinall S, Bouckenooghe A, Neuzil KM, Steele AD. 2012. Efficacy and immunogenicity of two or three dose rotavirus-vaccine regimen in South African children over two consecutive rotavirus-seasons: A randomized, double-blind, placebo-controlled trial. *Vaccine* 30 Suppl 1:A44–A51.
338. Anh DD, Carlos CC, Thiem DV, Hutagalung Y, Gatchalian S, Bock HL, Smolenov I, Suryakiran PV, Han HH. 2011. Immunogenicity, reactogenicity and safety of the human rotavirus vaccine RIX4414 (Rotarix) oral suspension (liquid formulation) when coadministered with expanded program on immunization (EPI) vaccines in Vietnam and the Philippines in 2006–2007. *Vaccine* 29:2029–2036.
339. Perez-Schael I, Salinas B, Tomat M, Linhares AC, Guerrero ML, Ruiz-Palacios GM, Bouckenooghe A, Yarzabal JP. 2007. Efficacy of the human rotavirus vaccine RIX4414 in malnourished children. *J Infect Dis* 196:537–540.
340. Zaman K, Dang DA, Victor JC, Shin S, Yunus M, Dallas MJ, Podder G, et al. 2010. Efficacy of pentavalent rotavirus vaccine against severe rotavirus gastroenteritis in infants in developing countries in Asia: A randomised, double-blind, placebo-controlled trial. *Lancet* 376:615–623.
341. Armah GE, Sow SO, Breiman RF, Dallas MJ, Tapia MD, Feikin DR, Binka FN, et al. 2010. Efficacy of pentavalent rotavirus vaccine against severe rotavirus gastroenteritis in infants in developing countries in sub-Saharan Africa: A randomised, double-blind, placebo-controlled trial. Lancet 376:606–614.
342. Angel J, Franco MA, Greenberg HB. 2012. Rotavirus immune responses and correlates of protection. *Curr Opin Virol* 2:419–425.
343. Tate JE, Patel MM, Steele AD, Gentsch JR, Payne DC, Cortese MM, Nakagomi O, et al. 2010. Global impact of rotavirus vaccines. *Expert Rev Vaccines* 9:395–407.
344. Tregnaghi MW, Abate HJ, Valencia A, Lopez P, Da Silveira TR, Rivera L, Rivera Medina DM, et al. 2011. Human rotavirus vaccine is highly efficacious when coadministered with routine expanded program of immunization vaccines including oral poliovirus vaccine in Latin America. *Pediatr Infect Dis J* 30:e103–108.
345. Zaman K, Sack DA, Yunus M, Arifeen SE, Podder G, Azim T, Luby S, et al. 2009. Successful coadministration of a human rotavirus and oral poliovirus vaccines in Bangladeshi infants in a 2-dose schedule at 12 and 16 weeks of age. *Vaccine* 27:1333–1339.
346. Steele AD, De Vos B, Tumbo J, Reynders J, Scholtz F, Bos P, de Beer MC, Van der Merwe CF, Delem A. 2010. Coadministration study in South African infants of a live-attenuated oral human rotavirus vaccine (RIX4414) and poliovirus vaccines. *Vaccine* 28:6542–6548.
347. Patel M, Steele AD, Parashar UD. 2012. Influence of oral polio vaccines on performance of the monovalent and pentavalent rotavirus vaccines. *Vaccine* 30 Suppl 1:A30–A35.
348. Fu C, He Q, Xu J, Xie H, Ding P, Hu W, Dong Z, Liu X, Wang M. 2012. Effectiveness of the Lanzhou lamb rotavirus vaccine against gastroenteritis among children. *Vaccine* 31:154–158.
349. Chandran A, Fitzwater S, Zhen A, Santosham M. 2010. Prevention of rotavirus gastroenteritis in infants and children: Rotavirus vaccine safety, efficacy, and potential impact of vaccines. *Biologics* 4:213–229.
350. Fu C, Tate JE, Jiang B. 2010. Effectiveness of Lanzhou lamb rotavirus vaccine against hospitalized gastroenteritis: Further analysis and update. *Hum Vaccin* 6:953.
351. Fu C, Wang M, Liang J, He T, Wang D, Xu J. 2007. Effectiveness of Lanzhou lamb rotavirus vaccine against rotavirus gastroenteritis requiring hospitalization: A matched case-control study. *Vaccine* 25:8756–8761.
352. Dang DA, Nguyen VT, Vu DT, Nguyen TH, Nguyen DM, Yuhuan W, Baoming J, Nguyen DH, Le TL. 2012. A dose-escalation safety and immunogenicity study of a new live attenuated human rotavirus vaccine (Rotavin-M1) in Vietnamese children. *Vaccine* 30 Suppl 1:A114–A121.
353. Danchin M, Kirkwood CD, Lee KJ, Bishop RF, Watts E, Justice FA, Clifford V, Cowley D, Buttery JP, Bines JE. 2013. Phase I trial of RV3-BB rotavirus vaccine: A human neonatal rotavirus vaccine. *Vaccine* 31:2610–2616.

354. Bhandari N, Sharma P, Taneja S, Kumar T, Rongsen-Chandola T, Appaiahgari MB, Mishra A, Singh S, Vrati S. 2009. A dose-escalation safety and immunogenicity study of live attenuated oral rotavirus vaccine 116E in infants: A randomized, double-blind, placebo-controlled trial. *J Infect Dis* 200:421–429.
355. Bhandari N, Rongsen-Chandola T, Bavdekar A, John J, Antony K, Taneja S, Goyal N, et al. 2014. Efficacy of a monovalent human-bovine (116E) rotavirus vaccine in Indian children in the second year of life. *Vaccine* 32 Suppl 1:A110–A116.
356. Vesikari T, Karvonen AV, Majuri J, Zeng S, Pang X, Kohberger R, Forrest BD, Hoshino Y, Chanock RM, Kapikian AZ. 2006. Safety, efficacy, and immunogenicity of 2 doses of bovine-human (UK) and rhesus-rhesus-human rotavirus reassortant tetravalent vaccines in Finnish children. *J Infect Dis* 194:370–376.
357. Kapikian AZ, Simonsen L, Vesikari T, Hoshino Y, Morens DM, Chanock RM, La Montagne JR, Murphy BR. 2005. A hexavalent human rotavirus-bovine rotavirus (UK) reassortant vaccine designed for use in developing countries and delivered in a schedule with the potential to eliminate the risk of intussusception. *J Infect Dis* 192 Suppl 1:S22–S29.
358. Azevedo MS, Gonzalez AM, Yuan L, Jeong KI, Iosef C, Van Nguyen T, Lovgren-Bengtsson K, Morein B, Saif LJ. 2010. An oral versus intranasal prime/boost regimen using attenuated human rotavirus or VP2 and VP6 virus-like particles with immunostimulating complexes influences protection and antibody-secreting cell responses to rotavirus in a neonatal gnotobiotic pig model. *Clin Vaccine Immunol* 17:420–428.
359. Blazevic V, Lappalainen S, Nurminen K, Huhti L, Vesikari T. 2011. Norovirus VLPs and rotavirus VP6 protein as combined vaccine for childhood gastroenteritis. *Vaccine* 29:8126–8133.
360. Tokuhara D, Alvarez B, Mejima M, Hiroiwa T, Takahashi Y, Kurokawa S, Kuroda M, et al. 2013. Rice-based oral antibody fragment prophylaxis and therapy against rotavirus infection. *J Clin Invest* 123:3829–3838.
361. Li T, Lin H, Zhang Y, Li M, Wang D, Che Y, Zhu Y, et al. 2014. Improved characteristics and protective efficacy in an animal model of *E. coli*-derived recombinant double-layered rotavirus virus-like particles. *Vaccine* 32:1921–1931.
362. Ward RL, McNeal MM. 2010. VP6: A candidate rotavirus vaccine. *J Infect Dis* 202 Suppl:S101–S107.
363. Vega CG, Bok M, Vlasova AN, Chattha KS, Gomez-Sebastian S, Nunez C, Alvarado C, et al. 2013. Recombinant monovalent llama-derived antibody fragments (VHH) to rotavirus VP6 protect neonatal gnotobiotic piglets against human rotavirus-induced diarrhea. *PLoS Pathog* 9:e1003334.
364. Zeller M, Rahman M, Heylen E, De Coster S, De Vos S, Arijs I, Novo L, Verstappen N, Van Ranst M, Matthijnssens J. 2010. Rotavirus incidence and genotype distribution before and after national rotavirus vaccine introduction in Belgium. *Vaccine* 28:7507–7513.
365. Hanquet G, Ducoffre G, Vergison A, Neels P, Sabbe M, Van Damme P, Van Herck K. 2011. Impact of rotavirus vaccination on laboratory confirmed cases in Belgium. *Vaccine* 29:4698–4703.
366. Braeckman T, Van Herck K, Raes M, Vergison A, Sabbe M, Van Damme P. 2011. Rotavirus vaccines in Belgium: Policy and impact. *Pediatr Infect Dis J* 30:S21–S24.
367. Raes M, Strens D, Vergison A, Verghote M, Standaert B. 2011. Reduction in pediatric rotavirus-related hospitalizations after universal rotavirus vaccination in Belgium. *Pediatr Infect Dis J* 30:e120–125.
368. Gagneur A, Nowak E, Lemaitre T, Segura JF, Delaperriere N, Abalea L, Poulhazan E, et al. 2011. Impact of rotavirus vaccination on hospitalizations for rotavirus diarrhea: the IVANHOE study. *Vaccine* 29:3753–3759.
369. Paulke-Korinek M, Rendi-Wagner P, Kundi M, Kronik R, Kollaritsch H. 2010. Universal mass vaccination against rotavirus gastroenteritis: Impact on hospitalization rates in Austrian children. *Pediatr Infect Dis J* 29:319–323.
370. Paulke-Korinek M, Kundi M, Rendi-Wagner P, de Martin A, Eder G, Schmidle-Loss B, Vecsei A, Kollaritsch H. 2011. Herd immunity after two years of the universal mass vaccination program against rotavirus gastroenteritis in Austria. *Vaccine* 29:2791–2796.
371. Trimis G, Koutsoumbari I, Kottaridi C, Palaiologou N, Assimakopoulou E, Spathis A, Lebessi E, et al. 2011. Hospital-based surveillance of rotavirus gastroenteritis in the era of limited vaccine uptake through the private sector. *Vaccine* 29:7292–7295.
372. Vesikari T, Karvonen A, Ferrante SA, Ciarlet M. 2010. Efficacy of the pentavalent rotavirus vaccine, RotaTeq(R), in Finnish infants up to 3 years of age: The Finnish Extension Study. *Eur J Pediatr* 169:1379–1386.

373. Vesikari T, Uhari M, Renko M, Hemming M, Salminen M, Torcel-Pagnon L, Bricout H, Simondon F. 2013. Impact and effectiveness of RotaTeq(r) vaccine based on 3 years of surveillance following introduction of a rotavirus immunization program in Finland. *Pediatr Infect Dis J* 32:1365–1373.
374. Giaquinto C, Jackson AE, Vesikari T. 2012. Report of the second European expert meeting on rotavirus vaccination. *Vaccine* 30:2237–2244.
375. Anderson EJ, Rupp A, Shulman ST, Wang D, Zheng X, Noskin GA. 2011. Impact of rotavirus vaccination on hospital-acquired rotavirus gastroenteritis in children. *Pediatrics* 127:e264–e270.
376. Boom JA, Tate JE, Sahni LC, Rench MA, Hull JJ, Gentsch JR, Patel MM, Baker CJ, Parashar UD. 2010. Effectiveness of pentavalent rotavirus vaccine in a large urban population in the United States. *Pediatrics* 125:e199–e207.
377. Boom JA, Tate JE, Sahni LC, Rench MA, Quaye O, Mijatovic-Rustempasic S, Patel MM, Baker CJ, Parashar UD. 2010. Sustained protection from pentavalent rotavirus vaccination during the second year of life at a large, urban United States pediatric hospital. *Pediatr Infect Dis J* 29:1133–1135.
378. Desai SN, Esposito DB, Shapiro ED, Dennehy PH, Vazquez M. 2010. Effectiveness of rotavirus vaccine in preventing hospitalization due to rotavirus gastroenteritis in young children in Connecticut, USA. *Vaccine* 28:7501–7506.
379. Eberly MD, Gorman GH, Eide MB, Olsen CH, Rajnik M. 2011. The effect of rotavirus immunization on rotavirus gastroenteritis hospitalization rates in military dependents. *Vaccine* 29:650–659.
380. Guh AY, Hadler JL. 2011. Use of the state immunization information system to assess rotavirus vaccine effectiveness in Connecticut, 2006–2008. *Vaccine* 29:6155–6158.
381. Staat MA, Payne DC, Donauer S, Weinberg GA, Edwards KM, Szilagyi PG, Griffin MR, et al. 2011. Effectiveness of pentavalent rotavirus vaccine against severe disease. *Pediatrics* 128:e267–e275.
382. Wang FT, Mast TC, Glass RJ, Loughlin J, Seeger JD. 2010. Effectiveness of the pentavalent rotavirus vaccine in preventing gastroenteritis in the United States. *Pediatrics* 125:e208–213.
383. Begue RE, Perrin K. 2010. Reduction in gastroenteritis with the use of pentavalent rotavirus vaccine in a primary practice. *Pediatrics* 126:e40–e45.
384. Chang HG, Smith PF, Tserenpuntsag B, Markey K, Parashar U, Morse DL. 2010. Reduction in hospitalizations for diarrhea and rotavirus infections in New York state following introduction of rotavirus vaccine. *Vaccine* 28:754–758.
385. Clark HF, Lawley D, Matthijnssens J, DiNubile MJ, Hodinka RL. 2010. Sustained decline in cases of rotavirus gastroenteritis presenting to the Children's Hospital of Philadelphia in the new rotavirus vaccine era. *Pediatr Infect Dis J* 29:699–702.
386. Cortese MM, Tate JE, Simonsen L, Edelman L, Parashar UD. 2010. Reduction in gastroenteritis in United States children and correlation with early rotavirus vaccine uptake from national medical claims databases. *Pediatr Infect Dis J* 29:489–494.
387. Cortes JE, Curns AT, Tate JE, Cortese MM, Patel MM, Zhou F, Parashar UD. 2011. Rotavirus vaccine and health care utilization for diarrhea in US children. *N Engl J Med* 365:1108–1117.
388. Curns AT, Steiner CA, Barrett M, Hunter K, Wilson E, Parashar UD. 2010. Reduction in acute gastroenteritis hospitalizations among US children after introduction of rotavirus vaccine: Analysis of hospital discharge data from 18 US states. *J Infect Dis* 201:1617–1624.
389. Custodio H, Masnita-Iusan C, Wludyka P, Rathore MH. 2010. Change in rotavirus epidemiology in northeast Florida after the introduction of rotavirus vaccine. *Pediatr Infect Dis J* 29:766–767.
390. Payne DC, Staat MA, Edwards KM, Szilagyi PG, Weinberg GA, Hall CB, Chappell J, et al. 2011. Direct and indirect effects of rotavirus vaccination upon childhood hospitalizations in 3 US Counties, 2006–2009. *Clin Infect Dis* 53:245–253.
391. Tate JE, Panozzo CA, Payne DC, Patel MM, Cortese MM, Fowlkes AL, Parashar UD. 2009. Decline and change in seasonality of US rotavirus activity after the introduction of rotavirus vaccine. *Pediatrics* 124:465–471.
392. Tate JE, Cortese MM, Payne DC, Curns AT, Yen C, Esposito DH, Cortes JE, et al. 2011. Uptake, impact, and effectiveness of rotavirus vaccination in the United States: Review of the first 3 years of postlicensure data. *Pediatr Infect Dis J* 30:S56–S60.
393. Tate JE, Mutuc JD, Panozzo CA, Payne DC, Cortese MM, Cortes JE, Yen C, et al. 2011. Sustained decline in rotavirus detections in the United States following the introduction of rotavirus vaccine in 2006. *Pediatr Infect Dis J* 30:S30–S34.
394. Staat MA, Rice MA, Donauer S, Payne DC, Bresee JS, Mast TC, Curns AT, et al. 2010. Estimating the rotavirus hospitalization disease burden and trends, using capture-recapture methods. *Pediatr Infect Dis J* 29:1083–1086.

395. Yen C, Tate JE, Wenk JD, Harris JM, II, Parashar UD. 2011. Diarrhea-associated hospitalizations among US children over 2 rotavirus seasons after vaccine introduction. *Pediatrics* 127:e9–e15.
396. Donauer S, Payne DC, Edwards KM, Szilagyi PG, Hornung RW, Weinberg GA, Chappell J, Hall CB, Parashar UD, Staat MA. 2013. Determining the effectiveness of the pentavalent rotavirus vaccine against rotavirus hospitalizations and emergency department visits using two study designs. *Vaccine* 31:2692–2697.
397. Cortese MM, Immergluck LC, Held M, Jain S, Chan T, Grizas AP, Khizer S, et al. 2013. Effectiveness of monovalent and pentavalent rotavirus vaccine. *Pediatrics* 132(1):e25–e33. doi:10.1542/peds.2012-3804.
398. Payne DC, Boom JA, Staat MA, Edwards KM, Szilagyi PG, Klein EJ, Selvarangan R, et al. 2013. Effectiveness of pentavalent and monovalent rotavirus vaccines in concurrent use among US children <5 years of age, 2009–2011. *Clin Infect Dis* 57:13–20.
399. Kilgore A, Donauer S, Edwards KM, Weinberg GA, Payne DC, Szilagyi PG, Rice M, et al. 2013. Rotavirus-associated hospitalization and emergency department costs and rotavirus vaccine program impact. *Vaccine* 31:4164–4171.
400. Mast TC, Khawaja S, Espinoza F, Paniagua M, Del Carmen LP, Cardellino A, Sanchez E. 2011. Case-control study of the effectiveness of vaccination with pentavalent rotavirus vaccine in Nicaragua. *Pediatr Infect Dis J* 30:e209–e215.
401. Patel M, Pedreira C, De Oliveira LH, Umana J, Tate J, Lopman B, Sanchez E, et al. 2012. Duration of protection of pentavalent rotavirus vaccination in Nicaragua. *Pediatrics* 130:e365–e372.
402. Orozco M, Vasquez J, Pedreira C, De Oliveira LH, Amador JJ, Malespin O, Andrus J, Tate J, Parashar U, Patel M. 2009. Uptake of rotavirus vaccine and national trends of acute gastroenteritis among children in Nicaragua. *J Infect Dis* 200:S125–S130.
403. Patel M, Pedreira C, De Oliveira LH, Tate J, Orozco M, Mercado J, Gonzalez A, et al. 2009. Association between pentavalent rotavirus vaccine and severe rotavirus diarrhea among children in Nicaragua. *JAMA* 301:2243–2251.
404. Becker-Dreps S, Paniagua M, Dominik R, Cao H, Shah NK, Morgan DR, Moreno G, Espinoza F. 2011. Changes in childhood diarrhea incidence in nicaragua following 3 years of universal infant rotavirus immunization. *Pediatr Infect Dis J* 30:243–247.
405. Ichihara MY, Rodrigues LC, Teles Santos CA, Teixeira MD, De Jesus SR, Alvim De Matos SM, Gagliardi Leite JP, Barreto ML. 2014. Effectiveness of rotavirus vaccine against hospitalized rotavirus diarrhea: A case-control study. *Vaccine* 32(23):2740–2747. doi:10.1016/j.vaccine.2014.01.007.
406. Lanzieri TM, Costa I, Shafi FA, Cunha MH, Ortega-Barria E, Linhares AC, Colindres RE. 2010. Trends in hospitalizations from all-cause gastroenteritis in children younger than 5 years of age in Brazil before and after human rotavirus vaccine introduction, 1998–2007. *Pediatr Infect Dis J* 29:673–675.
407. Lanzieri TM, Linhares AC, Costa I, Kolhe DA, Cunha MH, Ortega-Barria E, Colindres RE. 2011. Impact of rotavirus vaccination on childhood deaths from diarrhea in Brazil. *Int J Infect Dis* 15:e206–210.
408. Safadi MA, Berezin EN, Munford V, Almeida FJ, de Moraes JC, Pinheiro CF, Racz ML. 2010. Hospital-based surveillance to evaluate the impact of rotavirus vaccination in Sao Paulo, Brazil. *Pediatr Infect Dis J* 29:1019–1022.
409. Justino MC, Linhares AC, Lanzieri TM, Miranda Y, Mascarenhas JD, Abreu E, Guerra SF, et al. 2011. Effectiveness of the monovalent G1P[8] human rotavirus vaccine against hospitalization for severe G2P[4] rotavirus gastroenteritis in Belem, Brazil. *Pediatr Infect Dis J* 30:396–401.
410. Gurgel RG, Bohland AK, Vieira SC, Oliveira DM, Fontes PB, Barros VF, Ramos MF, et al. 2009. Incidence of rotavirus and all-cause diarrhea in northeast Brazil following the introduction of a national vaccination program. *Gastroenterology* 137:1970–1975.
411. Gurgel RQ, Ilozue C, Correia JB, Centenari C, Oliveira SM, Cuevas LE. 2011. Impact of rotavirus vaccination on diarrhoea mortality and hospital admissions in Brazil. *Trop Med Health* 16:1180–1184.
412. Assis AS, Valle DA, Antunes GR, Tibirica SH, Assis RM, Leite JP, Carvalho IP, Rosa e Silva ML. 2013. Rotavirus epidemiology before and after vaccine introduction. *J Pediatr (Rio J)* 89:470–476.
413. Quintanar-Solares M, Yen C, Richardson V, Esparza-Aguilar M, Parashar UD, Patel MM. 2011. Impact of rotavirus vaccination on diarrhea-related hospitalizations among children <5 years of age in Mexico. *Pediatr Infect Dis J* 30:S11–S15.
414. Richardson V, Hernandez-Pichardo J, Quintanar-Solares M, Esparza-Aguilar M, Johnson B, Gomez-Altamirano CM, Parashar U, Patel M. 2010. Effect of rotavirus vaccination on death from childhood diarrhea in Mexico. *N Engl J Med* 362:299–305.
415. Richardson V, Parashar U, Patel M. 2011. Childhood diarrhea deaths after rotavirus vaccination in Mexico. *N Engl J Med* 365:772–773.

416. Molto Y, Cortes JE, De Oliveira LH, Mike A, Solis I, Suman O, Coronado L, Patel MM, Parashar UD, Cortese MM. 2011. Reduction of diarrhea-associated hospitalizations among children aged <5 Years in Panama following the introduction of rotavirus vaccine. *Pediatr Infect Dis J* 30:S16–20.
417. Yen C, Armero Guardado JA, Alberto P, Rodriguez Araujo DS, Mena C, Cuellar E, Nolasco JB, et al. 2011. Decline in rotavirus hospitalizations and health care visits for childhood diarrhea following rotavirus vaccination in El Salvador. *Pediatr Infect Dis J* 30:S6–S10.
418. de Palma O, Cruz L, Ramos H, de Baires A, Villatoro N, Pastor D, de Oliveira LH, et al. 2010. Effectiveness of rotavirus vaccination against childhood diarrhoea in El Salvador: Case-control study. *BMJ* 340:c2825.
419. Desai R, Oliveira LH, Parashar UD, Lopman B, Tate JE, Patel MM. 2011. Reduction in morbidity and mortality from childhood diarrhoeal disease after species A rotavirus vaccine introduction in Latin America: A review. *Mem Inst Oswaldo Cruz* 106:907–911.
420. De Oliveira LH, Giglio N, Ciapponi A, Garcia Marti S, Kuperman M, Sanwogou NJ, Ruiz-Matus C, Marinho de Sousa MF. 2013. Temporal trends in diarrhea-related hospitalizations and deaths in children under age 5 before and after the introduction of the rotavirus vaccine in four Latin American countries. *Vaccine* 31 Suppl 3:C99–C108.
421. Dey A, Wang H, Menzies R, Macartney K. 2012. Changes in hospitalisations for acute gastroenteritis in Australia after the national rotavirus vaccination program. *Med J Aust* 197:453–457.
422. Skoff TH, Cohn AC, Clark TA, Messonnier NE, Martin SW. 2012. Early impact of the US Tdap vaccination program on pertussis trends. *Arch Pediatr Adolesc Med* 166:344–349.
423. Guris D, Jumaan AO, Mascola L, Watson BM, Zhang JX, Chaves SS, Gargiullo P, Perella D, Civen R, Seward JF. 2008. Changing varicella epidemiology in active surveillance sites—United States, 1995–2005. *J Infect Dis* 197 Suppl 2:S71–S75.
424. Goodson JL, Masresha BG, Wannemuehler K, Uzicanin A, Cochi S. 2011. Changing epidemiology of measles in Africa. *J Infect Dis* 204 Suppl 1:S205–S214.
425. Msimang VM, Page N, Groome MJ, Moyes J, Cortese MM, Seheri M, Kahn K, Chagan M, Madhi SA, Cohen C. 2013. Impact of rotavirus vaccine on childhood diarrheal hospitalization after introduction into the South African public immunization program. *Pediatr Infect Dis J* 32:1359–1364.
426. Dennehy PH. 2013. Treatment and prevention of rotavirus infection in children. *Curr Infect Dis Rep* 15:242–250.
427. Gastanaduy PA, Curns AT, Parashar UD, Lopman BA. 2013. Gastroenteritis hospitalizations in older children and adults in the United States before and after implementation of infant rotavirus vaccination. *JAMA* 310:851–853.
428. Hemming M, Räsänen S, Huhti L, Paloniemi M, Salminen M, Vesikari T. 2013. Major reduction of rotavirus, but not norovirus, gastroenteritis in children seen in hospital after the introduction of RotaTeq vaccine into the National Immunization Programme in Finland. *Eur J Pediatr* 172(6):739–746. doi:10.1007/s00431-013-1945–3:1–8.
429. Anderson EJ, Shippee DB, Weinrobe MH, Davila MD, Katz BZ, Reddy S, Cuyugan MGKP, et al. 2013. Indirect protection of adults from rotavirus by pediatric rotavirus vaccination. *Clin Infect Dis* 56:755–760.
430. Perdue DG, Bulkow LR, Gellin BG, Davidson M, Petersen KM, Singleton RJ, Parkinson AJ. 2000. Invasive Haemophilus influenzae disease in Alaskan residents aged 10 years and older before and after infant vaccination programs. *JAMA* 283:3089–3094.
431. Kyaw MH, Lynfield R, Schaffner W, Craig AS, Hadler J, Reingold A, Thomas AR, et al. 2006. Effect of introduction of the pneumococcal conjugate vaccine on drug-resistant Streptococcus pneumoniae. *N Engl J Med* 354:1455–1463.
432. Fine PE. 1993. Herd immunity: History, theory, practice. *Epidemiol Rev* 15:265–302.
433. Kim TH, Johnstone J, Loeb M. 2011. Vaccine herd effect. *Scand J Infect Dis* 43:683–689.
434. Curns AT, Panozzo CA, Tate JE, Payne DC, Patel MM, Cortese MM, Parashar UD. 2011. Remarkable postvaccination spatiotemporal changes in United States rotavirus activity. *Pediatr Infect Dis J* 30:S54–S55.
435. Tate JE, Haynes A, Payne DC, Cortese MM, Lopman BA, Patel MM, Parashar UD. 2013. Trends in national rotavirus activity before and after introduction of rotavirus vaccine into the national immunization program in the United States, 2000 to 2012. *Pediatr Infect Dis J* 32:741–744.
436. Chen SY, Tsai CN, Chen CL, Chao HC, Lee YS, Lai MW, Chen CC, Huang WL, Chiu CH. 2013. Severe viral gastroenteritis in children after suboptimal rotavirus immunization in Taiwan. *Pediatr Infect Dis J* 32:1335–1339.
437. Buttery JP, Standish J, Bines JE. 2014. Intussusception and rotavirus vaccines: Consensus on benefits outweighing recognized risk. *Pediatr Infect Dis J* 33:772–773.

438. Jiang J, Jiang B, Parashar U, Nguyen T, Bines J, Patel MM. 2013. Childhood intussusception: A literature review. *PLoS One* 8:e68482.
439. Lynch M, Shieh WJ, Bresee JS, Tatti KM, Gentsch JR, Jones T, Jiang B, et al. 2006. Intussusception after administration of the rhesus tetravalent rotavirus vaccine (Rotashield): The search for a pathogenic mechanism. *Pediatrics* 117:e827–e832.
440. Belongia EA, Irving SA, Shui IM, Kulldorff M, Lewis E, Yin R, Lieu TA, et al. 2010. Real-time surveillance to assess risk of intussusception and other adverse events after pentavalent, bovine-derived rotavirus vaccine. *Pediatr Infect Dis J* 29:1–5.
441. Buttery JP, Danchin MH, Lee KJ, Carlin JB, McIntyre PB, Elliott EJ, Booy R, Bines JE. 2011. Intussusception following rotavirus vaccine administration: Post-marketing surveillance in the National Immunization Program in Australia. *Vaccine* 29:3061–3066.
442. Carlin JB, Macartney KK, Lee KJ, Quinn HE, Buttery J, Lopert R, Bines J, McIntyre PB. 2013. Intussusception risk and disease prevention associated with rotavirus vaccines in Australia's National Immunization Program. *Clin Infect Dis* 57:1427–1434.
443. Haber P, Patel M, Pan Y, Baggs J, Haber M, Museru O, Yue X, Lewis P, Destefano F, Parashar UD. 2013. Intussusception after rotavirus vaccines reported to US VAERS, 2006–2012. *Pediatrics* 131:1042–1049.
444. Yih WK, Lieu TA, Kulldorff M, Martin D, McMahill-Walraven CN, Platt R, Selvam N, Selvan M, Lee GM, Nguyen M. 2014. Intussusception risk after rotavirus vaccination in U.S. infants. *N Engl J Med* 370:503–512.
445. Patel MM, Lopez-Collada VR, Bulhoes MM, De Oliveira LH, Bautista Marquez A, Flannery B, Esparza-Aguilar M, et al. 2011. Intussusception risk and health benefits of rotavirus vaccination in Mexico and Brazil. *N Engl J Med* 364:2283–2292.

11 Emerging Foodborne Pathogenic Kobuvirus, Picobirnavirus, and Torovirus

Pattara Khamrin, Niwat Maneekarn, Shoko Okitsu, and Hiroshi Ushijima

CONTENTS

INTRODUCTION

Foodborne and waterborne acute gastroenteritis outbreaks are increasing worldwide. The majority of acute gastroenteritis in humans is caused by viral infections. Among these, caliciviruses (norovirus and sapovirus), rotavirus, enteric adenovirus, and astrovirus have been reported as the most important etiologic viral agents [1,2]. In recent years, several novel viruses discovered in both humans and animals have been potentially linked with gastroenteritis. These novel viruses include kobuvirus, saffold virus, cosavirus, torovirus, picobirnavirus, and pestivirus [3–9]. However, some of these viruses are yet to be associated with clinical disease.

KOBUVIRUSES (AICHIVIRUS A, AICHIVIRUS B, AND AICHIVIRUS C)

Members of the *Kobuvirus* genus (*Picornaviridae* family) include three different species. The viruses were initially named according to the basis of their host tropisms. The picornaviruses isolated from humans were termed *Aichi virus*, viruses from cattle were termed *bovine*

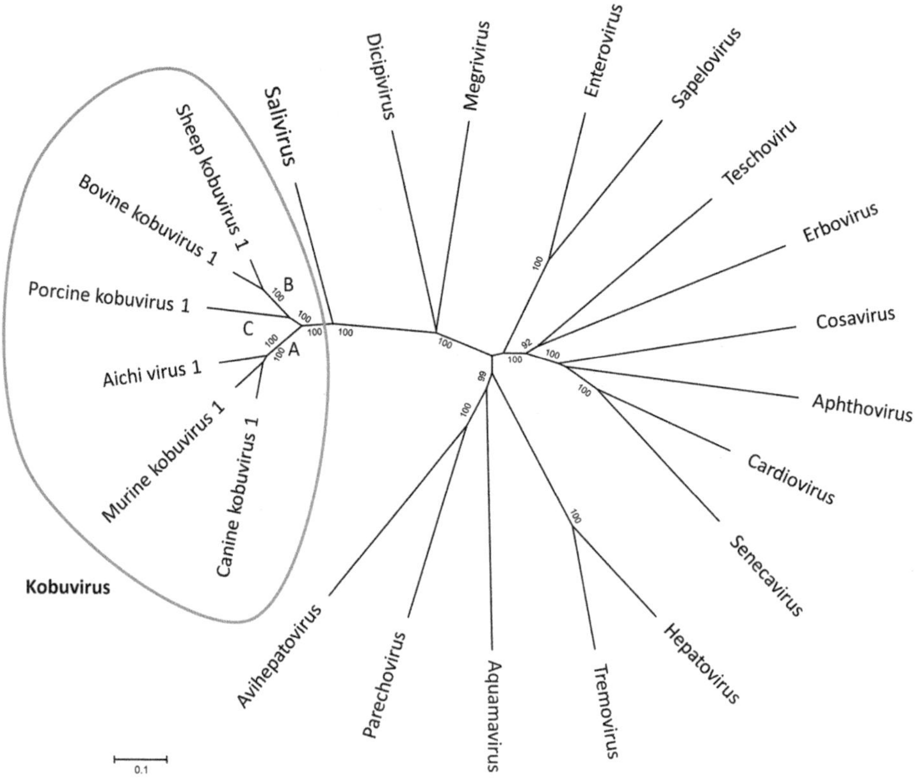

FIGURE 11.1 Phylogenetic relationship between kobuvirus A, B, and C and other picornaviruses, based on nucleotide sequences of the complete protein coding region. The scale bar at the bottom left indicates the number of amino acid substitutions per site, and bootstrap values (>80) are indicated for the corresponding nodes.

kobuvirus, and viruses from pigs were termed *porcine kobuvirus*. The International Committee on Taxonomy of Viruses (ICTV) has renamed these viruses, however, and Aichi virus, bovine kobuvirus, and porcine kobuvirus are now termed *Aichivirus A*, *Aichivirus B*, and *Aichivirus C*, respectively (Figure 11.1). In addition to these three host species, kobuviruses have also been detected in other animal species, including mice, dogs, and sheep. Murine and canine kobuviruses are genetically related to Aichivirus A, whereas ovine kobuvirus is similar to Aichivirus B (Figure 11.1).

The prototype strain in the *Kobuvirus* genus was found in Aichi prefecture, Japan, in 1989 [10]. Spherical virus-like particles of approximately 30 nm in diameter were detected using electron microscopy (EM) in a fecal specimen from a patient with gastroenteritis and linked with the consumption of contaminated oysters [11]. The word "kobu" means "bump" or "knob" in Japanese, because the prominent morphological characteristic of the virus particle is that it appears to be "bumpy" under EM [12]. Aichivirus B was first identified in contaminated culture medium from HeLa cells [13]. Morphologically, Aichivirus B resembles Aichivirus A, although genetically it is relatively distinct [13]. Therefore, Aichivirus B was classified as another species in the kobuvirus genus. Aichivirus C was discovered in pigs by two different research groups in Hungary and China [14,15]. The complete nucleotide sequences of porcine kobuvirus strains from Hungary and China demonstrated that they were considerably distinct from Aichivirus A and Aichivirus B and were therefore described as a new species [16,17].

CLASSIFICATION OF KOBUVIRUSES

The *Picornaviridae* family is genetically highly diverse and currently comprises 17 genera, many of which consist of several species, subspecies, and genotypes [18].

The kobuvirus genome is a single-stranded RNA molecule about 8.2–8.3 kb in length and contains a large open reading frame (ORF) encoding a single polyprotein [13,16,17,19]. The kobuvirus genome can be divided into three distinct functional protein–encoding regions termed *P1* (structural proteins), *P2*, and *P3* (nonstructural proteins). The organization of the kobuvirus genome consists of VPg (a viral protein that attaches to the 5′UTR), the 5′UTR and a genome that encodes a leader protein, three structural viral proteins (VP0, VP3, and VP1), and seven nonstructural proteins (2A–2C and 3A–3D), followed by a 3′UTR and poly(A) tail (Figure 11.2).

Kobuvirus strains have been isolated from various host species, and genetic variability between strains has allowed for their classification into three different species, Aichivirus A, B, and C (Figure 11.1). The current taxonomy of kobuvirus demonstrates that Aichivirus A comprises three distinct members, including Aichi virus 1 (found in humans), canine kobuvirus 1, and murine kobuvirus 1 [11,20,21]. Aichivirus B consists of two members, bovine kobuvirus 1 and ovine kobuvirus 1 [13,22], while for Aichivirus C, only a single type of porcine kobuvirus 1 has been identified.

Based on nucleotide and deduced amino acid (aa) sequences, at least three genotypes (A, B, and C) have been described for Aichi virus 1 [23]. Several studies have reported that the 3CD junction or 3D region of Aichi virus 1 is more conserved than the VP1 region [3,16,19,24]. In viral terms, this is not unexpected as nonstructural proteins are not under the same immune selection pressures as capsid proteins. The VP1 region codes for the structural, antigenic viral protein and is suitable for distinguishing subtypes or genotypes, whereas the more conserved 3CD junction or 3D sequences may not provide sufficient sequence diversity for subtyping or genotyping of Aichi virus 1.

The prototype strain of Aichi virus 1 (A846/88) is cytopathic in African green monkey kidney epithelial cell lines such as BS-C-1 and Vero cells. However, bovine kobuvirus and porcine kobuvirus cannot be readily propagated in cell culture. Therefore, the target cells, specific receptors, viral entry processes, and a clear understanding of viral replication are yet to be elucidated for these animal viruses. One evolutionary mechanism used by Aichi virus 1 to generate genetic diversity is thought to be recombination, as mosaic genomes of Aichi virus 1 have been found in several strains; recombination is a common trait in picornaviruses.

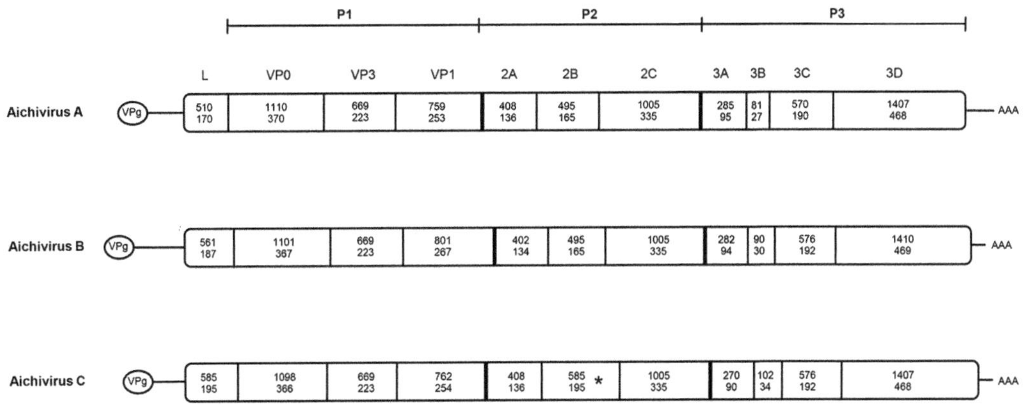

FIGURE 11.2 Genome organization of kobuviruses A (Aichi virus), B (bovine kobuvirus), and C (porcine kobuvirus). P1 represents the region encoding viral structural proteins; P2 and P3 represent regions encoding nonstructural proteins. Protein names are given at the top. The upper and lower numbers in each section represent nucleotide and amino acid lengths respectively. The virally encoded VPg is shown to attach to the 5'UTR (straight line). The 3'UTR is also depicted by a line, with AAA representing the poly(A) tail. * indicates that, within the 2B region of Aichivirus C, 2 copies of a 30 amino acid (90 nucleotides) long motif are observed.

VIRAL TRANSMISSION AND EPIDEMIOLOGY OF KOBUVIRUSES

Aichi virus 1 was first isolated in stool samples from acute gastroenteritis patients in Japan who had consumed raw oysters [10]. Subsequently, several outbreaks caused by Aichi virus 1 were linked with contaminated oysters or seafood consumption [12,23,26]. In addition, Aichi virus 1 has been found in untreated sewage water [26–28]. This presumably suggests that the kobuviruses are transmitted primarily via the fecal–oral route, either person-to-person or by ingestion of contaminated food.

Aichi virus 1 has been detected in numerous countries [23,29–34], and studies have suggested that transmission occurs via the fecal–oral route. Aichi virus 1 has been associated with a variety of clinical illnesses including diarrhea, vomiting, fever, purulent conjunctivitis, and respiratory symptoms [29]. Serological studies in Japan found that Aichi virus 1 antibody prevalence was higher in older adults. Approximately 30% of young adults were seropositive for Aichi virus 1, and this increased to 80% by middle age [11]. Studies in Spain, Germany, and Tunisia also confirmed that 70%, 76%, and 92% of the population across all age groups were seropositive against Aichi virus 1, respectively [30,35,36], indicating that it is a prevalent virus worldwide. For example, a report from China revealed that 1.8% of the stool specimens collected from hospitalized children in Shanghai were positive for Aichi virus 1 [37]. In Europe, the prevalence of Aichi virus 1 in stool specimens ranged from 0.5% to 1.5% [23,29–32].

In terms of animal carriage of kobuviruses, canine kobuvirus (canine kobuvirus 1) has been reported in both healthy and diarrheal domestic dogs [20,38], while mouse kobuvirus (murine kobuvirus 1) was found to be widely distributed in wild rodents [21].

The initial molecular epidemiology of Aichivirus B in the stool samples of healthy cattle in Japan found that 16.7% were positive [13]. Since then, Aichivirus B has been reported in several countries worldwide, including Thailand, Korea, Brazil, Belgium, Hungary, the Netherlands, and Italy [39–44], having a prevalence between 1% and 35%. Bovine-like kobuviruses have also been detected in domestic sheep in Hungary and black goats in Korea [22,45].

Aichivirus C has been detected in symptomatic (diarrheal) pigs from numerous countries including the United States, Hungary, the Czech Republic, the Netherlands, Brazil, China, Korea, Japan, and Thailand [46–57]. Aichivirus C has also been detected in healthy pigs, and a study has detected Aichivirus C genomes in pig serum samples [48].

DETECTION METHODS

The most reliable marker for the diagnosis of kobuvirus infection is the presence of kobuvirus RNA in stool specimens. Currently, reverse transcription-polymerase chain reaction (RT-PCR) is widely used as a tool for routine diagnosis of kobuvirus infections. RT-PCR detection of the three kobuvirus species can be performed using consensus kobuvirus-specific primers (Table 11.1), designed based on the conserved sequences of Aichivirus A from humans, Aichivirus B from cattle, and Aichivirus C from pigs. Similar to other picornaviruses, typing is usually performed by comparing the sequence of the capsid-encoding region (Table 11.2).

PICOBIRNAVIRUS

Discovery, Prevalence, and Genome

Picobirnavirus (PBV) was initially discovered through the unexpected detection of two genome segments by polyacrylamide gel electrophoresis (PAGE) of RNA extracted from fecal specimens [59]. The specimens were obtained from children that presented with gastroenteritis in Brazil [59]. At the same time, the virus was also detected in fecal specimens from rats [60]. Since this initial discovery in 1988, PBVs have been reported in stool samples from pediatric patients globally, including those from Brazil, Venezuela, France, Italy, Russia, India, Australia, Argentina, the United Kingdom, the

TABLE 11.1
Oligonucleotide Primers for the Detection of Kobuviruses

Primer Name	Region	PCR	Sequence 5′-3′	Sense	PCR Product Size (bp)	Species Specific	Reference
6261	3C–3D	First PCR	ACACTCCCACCTCCCGCCAGTA	+		Aichivirus A	[12]
6779	3C–3D	First PCR	GGAAGAGCTGGGTGTCAAGA	–	519	Aichivirus A	[12]
C94b	3C–3D	Second PCR	GACTTCCCCGGAGTCGTCGTCT	+		Aichivirus A	[12]
246k	3C–3D	Second PCR	GACATCCGGTTGACGTTGAC	–	223	Aichivirus A	[12]
C94b	3C	–	GACTTCCCCGGAGTCGTCGTCT	+		Aichivirus A	[12]
AiMP-R	3C	–	GCRGAGAATCCRCTCGTRCC	–	158	Aichivirus A	[25]
10f	3D	–	GATGCTCCTCGGTGGTCTCA	+		Aichivirus B	[13]
10r	3D	–	GTCGGGGTCCATCACAGGGT	–	631	Aichivirus B	[13]
UNIV-kobu-F	3D	–	TGGAYTACAARTGTTTTGATGC	+		Aichivirus A, B, C	[16]
UNIV-kobu-R	3D	–	ATGTTGTTRATGATGGTGTTGA	–	216	Aichivirus A, B, C	[16]

TABLE 11.2
Oligonucleotide Primers for the VP1 Amplification of Aichivirus 1

Primer Name	PCR	Sequence 5′-3′	Sense	PCR Product Size (bp)	Position[a]	Reference
Cap F	–	CAGGTGCCTACCAAGCAA AGAC	+		1104–1125	[24]
Cap1R	–	GGTGAACTCCTGGGACCAG	–	683	1786–1768	[24]
Cap2	–	CCTCGCCTACCCCACCGCC	+		1666–1684	[24]
Cap2R1	–	GAGACCGTGGAARGAGGAGTC	–	652	2317–2297	[24]
Cap3	–	CATAGAGGTCCCYTAYATCTC	+		2149–2169	[24]
Cap3R	–	CATACKGTGTATGTTCCGCGC	–	612	2760–2740	[24]
Cap4	–	CAGTGGCGYGGTGRACTCG	+		2618–2636	[24]
Cap4R	–	GCGATGTAYGTGAAGCACG	–	686	3303–3285	[24]
Cap E	–	CTAGTCGGACCCCACACCGC	+		2897–2915	[24]
CapER	–	GGATGGCCCAGTGGACGTAG	–	958	3854–3835	[24]
F2954	–	GGTGAATCCTTCAACGTACG	+		2954–2973	[23]
R3636	–	GCAAGAGAGCTGGAAGT	–	699	3652–3636	[23]
AiV-VP3-F1	First PCR	CACACCGCCCCTGCGTCRGCCCTCGT	+		2912–2937	[58]
AiV-VP1-R1	First PCR	GAGAGCTGGAAGTCRAAGGG	–	740	3651–3632	[58]
AiV-VP1-F2	Second PCR	CTCGATGCRCCMCAAGACACCGG	+		3023–3045	[58]
AiV-VP1-R2	Second PCR	CCTGACCAGTCCTCCCAWCCGAAGTA	–	530	3552–3527	[58]
AiV-VP1-F3	Second PCR	GTGCTTCACRTACATCGCYGCGG	+		3289–3311	[58]
AiV-VP1-R2	Second-PCR	CCTGACCAGTCCTCCCAWCCGAAGTA	–	264	3552–3527	[58]

[a] Positions are based on the Japanese Aichivirus 1 reference strain (accession number AB010145).

Netherlands, Hungary, Pakistan, and Thailand [61–72]. Literature on the subject reports a varied prevalence of PBVs in humans, depending on the detection method used. Using PAGE and silver staining, the frequency of virus detection ranged from 0.1% to 2.0% [68,73]. With molecular methods, the detection rates of PBV in humans have reached 20% [70]. Several studies have shown that PBV genogroup I is detected more frequently than genogroup II, for example [8].

Growing data from epidemiological studies show that PBVs have been detected in human feces and in a wide range of animal species, indicating that it may be a virus with zoonotic potential. As with other viruses such as influenza and norovirus, a close level of genetic relatedness between human and porcine PBV strains has been described in studies from Hungary, India, and Venezuela [73–75]. PBV has also been detected in a variety of animal species, including pigs, cows, horses, sheep, rabbits, rats, dogs, birds, chickens, and reptiles [8,73,76–80]. In addition, PBVs have been found in environmental samples, including sewage (at high levels) and natural surface waters, using molecular methods [81,82].

CLASSIFICATION

PBV has been classified as a new member of the *Picobirnaviridae* family by the ICTV. PBVs are named based on their characteristically small (pico) and bisegmented (bi) double-stranded RNA (rna) genome. The spherical, icosahedral virion is nonenveloped, 33–41 nm in diameter, and made up of a simple core capsid. The genome consists of two segmented ds-RNA molecules of 2.3–2.6

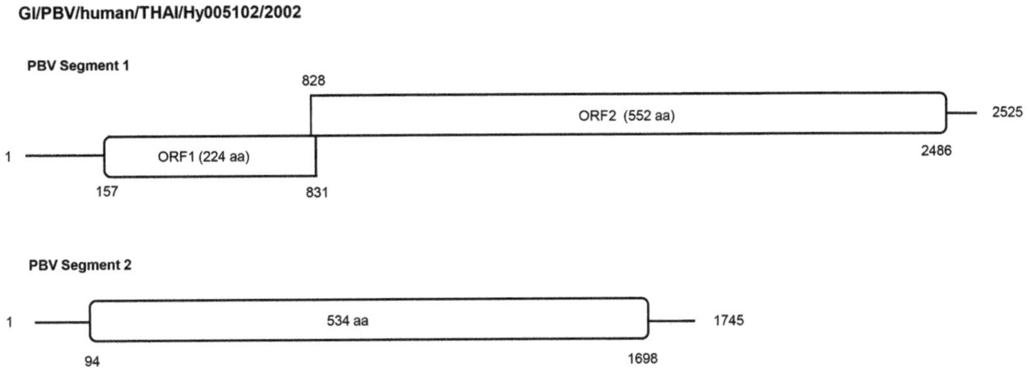

FIGURE 11.3 Genome organization of human picobirnavirus genogroup I, isolated in Thailand in 2002 (GI/PBV/human/THAI/Hy005102/2002, GenBank accession number NC_007026 for segment 1 and NC_007027 for segment 2). This two-segmented virus encodes three large viral proteins, as depicted by the open boxes. In segment 1, ORF1 encodes a protein of unknown function, and ORF2 encodes the viral capsid. Segment 2 encodes the viral RNA-dependent RNA polymerase. Nucleotide positions within the genome are shown outside the boxes (encoded sequence).

TABLE 11.3
Oligonucleotide Primers for Detection and Differentiation of Picobirnaviruses

Primer Name	Sequence 5′-3′	Sense	PCR Product Size (bp)	Position	Genogroup Specific	Reference
PicoB25	TGGTGTGGATGTTTC	+		665–679[a]		[84]
PicoB43	ARTGYTGGTCGAACTT	–	201	865–850[a]	Genogroup I	[84]
PicoB23	CGGTATGGATGTTTC	+		685–699[b]		[84]
PicoB24	AAGCGAGCCCATGTA	–	369	1053–1039b	Genogroup II	[84]

[a] Positions are based on the 1-CHN-97 (China) reference strain (accession number AF246939).
[b] Positions are based on the 4-GA-91 (United States) reference strain (accession number AF246940).

and 1.5–1.9 kbp for segment 1 and segment 2, respectively (Figure 11.3) [61,83]. The RNA segment 1 has two ORFs, encoding 224 and 552 aa proteins, respectively. ORF1 encodes a protein of unknown function, whereas ORF2 encodes the capsid protein. The smaller genomic segment 2 contains a single ORF encoding the 534 aa viral RNA-dependent RNA polymerase (RdRp) (Figure 11.3).

Molecular and epidemiological studies carried out in several countries have shown that human PBVs are highly diverse and can be classified into two genogroups. Genogroup I contains the prototype strain GI/PBV/human/China/1-CHN-97/1997 and genogroup II contains the prototype strain GII/PBV/human/USA/4-GA-91/1991. A nomenclature scheme for PBV has been proposed [85], and recommended that the genogroup specificity of PBV be classified as genogroup I, II, non-I, and non-II. Genogroup-specific primers targeting partial fragments of the RdRp gene of segment 2 revealed specific PCR fragments of 201 bp for genogroup I and 369 bp for genogroup II (Table 11.3) [84]. However, some stool samples that were positive for PBV by PAGE could not be amplified by these primers. As such, they should be classified as non-I and non-II PBVs as they do not belong to either of the genogroups. The current nomenclature system for PBV has been proposed to include the genogroup specificity, abbreviation of the genus (i.e., PBV), host of origin, country of first isolation, strain identifier, and year of isolation. For example, GI/PBV/human/THAI/Hy005102/2002 designates a GI PBV Hy005102 strain that was detected in a human in Thailand during 2002.

The site of viral primary replication for PBV has not been identified, and details of the replication cycle remain unclear as no animal model of infection currently exists.

VIRAL TRANSMISSION AND BURDEN OF DISEASE

The role of PBV as an etiological agent of gastroenteritis has not been clearly established, as this virus has been detected in the feces of subjects with and without diarrheal symptoms. Furthermore, mixed infections with PBV and other enteric agents have been commonly detected. However, most PBVs, particularly those of human origin, have been found in stools from acute gastroenteritis patients. The virus is thought to be transmitted through fecal–oral contact and the major route is likely to be waterborne infection, although data are lacking.

Although the association between PBV and diarrhea or other specific symptoms is not well understood, a potential link between PBV infection and diarrheal symptoms has been established in immunocompromised hosts [86–88]. Reports from the United States, Argentina, and Venezuela demonstrated that PBVs are common in immunocompromised patients and are more frequently detected in HIV-infected patients with diarrhea than in those without diarrhea [86–88]. Prolonged excretion of PBV has been observed for periods between 45 days and 7 months [86–88]. Chronic infections with enteric viruses in immunocompromised hosts are not uncommon, with chronic norovirus infections frequently reported, for example. These findings demonstrated the potential etiological role of PBV in diarrheal disease in immunocompromised patients. However, several studies have demonstrated that PBVs are often identified in mixed infections in patients with acute gastroenteritis along with other viral enteric agents (i.e., rotavirus, astrovirus, and norovirus) [61,67]. These data raise doubts about the true etiologic role of PBV in causing gastroenteritis, and further work is therefore needed. Although PBV may not act as the major causative agent of diarrhea, whether or not this virus might be a cofactor for increased pathogenesis by enteric viral infections remains unclear. In summary, this virus has not been well studied to date, and many questions remain in relation to its role in disease and ability to exacerbate gastroenteritis, particularly in patients already infected with an enteric virus.

TOROVIRUSES

History, Biology, Classification, and Pathogenesis

Toroviruses are gastrointestinal infectious agents that have been reported to infect humans and various animal species. Under EM, toroviruses are morphologically unique and typically appear as a collection of kidney-shaped particles with helical nucleocapsids, surrounded by an envelope containing spikes of approximately 100–170 nm in diameter. The term *torovirus* refers to the tubular nucleocapsid and comes from the Latin word "torus," denoting their circular convex shape.

Toroviruses, together with coronaviruses, belong to the *Coronaviridae* family of the *Nidovirales* order [89,90]. Toroviruses identified to date have been grouped on the basis of host tropism by the ICTV. Currently, four species are recognized in the torovirus genus; (i) equine torovirus (EToV), initially known as Berne virus; (ii) bovine torovirus (BToV), originally known as Breda virus; (iii) human torovirus (HToV); and (iv) porcine torovirus (PToV).

EToV was the first torovirus identified and was isolated in 1972 from the feces of a horse in Berne, Switzerland, hence the initial name Berne virus [91]. A morphologically related virus was later found during an outbreak of neonatal calf diarrhea in a farm from Breda, Iowa. This particular viral agent was initially designated as "Breda virus" and is now recognized as BToV [92]. Preliminary evidence of the existence of a human torovirus was subsequently reported [93]. Torovirus-like particles were detected in stool samples from patients with gastroenteritis using immune EM (IEM) [93]. Since then, toroviruses have been reported many times in the feces of humans and swine with diarrhea worldwide [7,89,94–96].

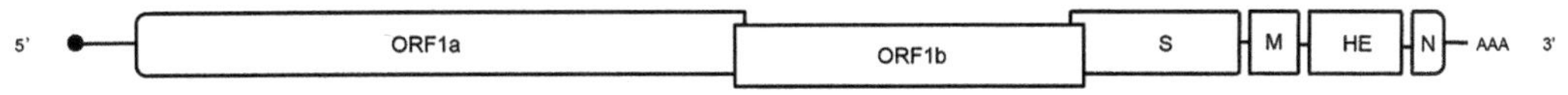

FIGURE 11.4 Torovirus genome organization. ORF1a and ORF1b encode the viral replication enzymes and proteins. The abbreviations S, M, HE, and N represent regions encoding the spike protein, membrane protein, hemagglutinin-esterase, and nucleocapsid protein, respectively. Untranslated regions are shown by lines at the beginning and end of the genome. The genome is flanked by a 5′ cap (black dot) and a 3′ poly(A) tail.

Torovirus genomes consist of a positive-sense, single-stranded RNA molecule of approximately 25–30 kb in length. The 5′ end of the genome contains two large overlapping ORFs (ORF1a and ORF1b), which encode the viral RNA replication proteins. The genome contains four additional ORFs (ORFs 2–5), which encode the spike (S), membrane (M), hemagglutinin-esterase (HE), and nucleocapsid (N) structural proteins, respectively (Figure 11.4) [90].

Since the first torovirus was identified in 1972, research on torovirus remains limited and several issues regarding torovirus pathogenesis remain unexplored. Few studies on torovirus pathogenesis have been performed because very few torovirus strains have ever been successfully propagated in cell culture. One isolate of EToV, which was detected in Berne, Switzerland (previously described in this section), was the first torovirus strain to be successfully propagated in immortalized cells [91]. Another BToV strain obtained from the intestine of a calf with diarrhea in 2004 was also successfully propagated in cell culture, using a human rectal adenocarcinoma cell line (HRT-18). In terms of *in vivo* models, an inoculation trial of BToV in cattle demonstrated that oral or intranasal inoculation of BToV in calves could cause moderate to severe diarrhea. Overall, these data support the relevance of BToV in causing diarrhea in cattle [92,97]. However, further research is needed to clearly define the role this virus has in terms of disease, both in humans and in livestock.

EPIDEMIOLOGY OF TOROVIRUSES

In 1984, torovirus-like particles resembling BToV were detected by EM in a stool specimen collected from a patient with acute gastroenteritis [93]. Since then, torovirus-like particles with BToV antisera crossreactivity have been detected in fecal samples of humans by IEM, and torovirus antigens have also been detected by enzyme-linked immunosorbent assay (ELISA) [98–100]. There are several reports describing the existence of torovirus in human stool specimens collected from several countries worldwide, such as the United States, Canada, the United Kingdom, France, the Netherlands, India, and Brazil [101]. One study found that torovirus was the most common viral pathogen associated with nosocomial viral gastroenteritis in a single pediatric ward in Canada [7]. Clearly, more studies are needed to define the role of this virus in human gastroenteritis.

PATHOGENESIS

According to the accumulated data on torovirus pathogenesis, the mechanism by which toroviruses induce diarrhea could involve the production of colonic lesions by inducing cytopathic effects (CPEs) and apoptosis in mucosa epithelial cells. Consequently, this results in a reduced ability to absorb water, producing a malabsorptive effect and, finally, inducing the classical symptoms of diarrhea.

Studies on the pathogenesis of BToV in experimentally infected calves demonstrate that BToV causes CPEs in the epithelial cells of the intestinal tract, and subsequently induces villous atrophy and crypt hyperplasia [92,97]. Furthermore, *in vitro* propagation of EToV has demonstrated that the virus induces apoptosis in infected cells at a late stage in the infection cycle. In addition, mucosal epithelial cell apoptosis was associated with gastroenteritis symptoms [102].

TRANSMISSION

Torovirus transmission is thought to occur via the fecal–oral route as toroviruses are found in the feces of infected hosts and in environmental sources [103]. However, it has been reported that toroviruses may also spread by airborne transmission, since BToV has been found in the respiratory tract of calves. Likewise, calves have been successfully infected with BToV by the intranasal route [97].

TOROVIRUSES IN ANIMALS

Heterologous neutralization assays against EToV demonstrated a high seroprevalence of torovirus antibodies in pigs in the Netherlands and Switzerland [104,105], and EToV-specific seroprevalence is high in equine populations in Germany [106]. In addition, neutralizing antibodies against EToV have been detected in sera from other host animals, such as cattle, goats, sheep, pigs, rabbits, and mice [104]. BToV was first detected in the United States in 1982 during an outbreak of diarrhea in cattle [92]. Studies have shown that BToV is distributed worldwide; it has been documented in the United States, Canada, Brazil, South Africa, Japan, South Korea, India, and in Europe, including the United Kingdom, Austria, Germany, Belgium, France, Switzerland, and Italy [89,98,107–113]. Subsequently, molecular techniques have been used to detect PToV in swine fecal specimens from Hungary, Italy, Spain, and South Korea [95,96,114–116].

DETECTION METHODS

For the diagnosis of torovirus infection, several methods are available. These include EM, IEM, enzyme immunoassay (EIA), RT-PCR, real-time RT-PCR, and nucleic acid sequence analysis [95,96,104,105,114–117]. Of these methods, RT-PCR is probably the most widely used for the detection and genotype identification of torovirus infections. This technique has replaced traditional immunological tests and has become the standard method for molecular diagnosis of torovirus infections. Different sets of primers specific to the four different torovirus species, EToV, BToV, HToV, and PToV, have been designed to identify the viruses. Primers used for the detection of genomic RNA from the four species of torovirus are summarized in Table 11.4.

In summary, toroviruses are described as enteric pathogens that cause diarrhea in several mammalian host species including humans, although additional confirmatory data are needed. Toroviruses can be detected in the feces several hours before the onset of clinical symptoms, during the presence of diarrhea, and for several days after infection [92].

CONCLUSION AND FUTURE PERSPECTIVES

Acute gastroenteritis is one of the most common illnesses in humans worldwide. An estimated 25%–30% of all deaths among children younger than 5 years of age are caused by viral infections, particularly by norovirus, rotavirus, sapovirus, and astrovirus [1]. In recent years, several novel viruses have been discovered, mostly due to the advent of advanced molecular screening methods [9]. Kobuvirus, picobirnavirus, and torovirus have also been considered as etiological agents of diarrhea in humans [12,59,93]. The standard laboratory methods for the diagnosis of viral infections are based mainly on viral isolation in cell culture. However, some viruses cannot be isolated using current cell culture systems. For epidemiological studies, the application of RT-PCR and sequencing techniques has become standard for the detection and characterization of these viral pathogens [16,84,118]. While molecular techniques are replacing traditional viral culture with the detection of viral genomes by RT-PCR and sequencing, new molecular techniques are also evolving rapidly. Advances in metagenomic analysis and next-generation sequencing provide new tools for viral discovery approaches. The development of these new technologies has increased the

TABLE 11.4
Oligonucleotide Primers for the Detection of Toroviruses

Primer Name	Region	PCR	Sequence 5′-3′	Sense	PCR Product Size (bp)	Species Specific	Reference
Sense	N-3′UTR	–	TAATGGCACTGAAGACTC	+		EToV, HToV	[118]
Antisense	N-3′UTR	–	ACATAACATCTTACATGG	–	219	EToV, HToV	[118]
293	3′UTR	–	AGCAACCTTGAGGTTGGGTCTGT	+		PToV	[105]
294	3′UTR	–	CTTACATGGAGACACTCAACCA	–	136	PToV	[105]
593	N	–	GTCAGAATAGATCACGCATT	+		PToV	[105]
620	N	–	CGCCAAACTCTGCAACTCAGGTGGA	–	185	PToV	[105]
S5	S	–	GTGTTAAGTTTGTGCAAAAAT	+		BToV	[107]
S7	S	–	TGCATGAACTCTATATGGTGT	–	741	BToV	[107]
1435	M	–	TCTTTGAAGATTGCCAAAA	+		BToV	[116]
1434	M	–	CATCTTCTAAAGATAAGTGG	–	740	BToV	[116]
1344	N	–	GAGAAAGAGCCAAGATGAATT	+		BToV	[116]
294	N	–	CTTACATGGAGACACTCAACCA	–	664	BToV	[116]
ToVM 5′	M	–	AGTATGACCTTTACTGGCTA	+		BToV, PToV	[114]
ToVM 3′	M	–	TAATCTGCAACACCTTG	–	410	BToV, PToV	[114]
–	N	First PCR	ATGAATTCTATGCTTAATCCA	+		BToV	[112]
–	N	First PCR	AATTCAAAGCCACTTTTATTG	–	471	BToV	[112]
–	N	Second PCR	CAAATGCTATGCCATTTCAGC	+		BToV	[112]
–	N	Second PCR	TGGAAACTTCAACAGTGGCAT	–	395	BToV	[112]
–	M	First PCR	TGTTTGAGACCAATTATTGGC	+		BToV	[112]
–	M	First PCR	TACTCAAACTTAACACTAGAC	–	682	BToV	[112]
–	M	Second PCR	CCAAACCCATTTACTGCTCAA	+		BToV	[112]
–	M	Second PCR	GTATAATCTGCAACACCTTGC	–	637	BToV	[112]
–	S	First PCR	GTGTTAAGTTTGTGCAAAAAT	+		BToV	[112]
–	S	First PCR	TGCATGAACTCTATATGGTGT	–	722	BToV	[112]
–	S	Second PCR	CAGAGGTGCCGTTGTTGTGTC	+		BToV	[112]
–	S	Second PCR	ACATAGAGCGGTGTCTGTTGA	–	616	BToV	[112]

(Continued)

TABLE 11.4 (CONTINUED)
Oligonucleotide Primers for the Detection of Toroviruses

Primer Name	Region	PCR	Sequence 5′-3′	Sense	PCR Product Size (bp)	Species Specific	Reference
–	HE	First PCR	CGGCAACACCAGTAACACCAT	+		BToV	[119]
–	HE	First PCR	TAACTAAAACTAATAACACC	–	1172	BToV	[119]
–	HE	Second PCR	GATTGGTGTGGGTTTGGTGA	+		BToV	[119]
–	HE	Second PCR	ATATGCAGAGGAGGTTACATC	–	1094	BToV	[119]
–	M	First PCR	TTCTTACTACACTTTTTGGA	+		BToV	[120]
–	M	First PCR	ACTCAAACTTAACACTAGAC	–	603	BToV	[120]
–	M	Second PCR	TATGTACTATGTTTCCAGCT	+		BToV	[120]
–	M	Second PCR	CCAACACAAATCCGCAACGC	–	409	BToV	[120]
–	N	First PCR	GCCTTTTCCAGACCAGGCCC	+		PToV	[96]
–	N	First PCR	GCAAACCATTGTCCATTAACAC	–	555	PToV	[96]
–	N	Second PCR	ATCTTTGGCAATTGCTTA	+		PToV	[96]
–	N	Second PCR	ACCACGAATAGCAATT	–	175	PToV	[96]

understanding of virus biology and the identification of viral contamination sources to assist in addressing public health concerns related to kobuvirus, picobirnavirus, and torovirus infections.

REFERENCES

1. Clark B, McKendrick M. 2004. A review of viral gastroenteritis. *Curr Opin Infect Dis* 17:461–469.
2. Dennehy PH. 2011. Viral gastroenteritis in children. *Pediatr Infect Dis J* 30:63–64.
3. Reuter G, Boros Á, Pankovics P. 2011. Kobuviruses: A comprehensive review. *Rev Med Virol* 21:32–41.
4. Jones MS, Lukashov VV, Ganac RD, Schnurr DP. 2007. Discovery of a novel human picornavirus in a stool sample from a pediatric patient presenting with fever of unknown origin. *J Clin Microbiol* 45:2144–2150.
5. Kapoor A, Victoria J, Simmonds P, Slikas E, Chieochansin T, Naeem A, Shaukat S, Sharif S, Alam MM, Angez M. 2008. A highly prevalent and genetically diversified Picornaviridae genus in South Asian children. *Proc Natl Acad Sci USA* 105:20482–20487.
6. Wilhelmi I, Roman E, Sanchez-Fauquier A. 2003. Viruses causing gastroenteritis. *Clin Microbiol Infect* 9:247–262.
7. Gubbay J, Al-Rezqi A, Hawkes M, Williams L, Richardson S, Matlow A. 2012. The role of torovirus in nosocomial viral gastroenteritis at a large tertiary pediatric centre. *Can J Infect Dis Med Microbiol* 23:78.
8. Ganesh B, Bányai K, Martella V, Jakab F, Masachessi G, Kobayashi N. 2012. Picobirnavirus infections: Viral persistence and zoonotic potential. *Rev Med Virol* 22:245–256.
9. Tang P, Chiu C. 2010. Metagenomics for the discovery of novel human viruses. *Future Microbiol* 5:177–189.
10. Yamashita T, Kobayashi S, Sakac K, Nakata S, Chiba S, Ishihara Y, Isomura S. 1991. Isolation of cytopathic small round viruses with BS-Cl cells from patients with gastroenteritis. *J Infect Dis* 164:954–957.
11. Yamashita T, Sakae K, Ishihara Y, Isomura S, Utagawa E. 1993. Prevalence of newly isolated, cytopathic small round virus (Aichi strain) in Japan. *J Clin Microbiol* 31:2938–2943.
12. Yamashita T, Sugiyama M, Tsuzuki H, Sakae K, Suzuki Y, Miyazaki Y. 2000. Application of a reverse transcription-PCR for identification and differentiation of Aichi virus, a new member of the Picornavirus family associated with gastroenteritis in humans. *J Clin Microbiol* 38:2955–2961.
13. Yamashita T, Ito M, Kabashima Y, Tsuzuki H, Fujiura A, Sakae K. 2003. Isolation and characterization of a new species of kobuvirus associated with cattle. *J Gen Virol* 84:3069–3077.
14. Reuter G, Boldizsár Á, Kiss I, Pankovics P. 2008. Candidate new species of kobuvirus in porcine hosts. *Emerg Infect Dis* 14:1968.
15. Yu J-m, Jin M, Zhang Q, Li H-y, Li D-d, Xu Z-q, Li J-s, Cui S-x, Yang S-h, Liu N. 2009. Candidate porcine kobuvirus, China. *Emerg Infect Dis* 15:823.
16. Reuter G, Boldizsár Á, Pankovics P. 2009. Complete nucleotide and amino acid sequences and genetic organization of porcine kobuvirus, a member of a new species in the genus kobuvirus, family Picornaviridae. *Arch Virol* 154:101–108.
17. Yu JM, Xu ZQ, Li BW, Zhang Q, Cui SX, Jin M, Duan ZJ. 2011. Analysis and characterization of the complete genome of a member of a new species of kobuvirus associated with swine. *Arch Virol* 156:747–751.
18. Adams M, King A, Carstens E. 2013. Ratification vote on taxonomic proposals to the International Committee on Taxonomy of Viruses. *Arch Virol* 158:2023–2030.
19. Yamashita T, Sakae K, Tsuzuki H, Suzuki Y, Ishikawa N, Takeda N, Miyamura T, Yamazaki S. 1998. Complete nucleotide sequence and genetic organization of Aichi virus, a distinct member of the Picornaviridae associated with acute gastroenteritis in humans. *J Virol* 72:8408–8412.
20. Li L, Pesavento PA, Shan T, Leutenegger CM, Wang C, Delwart E. 2011. Viruses in diarrhoeic dogs include novel kobuviruses and sapoviruses. *J Gen Virol* 92:2534–2541.
21. Phan TG, Kapusinszky B, Wang C, Rose RK, Lipton HL, Delwart EL. 2011. The fecal viral flora of wild rodents. *PLoS Path* 7:e1002218.
22. Reuter G, Boros Á, Pankovics P, Egyed L. 2010. Kobuvirus in domestic sheep, Hungary. *Emerg Infect Dis* 16:869.
23. Ambert-Balay K, Lorrot M, Bon F, Giraudon H, Kaplon J, Wolfer M, Lebon P, Gendrel D, Pothier P. 2008. Prevalence and genetic diversity of Aichi virus strains in stool samples from community and hospitalized patients. *J Clin Microbiol* 46:1252–1258.

24. Pham NTK, Trinh QD, Khamrin P, Nguyen TA, Dey SK, Phan TG, Hoang LP, Maneekarn N, Okitsu S, Mizuguchi M. 2008. Sequence analysis of the capsid gene of Aichi viruses detected from Japan, Bangladesh, Thailand, and Vietnam. *J Med Virol* 80:1222–1227.
25. Khamrin P, Okame M, Thongprachum A, Nantachit N, Nishimura S, Okitsu S, Maneekarn N, Ushijima H. 2011. A single-tube multiplex PCR for rapid detection in feces of 10 viruses causing diarrhea. *J Virol Methods* 173:390–393.
26. Le Guyader FS, Le Saux J-C, Ambert-Balay K, Krol J, Serais O, Parnaudeau S, Giraudon H, Delmas G, Pommepuy M, Pothier P. 2008. Aichi virus, norovirus, astrovirus, enterovirus, and rotavirus involved in clinical cases from a French oyster-related gastroenteritis outbreak. *J Clin Microbiol* 46:4011–4017.
27. Di Martino B, Di Profio F, Ceci C, Di Felice E, Marsilio F. 2013. Molecular detection of Aichi virus in raw sewage in Italy. *Arch Virol* 158:2001–2005.
28. Alcalá A, Vizzi E, Rodríguez-Díaz J, Zambrano JL, Betancourt W, Liprandi F. 2010. Molecular detection and characterization of Aichi viruses in sewage-polluted waters of Venezuela. *Appl Environ Microbiol* 76:4113–4115.
29. Reuter G, Boldizsár Á, Papp G, Pankovics P. 2009. Detection of Aichi virus shedding in a child with enteric and extraintestinal symptoms in Hungary. *Arch Virol* 154:1529–1532.
30. Oh D-Y, Silva P, Hauroeder B, Diedrich S, Cardoso D, Schreier E. 2006. Molecular characterization of the first Aichi viruses isolated in Europe and in South America. *Arch Virol* 151:1199–1206.
31. Kaikkonen S, Räsänen S, Rämet M, Vesikari T. 2010. Aichi virus infection in children with acute gastroenteritis in Finland. *Epidemiol Infect* 138:1166–1171.
32. Jonsson N, Wahlström K, Svensson L, Serrander L, Lindberg AM. 2012. Aichi virus infection in elderly people in Sweden. *Arch Virol* 157:1365–1369.
33. Pham NTK, Khamrin P, Nguyen TA, Kanti DS, Phan TG, Okitsu S, Ushijima H. 2007. Isolation and molecular characterization of Aichi viruses from fecal specimens collected in Japan, Bangladesh, Thailand, and Vietnam. *J Clin Microbiol* 45:2287–2288.
34. Sdiri-Loulizi K, Hassine M, Gharbi-Khelifi H, Sakly N, Chouchane S, Guediche MN, Pothier P, Aouni M, Ambert-Balay K. 2009. Detection and genomic characterization of Aichi viruses in stool samples from children in Monastir, Tunisia. *J Clin Microbiol* 47:2275–2278.
35. Sdiri-Loulizi K, Hassine M, Bour J-B, Ambert-Balay K, Mastouri M, Aho L-S, Gharbi-Khelifi H, Aouni Z, Sakly N, Chouchane S. 2010. Aichi virus IgG seroprevalence in Tunisia parallels genomic detection and clinical presentation in children with gastroenteritis. *Clin Vaccine Immunol* 17:1111–1116.
36. Ribes JM, Montava R, Téllez-Castillo CJ, Fernández-Jiménez M, Buesa J. 2010. Seroprevalence of Aichi virus in a Spanish population from 2007 to 2008. *Clin Vaccine Immunol* 17:545–549.
37. Yang S, Zhang W, Shen Q, Yang Z, Zhu J, Cui L, Hua X. 2009. Aichi virus strains in children with gastroenteritis, China. *Emerg Infect Dis* 15:1703.
38. Kapoor A, Simmonds P, Dubovi EJ, Qaisar N, Henriquez JA, Medina J, Shields S, Lipkin WI. 2011. Characterization of a canine homolog of human Aichivirus. *J Virol* 85:11520–11525.
39. Di Martino B, Di Profio F, Di Felice E, Ceci C, Pistilli MG, Marsilio F. 2012. Molecular detection of bovine kobuviruses in Italy. *Arch Virol* 157:2393–2396.
40. Jeoung H-Y, Lim J-A, Jeong W, Oem J-K, An D-J. 2011. Three clusters of bovine kobuvirus isolated in Korea, 2008–2010. *Virus Genes* 42:402–406.
41. Park S-J, Kim H-K, Song D-S, Moon H-J, Park B-K. 2011. Molecular detection and genetic characterization of kobuviruses in fecal samples collected from diarrheic cattle in Korea. *Infect Genet Evol* 11:1178–1182.
42. Mauroy A, Scipioni A, Mathijs E, Thys C, Thiry E. 2009. Molecular detection of kobuviruses and recombinant noroviruses in cattle in continental Europe. *Arch Virol* 154:1841–1845.
43. Reuter G, Egyed L. 2009. Bovine kobuvirus in Europe. *Emerg Infect Dis* 15:822.
44. Khamrin P, Maneekarn N, Peerakome S, Okitsu S, Mizuguchi M, Ushijima H. 2008. Bovine kobuviruses from cattle with diarrhea. *Emerg Infect Dis* 14:985.
45. Lee M-H, Jeoung H-Y, Lim J-A, Song J-Y, Song D-S, An D-J. 2012. Kobuvirus in South Korean black goats. *Virus Genes* 45:186–189.
46. Sisay Z, Wang Q, Oka T, Saif L. 2013. Prevalence and molecular characterization of porcine enteric caliciviruses and first detection of porcine kobuviruses in U.S. swine. *Arch Virol* 158:1583–1588.
47. Verma H, Mor SK, Abdel-Glil MY, Goyal SM. 2013. Identification and molecular characterization of porcine kobuvirus in U.S. swine. *Virus Genes* 46:551–553.
48. Reuter G, Kecskeméti S, Pankovics P. 2010. Evolution of porcine kobuvirus infection, Hungary. *Emerg Infect Dis* 16:696.

49. Dufkova L, Scigalkova I, Moutelikova R, Malenovska H, Prodelalova J. 2013. Genetic diversity of porcine sapoviruses, kobuviruses, and astroviruses in asymptomatic pigs: An emerging new sapovirus GIII genotype. *Arch Virol* 158:549–558.
50. Barry AF, Ribeiro J, Alfieri AF, van der Poel WH, Alfieri AA. 2011. First detection of kobuvirus in farm animals in Brazil and the Netherlands. *Infect Genet Evol* 11:1811–1814.
51. Wang C, Lan D, Hua X. 2011. Porcine kobuvirus from pig stool specimens in Shanghai, China. *Virus Genes* 43:350–352.
52. Zhang Q, Hu R, Tang X, Wu C, He Q, Zhao Z, Chen H, Wu B. 2013. Occurrence and investigation of enteric viral infections in pigs with diarrhea in China. *Arch Virol* 158:1631–1636.
53. Ge F-F, Yang D-Q, Ju H-B, Wang J, Liu J, Liu P-H, Zhou J-P. 2013. Epidemiological survey of porcine epidemic diarrhea virus in swine farms in Shanghai, China. *Arch Virol* 158:2227–2231.
54. Park S-J, Kim H-K, Moon H-J, Song D-S, Rho S-M, Han J-Y, Nguyen V-G, Park B-K. 2010. Molecular detection of porcine kobuviruses in pigs in Korea and their association with diarrhea. *Arch Virol* 155:1803–1811.
55. An D-J, Jeoung H-Y, Jeong W, Lee HS, Park J-Y, Kim B. 2011. Porcine kobuvirus from pig stool in Korea. *Virus Genes* 42:208–211.
56. Khamrin P, Maneekarn N, Kongkaew A, Kongkaew S, Okitsu S, Ushijima H. 2009. Porcine kobuvirus in piglets, Thailand. *Emerg Infect Dis* 15:2075–2076.
57. Khamrin P, Maneekarn N, Hidaka S, Kishikawa S, Ushijima K, Okitsu S, Ushijima H. 2010. Molecular detection of kobuvirus sequences in stool samples collected from healthy pigs in Japan. *Infect Genet Evol* 10:950–954.
58. Lodder WJ, Rutjes SA, Takumi K, de Roda Husman AM. 2013. Aichi virus in sewage and surface water, the Netherlands. *Emerg Infect Dis* 19:1222.
59. Pereira H, Fialho A, Flewett T, Teixeira J, Andrade Z. 1988. Novel viruses in human faeces. *Lancet* 332:103–104.
60. Pereira H, Flewett T, Candeias J, Barth O. 1988. A virus with a bisegmented double-stranded RNA genome in rat (*Oryzomys nigripes*) intestines. *J Gen Virol* 69:2749–2754.
61. Wakuda M, Pongsuwanna Y, Taniguchi K. 2005. Complete nucleotide sequences of two RNA segments of human picobirnavirus. *J Virol Methods* 126:165–169.
62. Ludert J, Liprandi F. 1993. Identification of viruses with bi-and trisegmented double-stranded RNA genome in faeces of children with gastroenteritis. *Res Virol* 144:219–224.
63. Gallimore C, Appleton H, Green J, Brown D, Lewis D. 1995. Detection and characterisation of bisegmented double-stranded RNA viruses (picobirnaviruses) in human fecal specimens. *J Med Virol* 45:135–140.
64. Cascio A, Bosco M, Vizi E, Giammanco A, Ferraro D, Arista S. 1996. Identification of picobirnavirus from faeces of Italian children suffering from acute diarrhea. *Eur J Epidemiol* 12:545–547.
65. Novikova N, Epifanova N, Fedorova O, Golitsyna L, Kupriianova N. 2002. Detection of picobirnaviruses by electrophoresis of RNA in polyacrylamide gel. *Vopr Virusol* 48:41–43.
66. Bhattacharya R, Sahoo GC, Nayak MK, Saha DR, Sur D, Naik T, Bhattacharya S, Krishnan T. 2006. Molecular epidemiology of human picobirnaviruses among children of a slum community in Kolkata, India. *Infect Genet Evol* 6:453–458.
67. Bhattacharya R, Sahoo GC, Nayak MK, Rajendran K, Dutta P, Mitra U, Bhattacharya M, Naik T, Bhattacharya S, Krishnan T. 2007. Detection of genogroup I and II human picobirnaviruses showing small genomic RNA profile causing acute watery diarrhoea among children in Kolkata, India. *Infect Genet Evol* 7:229–238.
68. Giordano MO, Masachessi G, Martinez LC, Barril PA, Ferreyra LJ, Isa MB, Nates SV. 2008. Two instances of large genome profile picobirnavirus occurrence in Argentinian infants with diarrhea over a 26-year period (1977–2002). *J Infect* 56:371–375.
69. Victoria JG, Kapoor A, Li L, Blinkova O, Slikas B, Wang C, Naeem A, Zaidi S, Delwart E. 2009. Metagenomic analyses of viruses in stool samples from children with acute flaccid paralysis. *J Virol* 83:4642–4651.
70. van Leeuwen M, Williams MM, Koraka P, Simon JH, Smits SL, Osterhaus AD. 2010. Human picobirnaviruses identified by molecular screening of diarrhea samples. *J Clin Microbiol* 48:1787–1794.
71. Flahault A, Hanslik T. 2010. Epidemiology of viral gastroenteritis in France and Europe. *Bull Acad Natl Med* 194:1415–1424; discussion 1424–1415.
72. Bányai K, Jakab F, Reuter G, Bene J, Uj M, Melegh B, Szücs G. 2003. Sequence heterogeneity among human picobirnaviruses detected in a gastroenteritis outbreak. *Arch Virol* 148:2281–2291.

73. Ganesh B, Nataraju SM, Rajendran K, Ramamurthy T, Kanungo S, Manna B, Nagashima S, Sur D, Kobayashi N, Krishnan T. 2010. Detection of closely related Picobirnaviruses among diarrhoeic children in Kolkata: Evidence of zoonoses? *Infect Genet Evol* 10:511–516.
74. Bányai K, Martella V, Bogdán A, Forgách P, Jakab F, Meleg E, Bíró H, Melegh B, Szűcs G. 2008. Genogroup I picobirnaviruses in pigs: Evidence for genetic diversity and relatedness to human strains. *J Gen Virol* 89:534–539.
75. Carruyo GM, Mateu G, Martínez LC, Pujol FH, Nates SV, Liprandi F, Ludert JE. 2008. Molecular characterization of porcine picobirnaviruses and development of a specific reverse transcription-PCR assay. *J Clin Microbiol* 46:2402–2405.
76. Fregolente MCD, de Castro-Dias E, Martins SS, Spilki FR, Allegretti SM, Gatti MSV. 2009. Molecular characterization of picobirnaviruses from new hosts. *Virus Res* 143:134–136.
77. Martínez LC, Masachessi G, Carruyo G, Ferreyra LJ, Barril PA, Isa MB, Giordano MO, Ludert JE, Nates SV. 2010. Picobirnavirus causes persistent infection in pigs. *Infect Genet Evol* 10:984–988.
78. Malik YS, Chandrashekar K, Sharma K, Haq AA, Vaid N, Chakravarti S, Batra M, Singh R, Pandey A. 2011. Picobirnavirus detection in bovine and buffalo calves from foothills of Himalaya and Central India. *Trop Anim Health Prod* 43:1475–1478.
79. Wang Y, Bányai K, Tu X, Jiang B. 2012. Simian genogroup I picobirnaviruses: Prevalence, genetic diversity, and zoonotic potential. *J Clin Microbiol* 50:2779–2782.
80. Ganesh B, Banyai K, Masachessi G, Mladenova Z, Nagashima S, Ghosh S, Nataraju SM, Pativada M, Kumar R, Kobayashi N. 2011. Genogroup I picobirnavirus in diarrhoeic foals: Can the horse serve as a natural reservoir for human infection? *Vet Res* 42:52.
81. Hamza IA, Jurzik L, Überla K, Wilhelm M. 2011. Evaluation of pepper mild mottle virus, human picobirnavirus and Torque teno virus as indicators of fecal contamination in river water. *Water Res* 45:1358–1368.
82. Symonds EM, Griffin DW, Breitbart M. 2009. Eukaryotic viruses in wastewater samples from the United States. *Appl Environ Microbiol* 75:1402–1409.
83. Ghosh S, Kobayashi N, Nagashima S, Naik TN. 2009. Molecular characterization of full-length genomic segment 2 of a bovine picobirnavirus (PBV) strain: Evidence for high genetic diversity with genogroup I PBVs. *J Gen Virol* 90:2519–2524.
84. Rosen BI, Fang Z-Y, Glass RI, Monroe SS. 2000. Cloning of human picobirnavirus genomic segments and development of an RT-PCR detection assay. *Virology* 277:316–329.
85. Fregolente MCD, Gatti MSV. 2009. Nomenclature proposal for picobirnavirus. *Arch Virol* 154:1953–1954.
86. Grohmann GS, Glass RI, Pereira HG, Monroe SS, Hightower AW, Weber R, Bryan RT. 1993. Enteric viruses and diarrhea in HIV-infected patients. *N Engl J Med* 329:14–20.
87. Giordano MO, Martinez LC, Rinaldi D, Gúinard S, Naretto E, Casero R, Yacci MR, Depetris AR, Medeot SI, Nates SV. 1998. Detection of picobirnavirus in HIV-infected patients with diarrhea in Argentina. *J Acquir Immune Defic Syndr Hum Retrovirol* 18:380–383.
88. Ludert JE. 1998. Prevalence of enteric viruses in human immunodeficiency virus seropositive patients in Venezuela. *J Med Virol* 55:288–292.
89. Koopmans M, Horzinek MC. 1994. Toroviruses of animals and humans: A review. *Adv Virus Res* 43:233–273.
90. Snijder EJ, Horzinek MC. 1993. Toroviruses: Replication, evolution and comparison with other members of the coronavirus-like superfamily. *J Gen Virol* 74:2305–2305.
91. Weiss M, Steck F, Horzinek MC. 1983. Purification and partial characterization of a new enveloped RNA virus (Berne virus). *J Gen Virol* 64:1849–1858.
92. Woode G, Reed D, Runnels P, Herrig M, Hill H. 1982. Studies with an unclassified virus isolated from diarrheic calves. *Vet Microbiol* 7:221–240.
93. Beards G, Green J, Hall C, Flewett T, Lamouliatte F, Du Pasquier P. 1984. An enveloped virus in stools of children and adults with gastroenteritis that resembles the Breda virus of calves. *Lancet* 323:1050–1052.
94. Koopmans MP, Goosen ESM, AM LIMA A, Mcauliffe IT, Nataro JP, Barrett LJ, Glass RI, Guerrant RL. 1997. Association of torovirus with acute and persistent diarrhea in children. *Pediatr Infect Dis J* 16:504–507.
95. Pignatelli J, Grau-Roma L, Jimenez M, Segales J, Rodriguez D. 2010. Longitudinal serological and virological study on porcine torovirus (PToV) in piglets from Spanish farms. *Vet Microbiol* 146:260–268.
96. Shin D-J, Park S-I, Jeong Y-J, Hosmillo M, Kim H-H, Kim H-J, Kwon H-J, Kang M-I, Park S-J, Cho K-O. 2010. Detection and molecular characterization of porcine toroviruses in Korea. *Arch Virol* 155:417–422.

97. Pohlenz J, Cheville N, Woode G, Mokresh A. 1984. Cellular lesions in intestinal mucosa of gnotobiotic calves experimentally infected with a new unclassified bovine virus (Breda virus). *Vet Pathol* 21:407–417.
98. Krishnan T, Naik TN. 1997. Electronmicroscopic evidence of torovirus-like particles in children with diarrhoea. *Indian J Med Res* 105:108–110.
99. Beards GM, Brown DW, Green J, Flewett TH. 1986. Preliminary characterisation of torovirus-like particles of humans: Comparison with Berne virus of horses and Breda virus of calves. *J Med Virol* 20:67–78.
100. Koopmans M, Petric M, Glass RI, Monroe SS. 1993. Enzyme-linked immunosorbent assay reactivity of torovirus-like particles in fecal specimens from humans with diarrhea. *J Clin Microbiol* 31:2738–2744.
101. Hoet AE, Saif LJ. 2004. Bovine torovirus (Breda virus) revisited. *Anim Health Res Rev* 5:157–171.
102. Maestre AM, Garzón A, Rodríguez D. 2011. Equine torovirus (BEV) induces caspase-mediated apoptosis in infected cells. *PloS One* 6:e20972.
103. Bosch A. 2010. Human enteric viruses in the water environment: A minireview. *Int Microbiol* 1:191–196.
104. Weiss M, Steck F, Kaderli R, Horzinek M. 1984. Antibodies to Berne virus in horses and other animals. *Vet Microbiol* 9:523–531.
105. Kroneman A, Cornelissen LA, Horzinek MC, de Groot RJ, Egberink HF. 1998. Identification and characterization of a porcine torovirus. *J Virol* 72:3507–3511.
106. Liebermann H. 1989. New types of virus infections of domestic animals in the German Democratic Republic. 1. Serologic survey studies of the distribution of equine torovirus infections in the GDR. *Arch Exp Veterinarmed* 44:251–253.
107. Hoet AE, Nielsen PR, Hasoksuz M, Thomas C, Wittum TE, Saif LJ. 2003. Detection of bovine torovirus and other enteric pathogens in feces from diarrhea cases in cattle. *J Vet Diagn Invest* 15:205–212.
108. Duckmanton L, Carman S, Nagy E, Petric M. 1998. Detection of bovine torovirus in fecal specimens of calves with diarrhea from Ontario farms. *J Clin Microbiol* 36:1266–1270.
109. Vorster JH, Gerdes GH. 1993. Breda virus-like particles in calves in South Africa. *J S Afr Vet Assoc* 64:58.
110. Nogueira JS, Asano KM, de Souza SP, Brandao PE, Richtzenhain LJ. 2013. First detection and molecular diversity of Brazilian bovine torovirus (BToV) strains from young and adult cattle. *Res Vet Sci* 95:799–801.
111. Park SJ, Oh EH, Park SI, Kim HH, Jeong YJ, Lim GK, Hyun BH, Cho KO. 2008. Molecular epidemiology of bovine toroviruses circulating in South Korea. *Vet Microbiol* 126:364–371.
112. Ito T, Okada N, Fukuyama S. 2007. Epidemiological analysis of bovine torovirus in Japan. *Virus Res* 126:32–37.
113. Haschek B, Klein D, Benetka V, Herrera C, Sommerfeld-Stur I, Vilcek S, Moestl K, Baumgartner W. 2006. Detection of bovine torovirus in neonatal calf diarrhoea in Lower Austria and Styria (Austria). *J Vet Med B Infect Dis Vet Public Health* 53:160–165.
114. Pignatelli J, Jimenez M, Luque J, Rejas MT, Lavazza A, Rodriguez D. 2009. Molecular characterization of a new PToV strain: Evolutionary implications. *Virus Res* 143:33–43.
115. Matiz K, Kecskemeti S, Kiss I, Adam Z, Tanyi J, Nagy B. 2002. Torovirus detection in faecal specimens of calves and pigs in Hungary: Short communication. *Acta Vet Hung* 50:293–296.
116. Smits SL, Lavazza A, Matiz K, Horzinek MC, Koopmans MP, de Groot RJ. 2003. Phylogenetic and evolutionary relationships among torovirus field variants: Evidence for multiple intertypic recombination events. *J Virol* 77:9567–9577.
117. Hosmillo MD, Jeong YJ, Kim HJ, Collantes TM, Alfajaro MM, Park JG, Kim HH, et al. 2010. Development of universal SYBR Green real-time RT-PCR for the rapid detection and quantitation of bovine and porcine toroviruses. *J Virol Methods* 168:212–217.
118. Duckmanton L, Luan B, Devenish J, Tellier R, Petric M. 1997. Characterization of torovirus from human fecal specimens. *Virology* 239:158–168.
119. Ito T, Okada N, Okawa M, Fukuyama S, Shimizu M. 2009. Detection and characterization of bovine torovirus from the respiratory tract in Japanese cattle. *Vet Microbiol* 136:366–371.
120. Park SJ, Jeong C, Yoon SS, Choy HE, Saif LJ, Park SH, Kim YJ, et al. 2006. Detection and characterization of bovine coronaviruses in fecal specimens of adult cattle with diarrhea during the warmer seasons. *J Clin Microbiol* 44:3178–3188.

12 Prions

Akikazu Sakudo and Takashi Onodera

CONTENTS

INTRODUCTION

Prions, which are proteinaceous infectious particles, are the causative agents of prion diseases [1] or transmissible spongiform encephalopathies (TSEs) including kuru, Creutzfeldt–Jakob disease (CJD), Gerstmann–Sträussler–Scheinker syndrome (GSS), and fatal familial insomnia (FFI) in humans (Table 12.1). Kuru is a chronic or subacute degenerative neurological disease in humans that was originally discovered in the Fore people of Papua New Guinea as a result of the ritualistic cannibalism endemic within their culture. The elucidation of kuru etiology led to the discovery that worldwide presenile dementias, such as CJD and its variants, are characterized by similar cellular lesions. Furthermore, these diseases are transmissible and are caused by prion agents, which differ fundamentally from other infectious agents such as viruses or bacteria [1]. Among human prion diseases, approximately 85% of CJD is classified as sporadic CJD (sCJD), although the mechanisms of pathogenesis remain unclear. The incidence of CJD is approximately one case per million per year [1]. By contrast, less than 15% of cases of human prion diseases are hereditary [2]. These inherited human prion diseases, such as familial CJD (fCJD), GSS, and FFI, are caused by mutation of the prion protein gene (*PRNP*). Moreover, prions may be transmitted on the surfaces of medical instruments used in neurosurgery [3,4], leading to iatrogenic CJD (iCJD). There are also reports of prion infections derived from injections of human growth hormone, and pituitary gonadotropin, dura mater grafts, corneal transplants, as well as stereotactic electrodes [3].

Animal-associated prion diseases include scrapie in sheep and goats, bovine spongiform encephalopathy (BSE) in cattle, chronic wasting disease (CWD) in cervids, feline spongiform encephalopathy (FSE) in cats, transmissible mink encephalopathy (TME) in mink, and exotic ungulate spongiform encephalopathy in zoonotic animals (Table 12.2). BSE or "mad cow disease" is a striking example of a prion disease, because the prion agent is transmitted to cattle when fed rendered offal derived from the meat and bones of scrapie-infected sheep and probably from BSE-infected cattle to humans, causing variant CJD (vCJD).

The main component of the prion agent is an abnormal isoform of the prion protein (PrP^{Sc}) [5]. A key event in prion disease is the conversion of the host-encoded, cellular isoform of prion protein (PrP^{C}) to PrP^{Sc}. Furthermore, there is increasing evidence that the major component of the

TABLE 12.1
Prion Diseases in Humans

Diseases	Pathogenesis
Iatrogenic CJD (iCJD)	Transmission (human growth hormone, pituitary gonadotropin, dura mater, and corneal transplants, or stereotactic electrodes)
Sporadic CJD (sCJD)	Unknown
Familial CJD (fCJD)	Mutation of the human PrP gene (*PRNP*)
Variant CJD (vCJD)	Transmission
Gerstmann–Sträussler–Scheinker syndrome (GSS)	Mutation of *PRNP*
Fatal familial insomnia (FFI)	Mutation of *PRNP*

TABLE 12.2
Prion Diseases in Animals

Diseases	Affected Animals
Bovine spongiform encephalopathy (BSE)	Cattle
Chronic wasting disease (CWD)	Elk
	Mule deer
Exotic ungulate spongiform encephalopathy	Arabian oryx
	Eland
	Greater kudu
	Gemsbok
	Mouflon
	Nyala
Feline spongiform encephalopathy (FSE)	Albino tigers
	Cats
	Cheetah
	Puma
Scrapie	Goats
	Sheep
Transmissible mink encephalopathy (TME)	Mink

infectious agent is PrP^{Sc} or a prion protein (PrP)-folding intermediate [6]. PrP^{C} is a cell surface–anchored glycoprotein of unknown function [2] that is predominantly found in the central nervous system (CNS) of animals, including humans [7]. PrP^{Sc} is derived from PrP^{C} in a posttranslational process that appears to involve direct molecular interactions between these two proteins [8]. The crucial role of PrP^{C} expression in prion infection and PrP^{Sc} formation has been demonstrated in transgenic mice with an ablated murine *Prnp* [9]. The *Prnp*-knockout mice do not show any symptoms of prion diseases without prion protein (PrP) accumulation. Furthermore, studies using transgenic animals have shown the influence of specific amino acid residues or domains in PrP on prion susceptibility [10–15]. In addition, this experimental system demonstrated a species barrier defined by amino acid residues in *Prnp* [16].

DISCOVERY OF PRIONS AND PRION HYPOTHESIS

In the 1930s, a flock of Karakul sheep was imported from Germany to Iceland, resulting in the introduction of maedi-visna disease in the naïve Icelandic sheep. The striking feature of maedi-visna is the slow progression to disease after primary infection, often taking more than 10 years [17]. Although the disease occurs primarily in sheep, both experimental and natural transmission to goats has been observed. The slow, relentless, and progressive brain infections are almost always fatal. On this basis, parallels have been drawn between maedi-visna disease and scrapie. However, maedi-visna disease is caused by a lentivirus, a subfamily of retroviruses, whereas the scrapie agent proved to be far more enigmatic.

As early as 1967, scientists recognized that the scrapie agent might be an unconventional pathogen [18,19]. Most notable were the results of radiation inactivation studies, which suggested that the scrapie agent might not contain nucleic acid and that the infectious particles may be extremely small. In 1967, Griffith [20] proposed a hypothesis that a protein could acquire a disease-inducing conformation, which eventually led researchers to discover prions.

In a seminal study in 1981, scrapie-associated fibrils (SAFs) were identified in the brains of infected sheep [21]. Furthermore, the scrapie agent could be concentrated by centrifugation [21], which led to the development of improved bioassays as well as a fractionation procedure that allowed the isolation of an unusual protein from scrapie-infected tissue [22]. This protein was found to be insoluble and relatively resistant to proteases, and the corresponding gene was subsequently cloned. The gene, called *Prnp* in animals and *PRNP* in humans, is highly conserved between different mammalian species [5]. *Prnp/PRNP* is now known to be essential for the pathogenesis of common prion diseases.

Stanley Prusiner, who won the Nobel Prize for Physiology or Medicine in 1997, named the scrapie infectious agent *prion* (an anagram of "proin," from "proteinaceous infectious particle"). He proposed that an altered form of PrP caused a fatal encephalopathy, characteristic of scrapie diseases. Furthermore, the protein-only hypothesis put forward by Prusiner suggested that the essential pathogenic component is an altered conformation of the host-encoded PrP, called PrP^{Sc} ("PrP-scrapie"; also called PrP^{res} for "protease-resistant form") [5]. In the simplest case, PrP^{Sc} is proposed to have the property of converting cellular PrP into more copies of pathogenic PrP^{Sc}. This hypothesis is consistent with findings that mice lacking both copies of *Prnp* are resistant to prion infection [9].

CLASSIFICATION OF PRIONS

Various criteria can be used to distinguish prions from viruses. For example, prions are nonimmunogenic in contrast to viruses, which almost always provoke an immune response. Additionally, there is no evidence of nucleic acid within the infectious prion particle, whereas viruses possess a nucleic acid genome that serves as a template for the synthesis of viral progeny. Furthermore, the main component of the prion is PrP^{Sc}, which is encoded by a chromosomal gene, whereas viruses are composed of nucleic acid, proteins, and often other constituents.

Although it is broadly accepted that the main component of prions is PrP^{Sc}, it remains unclear whether PrP^{Sc} is the sole agent responsible for infection or if another molecular species in addition to PrP^{Sc} might also constitute an essential component of the infectious prion agent. Size fractionation of the prion agent has shown that particles of 17–27 nm in diameter (300–600 kDa) display maximal infectivity. These findings suggest that the infectious unit of a prion comprises particle sizes corresponding to 14–28 PrP molecules [23]. Furthermore, recent protein misfolding cyclic amplification (PMCA) analyses, an *in vitro* amplification method for PrP^{Sc}, have shown that metal ions [24], RNA [25], anionic phospholipids [26], polyanions [27], and phosphatidylethanolamine [28] are PrP^{Sc} propagation cofactors. However, cofactor preference depends on the chosen PMCA system, including the precise constituents of the PMCA buffer (e.g., detergents and other additives). Phosphatidylethanolamine is not only a cellular factor that plays a role in prion generation, but it is also an integral component of the infectious prion agent [29]. Thus, the latest evidence suggests that prions are mainly composed of PrP^{Sc}, but also include other biomolecules.

STRUCTURE OF PRION PROTEIN AND ITS CORRESPONDING GENE

Prnp is encoded on chromosome 2, 3, 13, and 13 of mice [30], rats [31,32], sheep [33], and cattle [34], respectively, while *PRNP* is encoded on chromosome 20 in humans [35]. The structure and sequence of *Prnp/PRNP* is highly conserved across a variety of mammalian species. Murine

Prnp spans 38 kb and has three exons, although only the third exon contains the prion protein coding sequence. A similar genomic structure of *Prnp* is found in rats (16 kb), sheep (31 kb), and cattle (35 kb). By contrast, human *PRNP* spans 35 kb and comprises only two exons, with Prnp located within the second exon and an intervening sequence between the first and second exons that resembles the second mammalian exon. The open reading frame (ORF) of *Prnp* is also highly conserved among mammalian species. Intriguingly, the entire ORF of all known mammalian *Prnp/PRNP* genes is always located in the last exon.

DIFFERENCES BETWEEN CELLULAR AND ABNORMAL PRION PROTEINS AND THEIR CONVERSION COFACTORS

PrP^{Sc} accumulation and PrP^{C} deficiency in the CNS is a hallmark of prion disease [36]. Although the sequences of PrP^{C} and PrP^{Sc} are identical, the corresponding secondary structures are distinct. Infrared spectroscopy [37], nuclear magnetic resonance (NMR) spectroscopy [38], and circular dichroism (CD) spectroscopy [39] all showed that PrP^{C} comprises a high proportion of α-helices (~40% of the protein) and relatively little β-sheets (3% of the protein), whereas PrP^{Sc} has a high β-sheets content (~40% of the protein) and an α-helix content of 30%. Unfortunately, it is not possible to determine the high-resolution structure of PrP^{Sc} using general NMR or x-ray analyses because the protein is insoluble.

The prion hypothesis states that a conformational change from PrP^{C} to PrP^{Sc} plays a crucial role in the disease process, suggesting that this transition is essential for prion propagation. Currently, there are two main models describing the conversion of PrP^{C} to PrP^{Sc}; namely, the refolding and seeding models [3]. The refolding model proposes that a single particle of endogenous PrP^{C} interacts with exogenously introduced PrP^{Sc} to generate more PrP^{Sc}. The seeding model proposes that PrP^{Sc} exists in equilibrium with PrP^{C}. In healthy individuals, this equilibrium favors PrP^{C}, thereby resulting in an absence of prion disease. By contrast, an oligomer or short polymers of PrP^{Sc} can interact with PrP to form an infectious unit. The seeding model is supported by the recent application of an *in vitro* amplification method, PMCA, where the destruction of PrP^{Sc} aggregates causes fragmentation of infectious seed in the sonication process. Thus, the PMCA procedure mimics the autocatalytic replication of PrP^{Sc} [40].

Some researchers have speculated that there is an additional chaperone protein (known as protein X) involved in prion propagation [41]. Furthermore, the nonlethal effect of PrP^{C} deficiency in mice raises the possibility of a hypothetical PrP^{C} functional homolog termed "π" [11]. Shadoo is a brain glycoprotein that displays similarity to the unstructured region of PrP. However, the generation of Shadoo/PrP double knockout mice suggests that Shadoo is not π [42]. Nonetheless, it is likely that the role of π could be played by other proteins.

CLINICAL FEATURES

Prion diseases are a group of neurodegenerative disorders affecting mammals (Tables 12.1 and 12.2), which are transmissible under certain circumstances. However, unlike other transmissible disorders, prion diseases can also be caused by mutations in the host *Prnp/PRNP*. The mechanism by which prions spread among sheep and goats that develop natural scrapie is unknown. CWD, TME, BSE, FSE, and exotic ungulate spongiform encephalopathy are all thought to occur after the consumption of prion-infected materials such as meat and bone meal (MBM). Typical symptoms of scrapie include hyperexcitability, pruritus, and myoclonus with rapid progression to tetraparesis, and ultimately death [3]. The clinical symptoms of BSE are insidious, resulting in changes in temperament, including aggressive behavioral changes and uncoordinated gait [43].

As is the case with animal prion diseases, kuru among the Fore people of Papua New Guinea is thought to have resulted from the consumption of brain tissue during ritualistic cannibalism of their dead relatives [44–47]. GSS, FFI, and fCJD are all dominant, inherited prion diseases that have been shown to be genetically linked to mutations in *PRNP*. These inherited prion diseases feature many different mutations in the PrP gene (Figure 12.1). However, iatrogenic CJD cases can be traced to the inoculation of prions via human pituitary–derived growth hormone, corneal transplants, dura mater grafts, or cerebral electrode implants, although the number of cases recorded to date is small [3,5]. Most cases of CJD are sCJD, probably the result of somatic mutation of the PrP gene or the spontaneous conversion of PrP^{C} into PrP^{Sc}, while about 10%–15% of CJD cases are caused by germline mutations in *PRNP*. The inherited form of CJD is fCJD. CJD is typically a rapid progressive disease, exhibiting cerebellar ataxia, dementia, myoclonus, visual disturbance, and periodic sharp waves on electroencephalography (EEG), but these symptoms are not associated with fever or humoral immune response such as leukocytosis [48].

PrP has several functional domains [50] including a signal peptide (SP) region at the N-terminus and an octapeptide repeat (OR) region in the N-terminal half, as well as hydrophobic regions (HRs) in the central part (HR1) and C-terminal (HR2) part (Figure 12.1). In addition, disulfide links (S–S), Asn-linked glycosylation (CHO), and a glycosylphosphatidylinositol (GPI) anchor are present.

Some of the mutations and polymorphisms are related either to inherited prion diseases or susceptibility to prion diseases. For instance, polymorphisms of the *PRNP* ORF have been shown to be major determinants of susceptibility to prion diseases in humans. Polymorphisms of human *PRNP* are limited to single amino acids substitutions (e.g., M129V, N171S, and E219K) and to the deletion of one octapeptide repeat in the OR region. Among these mutations, one repeat deletion and N171S are not thought to be related to the onset of, or susceptibility to, prion disease [51,52]. Most studies have focused on the M129V polymorphism, which is related to susceptibility to prion diseases

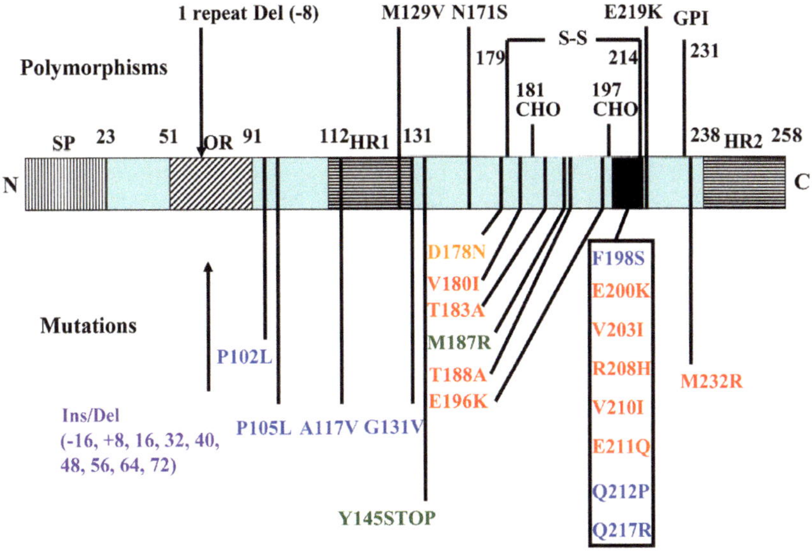

FIGURE 12.1 Polymorphisms and mutations of the prion protein gene (*PRNP*) in humans. The first 22 amino acid residues of the N-terminal region of human PrP are cleaved shortly after translation, and 24 residues from the C-terminal end are cleaved before the addition of the glycosylphosphatidylinositol (GPI) anchor. An octapeptide repeat region (OR) is found in the N-terminal portion of the protein. Polymorphisms (upper) and mutations (lower) of the human PrP gene are shown. GPI: glycosylphosphatidylinositol anchor site; CHO: glycosylation site; SP: signal peptide region; HR1: hydrophobic region in central part of PrP; HR2: hydrophobic region in C-terminal part of PrP; Red: mutations related to Creutzfeldt–Jakob disease (CJD); Blue: mutations related to Gerstmann–Sträussler–Scheinker syndrome (GSS); Purple: mutations related to GSS or CJD; Orange: mutations related to fatal familial insomnia (FFI) or CJD; Green: mutations related to other diseases (Y145STOP: PrP cerebral amyloidosis; M187R: inherited prion encephalopathy with curly PrP deposits). (Cited from figure 1 of Sakudo A. et al. 2010. Structure of the prion protein and its gene: An analysis using bioinformatics and computer simulation. *Curr Protein Pept Sci* 11:166–179 with permission from Bentham Science Publishers.)

including sporadic, iatrogenic, and variant CJD [53–55]. In addition, M129V is closely related to the clinical status and Western blotting patterns of PrP. Interestingly, E219K might have a protective role against the onset of sCJD [56] because E219K has never been found in patients with sCJD.

Different mutations combined with the M129V polymorphism in human *PRNP* generate various phenotypes. The main phenotypes are GSS, which is characterized by PrP plaques in the brain with slow disease progression; fCJD, which shows a CJD-like status; and FFI, which is associated with severe insomnia and neurodegeneration of the thalamus. Interestingly, mutations in PrP related to prion diseases cluster in the region between disulfide bonding sites. Moreover, representative mutations related to GSS, FFI, and other fCJDs have been analyzed. GSS is a group of syndromes showing inherited dementia with cerebellar abnormalities including amyloid plaques. The most common mutation associated with GSS is P102L(129M), that is, P102L with M at position 129. Patients with P102L(129M) show spongiform-related changes in the cerebrum and amyloid plaques in the cerebrum and cerebellum [57]. The P105L(129V) mutation (i.e., P105L with V at position 129) is found in patients with spastic and paralytic GSS, which has a less severe clinical onset than P102L-related GSS and does not show any spongiform in the cerebrum [51]. Y145STOP(129M), corresponding to a Y145STOP with M at position 129, also causes GSS with Alzheimer's neurofibrillary tangles (NFTs) in the cerebral cortex and amyloid deposits in the blood vessels [58]. D178N(129M) is associated with FFI, in which patients display either no visible abnormalities or only slight atrophy in the brain [59]. The most characteristic abnormalities are neurodegeneration and gliosis in the thalamus without spongiform and synaptic patterns of staining with anti-PrP antibody in the subcortical gray matter and the brain stem, cerebellum, and inferior olivary nucleus. Patients with D178N(129V) (i.e., D178N with V at position 129) show typical symptoms of CJD with spongiform-related changes and synaptic PrP deposits, but without amyloid deposits and neurodegeneration of the thalamus and olivary nucleus [60]. As for other mutations related to prion diseases, the deletion and insertion of repeats in the OR (−16, +8, 16, 32, 40, 48, 56, 64, and 72), V180I, V180I-M232R, E200K, V203I, R208H, V210I, E211Q, and M232R have been reported. All these deletions and insertions cause GSS or CJD, with affected individuals showing spongiform, neurodegeneration, gliosis, and amyloid deposits [61,62]. Patients with V180I(129M/V) and V180I(M232R) (i.e., V180I with either M or V at position 129 or R at position 232, respectively) show clinical symptoms of CJD, such as spongiform changes, without amyloid deposits in the cerebral cortex and thalamus [63]. There are numerous reports concerning mutation E200K, where patients display a typical CJD pathological status as well as gliosis [64].

vCJD was first identified by the CJD Surveillance Unit in Great Britain [65], followed by a report of a similar case in France [66]. The patients in these reports differed from typical cases of sCJD as they were relatively young at disease onset (vCJD: 19–39 years; sCJD: 55–70 years) and the duration of the illness was prolonged (vCJD: 7.5–22 months; sCJD: 2.5–6.6 months) [3]. These patients also displayed atypical clinical features, including psychiatric symptoms, ataxia, pain in the extremities, and myoclonus or chorea occurring in the late stage of illness. Dementia was not evident until the final stages of the illness, when it was often accompanied by cortical blindness and akinetic mutism. None of these patients showed changes in EEG readings that were characteristic of sCJD, and all were methionine homozygous at codon 129 without the pathogenic mutations in *PRNP* typically found in humans.

DIAGNOSIS

Conventional Techniques

The specific diagnosis of prion diseases relies on the detection of PrP^{Sc} (Table 12.3). PrP^{Sc} can be distinguished from PrP^{C} by its high β-sheet content [37], its partial resistance to protease digestion [67], and its tendency to form large aggregates both *in vivo* and *in vitro* [22]. The formation of multimeric aggregates can be assumed to be closely related to the infection process [68,69].

TABLE 12.3
Diagnostic Methods for Prion Infections

Methods	Detection Principle	Basic Procedure	References
Aptamer	PrP conformation	RNA aptamers that specifically recognize the conformation of PrP^{C} and/or PrP^{Sc}.	[70]
Bioassay	PK-resistant PrP, incubation time or infectivity titer	Transmission to mice.	[71]
Capillary gel electrophoresis (competitive assay)	PK-resistant PrP	Competition between fluorescein-labeled synthetic PrP peptide and PrP present in samples is assayed by separation of free peptide and antibody-peptide peaks using capillary electrophoresis.	[72,73]
Capillary gel electrophoresis (noncompetitive assay)	ProteinA-antibody-PrP complex	Complex with PrP is separated using fluorescein labeled proteinA[a] and anti-PrP antibody by capillary electrophoresis.	[74]
CDI	PrP conformation	Specific antibody binding to denatured and native forms of PrP.	[75]
Cell blot	PK-resistant PrP	Culture cells on cover slip, directly transfer to membrane, then detect the PK-resistant PrP using anti-PrP antibody.	[76]
Cell culture assay	PK-resistant PrP or infectivity titer	Transmission to cells.	[77]
DELFIA	Insoluble PrP	Quantitate the insoluble PrP extracted at two fixed concentrations of guanidine hydrochloride.	[78]
Electrochemical biosensor	Electrochemical signal induced by binding DNA aptamer and PrP	Detection of PrP using a DNA aptamer-bound carbon nanotube.	[79]
ELISA	PK-resistant PrP	Detect PrP^{res} adsorbed onto microtiter plates by anti-PrP antibody.	[80]
FCS	Aggregation of PrP	PrP aggregates labeled using anti-PrP antibody tagged with fluorescent dyes, resulting in a fluorescent signal measured by dual-color fluorescence intensity distribution analysis.	[81,82]
Filtration blotting	PK-resistant PrP	Detection of filtered/adsorbed PrP^{res} on PVDF membrane by anti-PrP antibody after treatment with PK + SDS/heating.	[83]
Flow microbead immunoassay	PK-resistant PrP	Detect PK-resistant PrP using a flow cytometer with anti-PrP antibody coupled to microbeads.	[84]
Fluorescent analysis using quantum dot	Increased fluorescence signal of quantum dots induced by sandwich binding of PrP	Sandwich detection of PrP using anti-PrP antibody-bound magnetic microbeads and quantum dots.	[85]
FT-IR spectroscopy	Alterations of spectral feature	Analyze FT-IR spectra using statistical analysis.	[86]
Histoblot	PK-resistant PrP	Cryosection is blotted onto a membrane prior to PK treatment and immunolabeling with anti-PrP antibody.	[72]

(*Continued*)

TABLE 12.3 (CONTINUED)
Diagnostic Methods for Prion Infections

Methods	Detection Principle	Basic Procedure	References
Immuno PCR	PK-resistant PrP	Biotinylated anti-PrP antibody is bound to streptavidin HRP with reporter DNA. Reporter DNA is then PCR amplified to detect PrP^{res}.	[87]
Immunohistochemistry	PK-resistant PrP	Immunostaining of tissue sections.	[88]
Mass spectrometry	PrP27-30 (Residual PrP after PK treatment)	Detection of MRM signal by mass spectroscopic analysis.	[89]
Mechanical resonance	Frequency sift induced by PrP binding	Sandwich detection of PrP using anti-PrP antibody-bound mechanical resonator and nanoparticles.	[90]
MUFS	Alterations to spectral characteristics	Analyze spectra of emission excited by ultraviolet radiation.	[91]
PET blot	PK-resistant PrP	Paraffin-embedded tissue section is collected on a membrane, and PrP^{res} is immunolabeled with anti-PrP antibody.	[92,93]
PMCA	PK-resistant PrP	Amplification of misfolded protein by cycles of incubation and sonication.	[94]
RT-QuiC	Formation of amyloid fibrils from recombinant PrP	ThT fluorescence readout of the amyloid seeding assay using conditions selected to minimize the formation of spontaneous (prion-independent) ThT-positive amyloid fibrils.	[95]
Slot blot	PK-resistant PrP	Filter the cell lysate through a nitrocellulose membrane and then detect PrP^{res} using an anti-PrP antibody.	[96]
SOFIA	PK-resistant PrP	Fluorescent detection of PrP^{res} using biotinylated anti-PrP antibody and streptavidin labelled with Rhodamine red X.	[97]
Surface plasmon resonance	Reaction rate of binding DNA aptamer with PrP	Detection of PrP using a DNA aptamer coated on the surface of an electrically conductive polypyrrole.	[98]
Surrogate marker	Change of expression level of 14-3-3 protein, erythroid-specific marker, or plasminogen	Detect expression level changes by two-dimensional gel electrophoresis, differential display reverse-transcriptase PCR, or Western blotting of surrogate marker proteins for prion diseases.	[99–101]
Western blotting	PK-resistant PrP	Detect PrP^{res} on the membrane.	[102]

Source: This table is modified and updated from table 1 in Sakudo A. et al. 2007. Recent developments in prion disease research: Diagnostic tools and *in vitro* cell culture models. *J Vet Med Sci* 69:329–337 with permission from The Japanese Society of Veterinary Science.

Note: CDI, conformation-dependent immunoassay; DELFIA, dissociation-enhanced lanthanide fluorescent immunoassay; ELISA, enzyme-linked immunosorbent assay; FCS, fluorescence correlation spectroscopy; FT-IR, Fourier transform infrared; MRM, multiple reaction monitoring; MUFS, multispectral ultraviolet fluorescence spectroscopy; PCR, polymerase chain reaction; PET, paraffin-embedded tissue; PK, proteinase K; PMCA, protein misfolding cyclic amplification; PrP, prion protein; PrP^{C}, cellular isoform of PrP; PrP^{res}, PK-resistant PrP; PrP^{Sc}, abnormal isoform of PrP; PVDF, polyvinylidene difluoride; RT-QuIC, real-time quaking-induced conversion; SDS, sodium dodecyl sulfate; SDS-PAGE, sodium dodecyl sulfate-polyacrylamide gel electrophoresis; SOFIA, surround optical fiber immunoassay; WB, Western blotting.

[a] Protein A binds antibody.

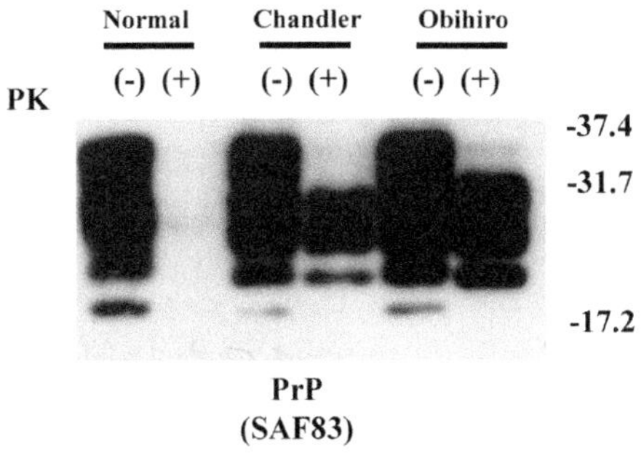

FIGURE 12.2 PrP^{Sc} in the brains of terminally diseased mice assessed by Western blotting. Prion was intracerebrally inoculated into C57BL/6J mice older than 8 weeks of age. Twenty microliters of inoculum containing 1% homogenate prepared from the brains of terminally diseased mice with mouse-adapted scrapie (Chandler and Obihiro strain) or uninfected mice were injected into the cerebral ventricular system of mice using a microsyringe. The brains of infected mice showing clinical symptoms (tremors and ataxia) were collected and used for further analysis. The brains of uninfected mice at a similar age were also collected. The mouse brains infected with prions (Chandler or Obihiro) or injected with normal brain homogenate (Normal) were treated with proteinase K (PK) (+). Untreated controls (−) are also included. The samples were subjected to Western blotting with the anti-PrP antibody SAF83 (SPI-Bio, Montigny-le-Bretonneux, France) to detect PK-resistant PrP (PrP^{res}), which included PrP^{Sc}. Bands corresponding to the molecular mass marker (kDa) are shown on the right. (Cited from Supplemental figure 2 of Sakudo and Onodera. 2011. Tissue- and cell type-specific modification of prion protein (PrP)-like protein Doppel, which affects PrP endoproteolysis. *Biochem Biophys Res Commun* 404:523–527. With permission from Elsevier.)

For the Western blotting procedure (Figure 12.2), sodium dodecyl sulfate (SDS)-polyacrylamide gel electrophoresis (PAGE) is used to separate proteinase K (PK)-treated proteins, which are then transferred to a membrane. PK-resistant PrP (PrP^{res}) on the membrane is detected using an anti-PrP antibody. Importantly, Western blotting analysis establishes not only the presence or absence of PrP^{res}, but also provides information on the molecular weight of the various peptides based on their electrophoretic mobility [105]. Western blotting for screening large numbers of cattle for BSE was originally considered to be impractical. However, Oesch and coworkers developed a rapid Western blotting procedure (named "Prionics Check" or "PWB"), which has since undergone modifications by the Swiss and British veterinary authorities, as well as the European Commission. Using this technique, samples of brain stem from officially confirmed BSE or scrapie cases and negative control samples from healthy New Zealand adult bovines, could be diagnosed with 100% specificity and sensitivity [106,107]. Furthermore, autolyzed samples could still be diagnosed correctly, leading to the development of a unique surveillance program in Switzerland. This surveillance program involved testing fallen stock for hidden cases of BSE. The routine application of the Prionics Check Western blotting technique has also revealed unrecognized BSE cases that could have entered the food chain. When peptides are modified with aberrant glycosylation, a different electrophoretic mobility is observed. Therefore, the mobility is influenced by the host genotype and also by prion strains [105,108]. Accordingly, a mobility index is used to differentiate sCJD subtypes (i.e., type 1 or type 2A), or iCJD and sCJD from vCJD, because of the different PrP^{Sc} glycoform profiles [109–114].

A modified Western blotting method has also been developed for prion detection. The cell blot procedure [76] involves growing cells on coverslips and then transferring them onto a nitrocellulose membrane, where PrP^{res} is then detected by Western blotting [76]. In addition, slot blotting [115] is a method whereby cell lysates are filtered through a nitrocellulose membrane using a slot-blot device before the PrP^{res} on the membrane is detected by Western blotting [115]. The sensitivity of the Western blotting technique can also be enhanced by centrifugation [116] or by the development of a suitable extraction method. An example of a modified extraction method is the sodium phosphotungstic acid (PTA) procedure [75,117], which precipitates PrP^{Sc} from solution. Guanidine hydrochloride has also been used to extract PrP^{Sc}. Western blotting has disadvantages in terms of processing time, the limited number of samples that can be processed, and the need for experienced personnel. These problems have been overcome, at least in part, by the enzyme-linked immunosorbent assay (ELISA).

Using the ELISA method, PrP^{res} is directly absorbed onto a microtiter plate or captured by anti-PrP antibody coated on a microtiter plate, and detected or "sandwiched" by other anti-PrP antibodies. Meanwhile, Grassi et al. chose to develop a conventional two-site immunometric assay (sandwich immunoassay) based on the use of two different monoclonal antibodies recognizing two distinct epitopes on the PrP molecule. Using this procedure, the first antibody (capture antibody) is immobilized on a solid phase (e.g., a microtiter plate) while a second antibody, covalently labeled with an enzyme (e.g., acetylcholine esterase [AchE]), is used as a tracer [118,119]. PrP within a sample is detected by measuring enzyme activity bound to the solid phase through the intermediary of capture antibody–PrP–tracer antibody reactions. A two-site immunometric assay was chosen because this procedure is demonstrably more sensitive and specific than other testing methods, as well as facilitating a higher rate of sample throughput [120]. Indeed, the two-site immunometric principle is commercially exploited in the Prionics Check EIA (Prionics/Roche, Switzerland), Roche Applied Science PrionScreen (Roche, Switzerland), Bio-Rad Platelia BSE purification kit and Bio-Rad BSE detection kit (Bio-Rad, France), FRELISA BSE kit (FUJIREBIO Inc., Japan), IDEXX HerdChek BSE antigen test kit (IDEXX Laboratories, Maine), and Institut Pourquier Speed'it BSE (Institut Pourquier, France) procedures. Modified versions of this method are also commercially available, such as VMRD CWD dbELISA (VMRD Inc., Washington), Enfer-TSE kit (Abbott Laboratories, Illinois), Prionics-Check LIA (Prionics, Switzerland), and CediTect BSE test (Eurofins Analytico Food, the Netherlands). Nonetheless, although the extensively employed ELISA method is sensitive, capable of high throughput, and does not require sophisticated techniques, the high frequency of false positive results remains a problem.

Another common method to diagnose prion diseases involves the immunohistochemical (IHC) analysis of brain sections [121,122]. IHC analysis employs light microscopy to analyze typical features of prion diseases, such as the accumulation of PrP amyloid plaques, astroglyosis, and neuronal cell loss. Despite the fact that vacuolation is sometimes used as an index of prion infection, many combinations of prion strains and host species result in the accumulation of PrP^{Sc} without vacuolation in brain sections after prion infection [123,124]. In addition, the region of the brain where PrP^{Sc} accumulates depends on the particular prion strain under investigation [125]. Histopathological analysis of organs/tissues other than brain have been studied for the diagnosis of prion diseases. These studies reveal that the tonsils are suitable for the diagnosis of prion infection in humans, deer, and sheep [117,126–131], while the appendix has been used for the preclinical diagnosis of vCJD in humans [117,132–135]. Prion strains can often be discriminated from the lesion profiles, which result from strain-specific patterns of vacuolation and PrP^{Sc} accumulation in the brain. In addition, IHC analysis has been further improved with modified histoblotting [72] or paraffin-embedded tissue (PET) blotting techniques [92,93]. These modifications can also contribute to increasing the sensitivity of detection of PrP^{Sc} in cryosections [72,92,93]. A sensitive PET blotting procedure that detects prion PrP^{Sc} deposits in formalin-fixed PET after blotting on a nitrocellulose membrane has been developed [92]. The sensitivity of this blotting procedure has been compared with that of histological and immunohistochemical methods and with the Western blot method used in prionics. Using the PET blot method, four clinically similar cases showed the same PrP^{Sc} deposition pattern observed in clinical BSE, whereas the results of histological examination and Western blotting were negative [92]. Therefore, PET is potentially a promising method that will enable the sensitive diagnosis of prion diseases.

FRONTIERS OF MOLECULAR TECHNIQUES FOR PRION DETECTION

Several methods have been developed that utilize anti-PrP antibody for the detection of PrP or PrP isoforms. Conformation-dependent immunoassay (CDI) detects conformational differences between PrP isoforms by evaluating the relative binding of antibodies to denatured and native proteins [75]. Indeed, CDI is commercially used for the CDI test of InPro Biotechnology Inc. (California). Moreover, the flow microbead immunoassay (FMI), which uses anti-PrP antibody–coupled microbeads and a flow cytometer, detected 7 pmol and 7 nmol recombinant PrP and PrP^{Sc} spiked in bovine MBM, respectively, at concentrations greater than 0.3% [84]. It should be emphasized that the novel PMCA technology cyclically amplifies the misfolding and aggregation process *in vitro* [40]. Therefore, PMCA is conceptually analogous to DNA amplification by polymerase chain reaction (PCR). Specifically, PMCA involves sequential cycles of incubation and sonication in the presence of seed PrP^{Sc} supplied with PrP^{C}. Importantly, the PMCA method not only amplifies PrP^{Sc}, but also measures prion infectivity titers [94]. Furthermore, this method is capable of detecting prions in blood, from not only terminally diseased hamsters, but also from prion-infected presymptomatic hamsters [136]. Five cycles of PMCA achieved 97% conversion of PrP^{C} to PrP^{Sc} in brain homogenate [137], which represents the highest level of sensitivity among the detection methods for prion proteins reported so far. Thus, PrP^{Sc} detection is possible at >10,000-fold dilution, although sensitivity could be increased even further if a modified Western blotting method was to be combined with PMCA. However, it should be noted that the protocol for PrP^{Sc} amplification depends on prion strain, species, and sample source (e.g., brain homogenate or blood). There are reports of the application of this method to detect vCJD, CWD, and hamster and mouse scrapie [137–140]. Recently, modified versions of PMCA, such as recombinant PrP-PMCA (rPrP-PMCA) and real-time quaking-induced conversion (RT-QuiC), have also been studied [141]. To date, PMCA has primarily been restricted to the study of prion conversion mechanisms and the diagnosis of prion diseases. Modified versions of this technology are anticipated to expand its use into the field of prion biology.

Capillary gel electrophoresis is an approach that takes advantage of competitive antibody binding between a fluorescein-labeled synthetic PrP peptide and PrP present in tissue samples [73,74]. The free peptide and antibody–peptide peaks are separated by capillary electrophoresis. The sensitivity of this method is extremely high; only 50 amol/L (10^{-18} mol/L) of fluorescent marker is required [73]. However, the specificity of the capillary gel electrophoresis approach remains unclear. For example, a recent report from Brown et al. [142] has shown the difficulty of reproducing results presented in a previous report by Schmerr et al. [74]. A novel noncompetitive capillary gel electrophoresis method for detecting prion protein has subsequently been developed [143]. Fluorescence correlation spectroscopy (FCS) is a highly sensitive method that detects single fluorescently labeled molecules in solution [69,144]. In this procedure, PrP^{Sc} can be labeled either using anti-PrP antibody or by conjugation with a labeled recombinant PrP.

Fourier transform infrared spectroscopy is a diagnostic method that incorporates the multivariate analysis of infrared spectra and can discriminate between prion-infected and uninfected animals [86]. A similar spectroscopy-based assay is also used in multispectral ultraviolet fluorescence spectroscopy (MUFS) [91]. During MUFS, the sample is sequentially illuminated with a monochromatic light of specific wavelengths in the ultraviolent spectrum, and the resulting fluorescence is measured over a finite bandwidth to generate a two-dimensional spectral signature. Multivariant techniques can then be employed to determine the presence of prion agent within the sample. The advantages of these methods are the lack of pretreatment steps to eliminate PrP^{C} (such as PK treatment) and that it is not necessary to use an anti-PrP antibody. However, to date, MUFS has not been adopted as a means of detecting prion infection.

In the case of ELISA, anti-PrP antibody is labeled either directly or indirectly with enzyme, allowing the PrP signal to be visualized upon the addition of an enzyme substrate. In the case of the more recently developed and more sensitive PrP detection method, dissociation-enhanced lanthanide fluorescent immunoassay, anti-PrP antibody is labeled with the rare earth element europium (lanthanide chelate). Surround optical fiber immunoassay (SOFIA) is an ultrasensitive detection method [97] based on sandwich ELISA, and involves the incubation of the sample with biotinylated anti-PrP antibody followed by the addition of streptavidin conjugated to Rhodamine (RedX). Immuno-PCR combines the specificity of immunological detection methods with the exponential amplification of PCR. In this technique, PK-resistant PrP is detected by a capture anti-PrP antibody bound to streptavidin-horseradish peroxidase attached to biotinylated reporter DNA [87]. Sensitive analysis is achieved by amplification of the reporter DNA using PCR.

New tools for detecting PrP^{Sc} have also been developed. Although some antibodies and aptamers are able to distinguish PrP^{Sc} from PrP^{C}, no extensive practical applications have been found for these. Antibodies with a conformational epitope of PrP^{Sc} were obtained by immunization with a Tyr-Tyr-Arg peptide [145]. Antibodies 15B3 and V5B2 can also recognize PrP^{Sc}-specific epitopes [146,147]. However, these antibodies have not yet been exploited for commercial applications and their use in research programs is somewhat limited. Interestingly, PrP^{C} and/or PrP^{Sc}-specific binding of RNA or DNA aptamers has also been reported [70,148,149].

Generally, the use of animal bioassays is considered the most sensitive way to assess the infectivity of a prion agent, although the previously described Western blot test in combination with PMCA is said to match or even surpass animal bioassays in sensitivity [150]. PrP-expressing transgenic mice inoculated with PMCA-amplified PrP^{Sc} exhibit a similar neuropathology to that found in prion diseases [151]. However, it remains to be established whether PrP^{Sc} is identical in these cases, and whether this is the only entity making up the prion. Therefore, the determination of infectivity with a reliable animal bioassay remains the gold standard, although these assays require a longer time to perform and often involve the use of numerous animals to confirm the findings. Furthermore, the assay may occasionally demand the use of additional sophisticated techniques [7]. It should also be noted that the volume of inoculum is critical for reducing variation in the results obtained from animal bioassays. Consequently, some animal bioassays for prion diagnosis are considered too time consuming and economically impractical for general use. Nonetheless, the transmission of prions to their natural host is very informative and helpful for understanding the clinical phenotype of prion diseases. Accordingly, cell culture systems that specifically and reliably detect prions have been developed for specific prion strains [103]. These *in vitro* cell culture model systems will be invaluable in future prion infectivity studies.

CONCLUSIONS AND FUTURE PERSPECTIVES

Researchers from the United Kingdom have estimated that the number of BSE-infected cattle entering the human food chain between 1985 and 1995 was approximately 446,000, prior to the introduction of a bovine offal ban at the end of 1989, followed by a further 283,000 cattle before the end of 1995 [152]. In total, the estimated number of BSE-infected cattle that entered the food chain over the period from 1974 to the end of 1995 is 903,000. However, there were only 16,412 confirmed cases of BSE in the United Kingdom, as most infected cattle were slaughtered before exhibiting signs of the disease. Consequently, there is a clear need to develop sensitive methods that are capable of identifying BSE at an early presymptomatic stage of the disease. In addition, the appearance of BSE epidemics in the United Kingdom and Europe, and the subsequent emergence of vCJD in young adults and teenagers, have raised concerns about the spread of prion disease through the food chain. There is thus an urgent need to carry out a risk assessment of prion transmission via this route. Moreover, as discussed previously, there is a potential risk of prion agents in blood products being transmitted by transfusion. Indeed, several experimental reports clearly demonstrate prion transmission via blood transfusion procedures [153,154]. A recent report on the analysis of appendix samples highlights the importance of introducing a blood test for prion diseases. This report concludes that among 32,441 appendix samples analyzed in the United Kingdom, 16 were positive for vCJD, suggesting a potential reservoir of 30,000 individuals infected with vCJD and in the presymptomatic stage of the disease [155]. Therefore, stringent procedures must be introduced to monitor the management of blood or blood products and the handling of surgical instruments, in order to limit the potential spread of infectious prion agents. Unfortunately, current technology aimed at prion removal by filtering blood is insufficient to prevent infection [156]. In addition, prions are exceptionally resistant pathogens that cannot be inactivated by conventional sterilization procedures. For example, treatment with alcohol, autoclaving (121°C for 20 min), and exposure to γ-ray irradiation, which are used to kill viruses and bacteria, are ineffective against prion agents. Therefore, to prevent prion-related iatrogenic diseases, appropriate procedures to inactivate prions should be introduced. Recently, the authors and others have written review papers addressing this issue [157,158]. Readers interested in learning about methods for prion inactivation should refer to these reviews as well as to the guidelines, Disinfection and Sterilization in Health Care Facilities, 2008, by the Centers for Disease Control and Prevention (CDC) [159].

As discussed previously, there is increasing evidence of human-to-human transmission via blood products. Therefore, researchers are devising tests to reliably identify blood donors that are asymptomatic carriers of prion diseases. Several highly sensitive methods for detecting prions have been developed, such as PMCA [94]. To date, a broad spectrum of PrP^{Sc} species has been successfully amplified using PMCA, including CWD, mouse-adapted scrapie, and BSE. PMCA has been used to detect PrP^{Sc} in blood from an asymptomatic prion-infected mouse. Further development of PMCA offers the prospect of a highly sensitive detection and quantification procedure. These novel methods include rPMCA, standard quaking-induced conversion (S-QuIC), amyloid seeding assay (ASA), RT-QuIC, and enhanced QuIC (eQuIC) [95]. In particular, RT-QuIC-based assays enhance the practicality, sensitivity, and quantitative attributes of the QuIC procedure, and will help to promote the detection of prion seeding activity in body fluids, including blood plasma. In conclusion, there have been significant advances in the routine identification of prion infection in clinical samples that merit further evaluation. However, additional improvements in these techniques are still required. In particular, the establishment of a more reliable and robust practical blood test for prion diseases is urgently needed.

ACKNOWLEDGMENTS

This work was supported by Grant-in-Aids for Scientific Research from the Ministry of Education, Science, Culture, and Technology of Japan (25450447, 24110717, 23780299, 22110514, 20780219, and 17780228), Grants-in-Aids from the Research Committee of Prion Disease and Slow Virus Infection, the Ministry of Health, Labour and Welfare of Japan, and the Japan Agency for Medical Research and Development (AMED).

REFERENCES

1. Gajdusek DC. 1977. Unconventional viruses and the origin and disappearance of kuru. *Science* 197:943–960.
2. Bucelli RC, Ances BM. 2013. Diagnosis and evaluation of a patient with rapidly progressive dementia. *Mo Med* 110:422–428.
3. Aguzzi A, Heppner FL. 2000. Pathogenesis of prion diseases: A progress report. *Cell Death Differ* 7:889–902.
4. Thomas JG, Chenoweth CE, Sullivan SE. 2013. Iatrogenic Creutzfeldt–Jakob disease via surgical instruments. *J Clin Neurosci* 20:1207–1212.
5. Prusiner SB. 1998. Prions. *Proc Natl Acad Sci USA* 95:13363–13383.
6. Gibbs CJ, Jr., Gajdusek DC, Asher DM, Alpers MP, Beck E, Daniel PM, Matthews WB. 1968. Creutzfeldt–Jakob disease (spongiform encephalopathy): Transmission to the chimpanzee. *Science* 161:388–389.
7. Aguzzi A, Polymenidou M. 2004. Mammalian prion biology: One century of evolving concepts. *Cell* 116:313–327.
8. Horiuchi M, Caughey B. 1999. Specific binding of normal prion protein to the scrapie form via a localized domain initiates its conversion to the protease-resistant state. *EMBO J* 18:3193–3203.
9. Büeler H, Aguzzi A, Sailer A, Greiner R-A, Autenried P, Aguet M, Weissmann C. 1993. Mice devoid of PrP are resistant to scrapie. *Cell* 73:1339–1347.
10. Scott M, Groth D, Foster D, Torchia M, Yang S-L, DeArmond SJ, Prusiner SB. 1993. Propagation of prions with artificial properties in transgenic mice expressing chimeric PrP genes. *Cell* 73:979–988.

11. Shmerling D, Hegyi I, Fischer M, Blättler T, Brandner S, Götz J, Rülicke T, Flechsig E, Cozzio A, von Mering C. 1998. Expression of amino-terminally truncated PrP in the mouse leading to ataxia and specific cerebellar lesions. *Cell* 93:203–214.
12. Barron RM, Thomson V, Jamieson E, Melton DW, Ironside J, Will R, Manson JC. 2001. Changing a single amino acid in the N-terminus of murine PrP alters TSE incubation time across three species barriers. *EMBO J* 20:5070–5078.
13. Perrier V, Kaneko K, Safar J, Vergara J, Tremblay P, DeArmond SJ, Cohen FE, Prusiner SB, Wallace AC. 2002. Dominant-negative inhibition of prion replication in transgenic mice. *Proc Natl Acad Sci USA* 99:13079–13084.
14. Chiesa R, Piccardo P, Quaglio E, Drisaldi B, Takao M, Ghetti B, Harris DA. 2003. Molecular distinction between pathogenic and infectious properties of the prion protein. *J Virol* 77:7611–7622.
15. Chesebro B, Trifilo M, Race R, Meade-White K, Teng C, LaCasse R, Raymond L, Favara C, Baron G, Priola S. 2005. Anchorless prion protein results in infectious amyloid disease without clinical scrapie. *Science* 308:1435–1439.
16. Sharma A, Bruce KL, Chen B, Gyoneva S, Behrens SH, Bommarius AS, Chernoff YO. 2016. Contributions of the prion protein sequence, strain, and environment to the species barrier. *J Biol Chem* 291:1277–1288.
17. Sigurðsson B. 1954. Rida, a chronic encephalitis of sheep: With general remarks on infections which develop slowly and some of their special characteristics. *Br Vet J* 110:341–354.
18. Alper T, Cramp W. 1967. Does the agent of scrapie replicate without nucleic acid? *Nature* 214:764–766.
19. Alper T, Haig D, Clarke M. 1966. The exceptionally small size of the scrapie agent. *Biochem Biophys Res Commun* 22:278–284.
20. Griffith JS. 1967. Self-replication and scrapie. *Nature* 215:1043.
21. Merz PA, Somerville R, Wisniewski H, Iqbal K. 1981. Abnormal fibrils from scrapie-infected brain. *Acta Neuropathol* 54:63–74.
22. Prusiner SB, McKinley MP, Bowman KA, Bolton DC, Bendheim PE, Groth DF, Glenner GG. 1983. Scrapie prions aggregate to form amyloid-like birefringent rods. *Cell* 35:349–358.
23. Silveira JR, Raymond GJ, Hughson AG, Race RE, Sim VL, Hayes SF, Caughey B. 2005. The most infectious prion protein particles. *Nature* 437:257–261.
24. Kim NH, Choi JK, Jeong BH, Kim JI, Kwon MS, Carp RI, Kim YS. 2005. Effect of transition metals (Mn, Cu, Fe) and deoxycholic acid (DA) on the conversion of PrP^{C} to PrP^{res}. *FASEB J* 19:783–785.
25. Deleault NR, Kascsak R, Geoghegan JC, Supattapone S. 2010. Species-dependent differences in cofactor utilization for formation of the protease-resistant prion protein *in vitro*. *Biochemistry* 49:3928–3934.
26. Wang F, Wang X, Yuan C-G, Ma J. 2010. Generating a prion with bacterially expressed recombinant prion protein. *Science* 327:1132–1135.
27. Deleault NR, Geoghegan JC, Nishina K, Kascsak R, Williamson RA, Supattapone S. 2005. Protease-resistant prion protein amplification reconstituted with partially purified substrates and synthetic polyanions. *J Biol Chem* 280:26873–26879.
28. Deleault NR, Piro JR, Walsh DJ, Wang F, Ma J, Geoghegan JC, Supattapone S. 2012. Isolation of phosphatidylethanolamine as a solitary cofactor for prion formation in the absence of nucleic acids. *Proc Natl Acad Sci USA* 109:8546–8551.
29. Deleault NR, Walsh DJ, Piro JR, Wang F, Wang X, Ma J, Rees JR, Supattapone S. 2012. Cofactor molecules maintain infectious conformation and restrict strain properties in purified prions. *Proc Natl Acad Sci USA* 109:E1938–E1946.
30. Baybutt H, Manson J. 1997. Characterisation of two promoters for prion protein (PrP) gene expression in neuronal cells. *Gene* 184:125–131.
31. Saeki K, Matsumoto Y, Hirota Y, Matsumoto Y, Onodera T. 1996. Three-exon structure of the gene encoding the rat prion protein and its expression in tissues. *Virus Genes* 12:15–20.
32. Saeki K, Matsumoto Y, Matsumoto Y, Onodera T. 1996. Identification of a promoter region in the rat prion protein gene. *Biochem Biophys Res Commun* 219:47–52.
33. O'Neill G, Donnelly K, Marshall E, Cairns D, Goldmann W, Hunter N. 2003. Characterization of ovine PrP gene promoter activity in N2a neuroblastoma and ovine foetal brain cell lines. *J Anim Breed Genet* 120:114–123.
34. Inoue S, Tanaka M, Horiuchi M, Ishiguro N, Shinagawa M. 1997. Characterization of the bovine prion protein gene: The expression requires interaction between the promoter and intron. *J Vet Med Sci* 59:175–183.
35. Mahal SP, Asante EA, Antoniou M, Collinge J. 2001. Isolation and functional characterisation of the promoter region of the human prion protein gene. *Gene* 268:105–114.

36. Sakudo A, Ikuta K. 2009. Prion protein functions and dysfunction in prion diseases. *Curr Med Chem* 16:380–389.
37. Pan K-M, Baldwin M, Nguyen J, Gasset M, Serban A, Groth D, Mehlhorn I, Huang Z, Fletterick RJ, Cohen FE. 1993. Conversion of alpha-helices into beta-sheets features in the formation of the scrapie prion proteins. *Proc Natl Acad Sci USA* 90:10962–10966.
38. Riek R, Hornemann S, Wider G, Billeter M, Glockshuber R, Wfithrich K. 1996. NMR structure of the mouse. *Nature* 382:11.
39. Safar J, Roller PP, Gajdusek DC, Gibbs CJ. 1993. Thermal stability and conformational transitions of scrapie amyloid (prion) protein correlate with infectivity. *Protein Sci* 2:2206–2216.
40. Soto C, Saborio GP, Anderes L. 2002. Cyclic amplification of protein misfolding: Application to prion-related disorders and beyond. *Trends Neurosci* 25:390–394.
41. Kaneko K, Zulianello L, Scott M, Cooper CM, Wallace AC, James TL, Cohen FE, Prusiner SB. 1997. Evidence for protein X binding to a discontinuous epitope on the cellular prion protein during scrapie prion propagation. *Proc Natl Acad Sci USA* 94:10069–10074.
42. Daude N, Wohlgemuth S, Brown R, Pitstick R, Gapeshina H, Yang J, Carlson GA, Westaway D. 2012. Knockout of the prion protein (PrP)-like *Sprn* gene does not produce embryonic lethality in combination with PrP^{C}-deficiency. *Proc Natl Acad Sci USA* 109:9035–9040.
43. McGill IS, Wells GA. 1993. Neuropathological findings in cattle with clinically suspect but histologically unconfirmed bovine spongiform encephalopathy (BSE). *J Comp Pathol* 108:241–260.
44. Gajdusek DC, Zigas V. 1959. Kuru: Clinical, pathological and epidemiological study of an acute progressive degenerative disease of the central nervous system among natives of the Eastern Highlands of New Guinea. *Am J Med* 26:442–469.
45. Gajdusek DC, Gibbs CJ. 1964. Attempts to demonstrate a transmissible agent in kuru, amyotrophic lateral sclerosis, and other subacute and chronic nervous system degenerations of man. *Nature* 204:257–259.
46. Hadlow WJ. 1959. Scrapie and kuru. *Lancet* 274:289–290.
47. Prusiner SB, Gajdusek C, Alpers MP. 1982. Kuru with incubation periods exceeding two decades. *Ann Neurol* 12:1–9.
48. Weber T, Aguzzi A. 1997. The spectrum of transmissible spongiform encephalopathies. *Intervirology* 40:198–212.
49. Sakudo A, Xue G, Kawashita N, Ano Y, Takagi T, Shintani H, Tanaka Y, Onodera T, Ikuta K. 2010. Structure of the prion protein and its gene: An analysis using bioinformatics and computer simulation. *Curr Protein Pept Sci* 11:166–179.
50. Sakudo A, Onodera T, Suganuma Y, Kobayashi T, Saeki K, Ikuta K. 2006. Recent advances in clarifying prion protein functions using knockout mice and derived cell lines. *Mini Rev Med Chem* 6:589–601.
51. Yamada M, Itoh Y, Inaba A, Wada Y, Takashima M, Satoh S, Kamata T, et al. 1999. An inherited prion disease with a PrP P105L mutation: Clinicopathologic and PrP heterogeneity. *Neurology* 53:181–188.
52. Samaia HB, Mari JJ, Vallada HP, Moura RP, Simpson AJ, Brentani RR. 1997. A prion-linked psychiatric disorder. *Nature* 390:241.
53. Alperovitch A, Zerr I, Pocchiari M, Mitrova E, de Pedro Cuesta J, Hegyi I, Collins S, Kretzschmar H, van Duijn C, Will RG. 1999. Codon 129 prion protein genotype and sporadic Creutzfeldt–Jakob disease. *Lancet* 353:1673–1674.
54. Brown P, Preece M, Brandel JP, Sato T, McShane L, Zerr I, Fletcher A, et al. 2000. Iatrogenic Creutzfeldt–Jakob disease at the millennium. *Neurology* 55:1075–1081.
55. The National CJD surveillance Unit, Western General Hospital. 2001. 10th annual report 2001, Creutzfeldt–Jakob disease surveillance in the U.K. http://www.cjd.ed.ac.uk/documents/report10.pdf. Accessed December 2015.
56. Shibuya S, Higuchi J, Shin RW, Tateishi J, Kitamoto T. 1998. Codon 219 Lys allele of *PRNP* is not found in sporadic Creutzfeldt–Jakob disease. *Ann Neurol* 43:826–828.
57. Hainfellner JA, Brantner-Inthaler S, Cervenakova L, Brown P, Kitamoto T, Tateishi J, Diringer H, et al. 1995. The original Gerstmann–Straussler–Scheinker family of Austria: Divergent clinicopathological phenotypes but constant PrP genotype. *Brain Pathol* 5:201–211.
58. Ghetti B, Piccardo P, Spillantini MG, Ichimiya Y, Porro M, Perini F, Kitamoto T, et al. 1996. Vascular variant of prion protein cerebral amyloidosis with tau-positive neurofibrillary tangles: The phenotype of the stop codon 145 mutation in *PRNP*. *Proc Natl Acad Sci USA* 93:744–748.
59. Gambetti P, Lugaresi E. 1998. Conclusions of the symposium. *Brain Pathol* 8:571–575.
60. Parchi P, Capellari S, Chin S, Schwarz HB, Schecter NP, Butts JD, Hudkins P, Burns DK, Powers JM, Gambetti P. 1999. A subtype of sporadic prion disease mimicking fatal familial insomnia. *Neurology* 52:1757–1763.

61. Goldfarb LG, Brown P, McCombie WR, Goldgaber D, Swergold GD, Wills PR, Cervenakova L, Baron H, Gibbs CJ, Jr., Gajdusek DC. 1991. Transmissible familial Creutzfeldt–Jakob disease associated with five, seven, and eight extra octapeptide coding repeats in the *PRNP* gene. *Proc Natl Acad Sci USA* 88:10926–10930.
62. Oda T, Kitamoto T, Tateishi J, Mitsuhashi T, Iwabuchi K, Haga C, Oguni E, et al. 1995. Prion disease with 144 base pair insertion in a Japanese family line. *Acta Neuropathol* 90:80–86.
63. Hitoshi S, Nagura H, Yamanouchi H, Kitamoto T. 1993. Double mutations at codon 180 and codon 232 of the *PRNP* gene in an apparently sporadic case of Creutzfeldt–Jakob disease. *J Neurol Sci* 120:208–212.
64. Seno H, Tashiro H, Ishino H, Inagaki T, Nagasaki M, Morikawa S. 2000. New haplotype of familial Creutzfeldt–Jakob disease with a codon 200 mutation and a codon 219 polymorphism of the prion protein gene in a Japanese family. *Acta Neuropathol* 99:125–130.
65. Will RG, Ironside J, Zeidler M, Estibeiro K, Cousens S, Smith P, Alperovitch A, Poser S, Pocchiari M, Hofman A. 1996. A new variant of Creutzfeldt–Jakob disease in the UK. *Lancet* 347:921–925.
66. Chazot G, Broussolle E, Lapras C, Blättler T, Aguzzi A, Kopp N. 1996. New variant of Creutzfeldt–Jakob disease in a 26-year-old French man. *Lancet* 347:1181.
67. Meyer RK, McKinley MP, Bowman KA, Braunfeld MB, Barry RA, Prusiner SB. 1986. Separation and properties of cellular and scrapie prion proteins. *Proc Natl Acad Sci USA* 83:2310–2314.
68. Eigen M. 1996. Prionics or the kinetic basis of prion diseases. *Biophys Chem* 63:A1–A18.
69. Caughey B, Kocisko DA, Raymond GJ, Lansbury PT. 1995. Aggregates of scrapie-associated prion protein induce the cell-free conversion of protease-sensitive prion protein to the protease-resistant state. *Chem Biol* 2:807–817.
70. Weiss S, Proske D, Neumann M, Groschup MH, Kretzschmar HA, Famulok M, Winnacker E-L. 1997. RNA aptamers specifically interact with the prion protein PrP. *J Virol* 71:8790–8797.
71. Prusiner SB, Cochran SP, Groth DF, Downey DE, Bowman KA, Martinez HM. 1982. Measurement of the scrapie agent using an incubation time interval assay. *Ann Neurol* 11:353–358.
72. Taraboulos A, Jendroska K, Serban D, Yang S-L, DeArmond SJ, Prusiner SB. 1992. Regional mapping of prion proteins in brain. *Proc Natl Acad Sci USA* 89:7620–7624.
73. Jackman R, Schmerr MJ. 2003. Analysis of the performance of antibody capture methods using fluorescent peptides with capillary zone electrophoresis with laser-induced fluorescence. *Electrophoresis* 24:892–896.
74. Schmerr MJ, Jenny AL, Bulgin MS, Miller JM, Hamir AN, Cutlip RC, Goodwin KR. 1999. Use of capillary electrophoresis and fluorescent labeled peptides to detect the abnormal prion protein in the blood of animals that are infected with a transmissible spongiform encephalopathy. *J Chromatogr A* 853:207–214.
75. Safar J, Wille H, Itri V, Groth D, Serban H, Torchia M, Cohen FE, Prusiner SB. 1998. Eight prion strains have PrPSc molecules with different conformations. *Nat Med* 4:1157–1165.
76. Bosque PJ, Prusiner SB. 2000. Cultured cell sublines highly susceptible to prion infection. *J Virol* 74:4377–4386.
77. Klöhn P-C, Stoltze L, Flechsig E, Enari M, Weissmann C. 2003. A quantitative, highly sensitive cell-based infectivity assay for mouse scrapie prions. *Proc Natl Acad Sci USA* 100:11666–11671.
78. Barnard G, Helmick B, Madden S, Gilbourne C, Patel R. 2000. The measurement of prion protein in bovine brain tissue using differential extraction and DELFIA® as a diagnostic test for BSE. *Luminescence* 15:357–362.
79. Miodek A, Castillo G, Hianik T, Korri-Youssoufi H. 2013. Electrochemical aptasensor of human cellular prion based on multiwalled carbon nanotubes modified with dendrimers: A platform for connecting redox markers and aptamers. *Anal Chem* 85:7704–7712.
80. Grathwohl K-UD, Horiuchia M, Ishiguro N, Shinagawa M. 1997. Sensitive enzyme-linked immunosorbent assay for detection of PrP^{Sc} in crude tissue extracts from scrapie-affected mice. *J Virol Methods* 64:205–216.
81. Bieschke J, Giese A, Schulz-Schaeffer W, Zerr I, Poser S, Eigen M, Kretzschmar H. 2000. Ultrasensitive detection of pathological prion protein aggregates by dual-color scanning for intensely fluorescent targets. *Proc Natl Acad Sci USA* 97:5468–5473.
82. Giese A, Bieschke J, Eigen M, Kretzschmar H. 2000. Putting prions into focus: Application of single molecule detection to the diagnosis of prion diseases. *Arch Virol Suppl* 16:161–171.
83. Kobayashi Y, Kohno N, Wanibe S, Hirayasu K, Uemori H, Tagawa Y, Yokoyama T, Shinagawa M. 2006. A solid-phase immunoassay of protease-resistant prion protein with filtration blotting involving sodium dodecyl sulfate. *Anal Biochem* 349:218–228.

84. Murayama Y, Yoshioka M, Horii H, Takata M, Miura K, Shinagawa M. 2006. Specific detection of prion antigenic determinants retained in bovine meat and bone meal by flow microbead immunoassay. *J Appl Microbiol* 101:369–376.
85. Xiao SJ, Hu PP, Wu XD, Zou YL, Chen LQ, Peng L, Ling J, et al. 2010. Sensitive discrimination and detection of prion disease-associated isoform with a dual-aptamer strategy by developing a sandwich structure of magnetic microparticles and quantum dots. *Anal Chem* 82:9736–9742.
86. Schmitt J, Beekes M, Brauer A, Udelhoven T, Lasch P, Naumann D. 2002. Identification of scrapie infection from blood serum by Fourier transform infrared spectroscopy. *Anal Chem* 74:3865–3868.
87. Barletta JM, Edelman DC, Highsmith WE, Constantine NT. 2005. Detection of ultra-low levels of pathologic prion protein in scrapie infected hamster brain homogenates using real-time immuno-PCR. *J Virol Methods* 127:154–164.
88. McBride P, Bruce M, Fraser H. 1988. Immunostaining of scrapie cerebral amyloid plaques with antisera raised to scrapie-associated fibrils (SAF). *Neuropathol Appl Neurobiol* 14:325–336.
89. Silva CJ, Onisko BC, Dynin I, Erickson ML, Requena JR, Carter JM. 2011. Utility of mass spectrometry in the diagnosis of prion diseases. *Anal Chem* 83:1609–1615.
90. Varshney M, Waggoner PS, Tan CP, Aubin K, Montagna RA, Craighead HG. 2008. Prion protein detection using nanomechanical resonator arrays and secondary mass labeling. *Anal Chem* 80:2141–2148.
91. Rubenstein R, Gray P, Wehlburg C, Wagner J, Tisone G. 1998. Detection and discrimination of PrP Sc by multispectral ultraviolet fluorescence. *Biochem Biophys Res Commun* 246:100–106.
92. Schulz-Schaeffer WJ, Tschöke S, Kranefuss N, Dröse W, Hause-Reitner D, Giese A, Groschup MH, Kretzschmar HA. 2000. The paraffin-embedded tissue blot detects PrP Sc early in the incubation time in prion diseases. *Am J Med* 156:51–56.
93. Ritchie D, Head M, Ironside J. 2004. Advances in the detection of prion protein in peripheral tissues of variant Creutzfeldt–Jakob disease patients using paraffin-embedded tissue blotting. *Neuropathol Appl Neurobiol* 30:360–368.
94. Castilla J, Saá P, Hetz C, Soto C. 2005. *In vitro* generation of infectious scrapie prions. *Cell* 121:195–206.
95. Orru CD, Wilham JM, Vascellari S, Hughson AG, Caughey B. 2012. New generation QuIC assays for prion seeding activity. *Prion* 6:147–152.
96. Gyllberg H, Lofgren K. 2008. Immunodetection of PrP^{Sc} using Western and Slot blotting techniques. *Methods Mol Biol* 459:35–48.
97. Chang B, Gray P, Piltch M, Bulgin MS, Sorensen-Melson S, Miller MW, Davies P, Brown DR, Coughlin DR, Rubenstein R. 2009. Surround optical fiber immunoassay (SOFIA): An ultra-sensitive assay for prion protein detection. *J Virol Methods* 159:15–22.
98. Miodek A, Poturnayova A, Snejdarkova M, Hianik T, Korri-Youssoufi H. 2013. Binding kinetics of human cellular prion detection by DNA aptamers immobilized on a conducting polypyrrole. *Anal Bioanal Chem* 405:2505–2514.
99. Fischer MB, Röckl C, Parizek P, Schwarz HP, Aguzzi A. 2000. Binding of disease-associated prion protein to plasminogen. *Nature* 408:479–483.
100. Kenney K, Brechtel C, Takahashi H, Kurohara K, Anderson P, Gibbs C. 2000. An enzyme-linked immunosorbent assay to quantify 14-3-3 proteins in the cerebrospinal fluid of suspected Creutzfeldt–Jakob disease patients. *Ann Neurol* 48:395–398.
101. Miele G, Manson J, Clinton M. 2001. A novel erythroid-specific marker of transmissible spongiform encephalopathies. *Nat Med* 7:361–364.
102. Inoue Y, Yamakawa Y, Sakudo A, Kinumi T, Nakamura Y, Matsumoto Y, Saeki K, Kamiyama T, Onodera T, Nishijima M. 2005. Infection route-independent accumulation of splenic abnormal prion protein. *Jpn J Infect Dis* 58:78–82.
103. Sakudo A, Nakamura I, Ikuta K, Onodera T. 2007. Recent developments in prion disease research: Diagnostic tools and *in vitro* cell culture models. *J Vet Med Sci* 69:329–337.
104. Sakudo A, Onodera T. 2011. Tissue- and cell type-specific modification of prion protein (PrP)-like protein Doppel, which affects PrP endoproteolysis. *Biochem Biophys Res Commun* 404:523–527.
105. Thuring C, Erkens J, Jacobs J, Bossers A, Van Keulen L, Garssen G, Van Zijderveld F, Ryder S, Groschup M, Sweeney T. 2004. Discrimination between scrapie and bovine spongiform encephalopathy in sheep by molecular size, immunoreactivity, and glycoprofile of prion protein. *J Clin Microbiol* 42:972–980.
106. Schaller O, Fatzer R, Stack M, Clark J, Cooley W, Biffiger K, Egli S, Doherr M, Vandevelde M, Heim D. 1999. Validation of a Western immunoblotting procedure for bovine PrP^{Sc} detection and its use as a rapid surveillance method for the diagnosis of bovine spongiform encephalopathy (BSE). *Acta Neuropathol* 98:437–443.

107. Schimmel H, Moynagh J. 1999. Preliminary report concerning the evaluation of test for the diagnosis of transmissible spongiform encephalopathy in bovines. http://ec.europa.eu/food/fs/bse/bse12_en.html. Accessed November, 2015.
108. Pan T, Colucci M, Wong B-S, Li R, Liu T, Petersen RB, Chen S, Gambetti P, Sy M-S. 2001. Novel differences between two human prion strains revealed by two-dimensional gel electrophoresis. *J Biol Chem* 276:37284–37288.
109. Collinge J, Sidle KC, Meads J, Ironside J, Hill AF. 1996. Molecular analysis of prion strain variation and the aetiology of "new variant" CJD. *Nature* 383:685–690.
110. Head MW, Bunn TJ, Bishop MT, McLoughlin V, Lowrie S, McKimmie CS, Williams MC, McCardle L, MacKenzie J, Knight R. 2004. Prion protein heterogeneity in sporadic but not variant Creutzfeldt–Jakob disease: UK cases 1991–2002. *Ann Neurol* 55:851–859.
111. Notari S, Capellari S, Giese A, Westner I, Baruzzi A, Ghetti B, Gambetti P, Kretzschmar HA, Parchi P. 2004. Effects of different experimental conditions on the PrP^{Sc} core generated by protease digestion: Implications for strain typing and molecular classification of CJD. *J Biol Chem* 279:16797–16804.
112. Parchi P, Giese A, Capellari S, Brown P, Schulz-Schaeffer W, Windl O, Zerr I, Budka H, Kopp N, Piccardo P. 1999. Classification of sporadic Creutzfeldt–Jakob disease based on molecular and phenotypic analysis of 300 subjects. *Ann Neurol* 46:224–233.
113. Schoch G, Seeger H, Bogousslavsky J, Tolnay M, Janzer RC, Aguzzi A, Glatzel M. 2005. Analysis of prion strains by PrP^{Sc} profiling in sporadic Creutzfeldt–Jakob disease. *PLoS Med* 3:e14.
114. Wadsworth JD, Hill AF, Beck JA, Collinge J. 2003. Molecular and clinical classification of human prion disease. *Br Med Bull* 66:241–254.
115. Winklhofer KF, Hartl FU, Tatzelt J. 2001. A sensitive filter retention assay for the detection of PrP^{Sc} and the screening of anti-prion compounds. *FEBS Lett* 503:41–45.
116. Fujita R, Urayama T, Ohmizu A, Yunoki M, Tsuji S, Ikuta K. 2006. Efficient detection of PrP^{Sc} (263K) in human plasma. *Biologicals* 34:187–189.
117. Wadsworth J, Joiner S, Hill A, Campbell T, Desbruslais M, Luthert P, Collinge J. 2001. Tissue distribution of protease resistant prion protein in variant Creutzfeldt–Jakob disease using a highly sensitive immunoblotting assay. *Lancet* 358:171–180.
118. Ekins RP, Jackson T. 1984. Nonisotopic immunoassays. An overview. *In* Bizollon CA (ed), *Monoclonal Antibodies and New-Trends in Immunoassay*, pp. 149–163. Elsevier, Amsterdam, the Netherlands.
119. Grassi J, Frobert Y, Pradelles P, Chercuitte F, Gruaz D, Dayer JM, Poubelle PE. 1989. Production of monoclonal antibodies against interleukin-1 alpha and -1 beta. Development of two enzyme immunometric assays (EIA) using acetylcholinesterase and their application to biological media. *J Immunol Methods* 123:193–210.
120. Moynagh J, Schimmel H. 1999. Tests for BSE evaluated. *Nature* 400:105–105.
121. Armstrong RA, Cairns NJ, Ironside J, Lantos PL. 2002. Quantification of vacuolation ("spongiform change"), surviving neurones and prion protein deposition in eleven cases of variant Creutzfeldt–Jakob disease. *Neuropathol Appl Neurobiol* 28:129–135.
122. Ironside J, McCardle L, Horsburgh A, Lim Z, Head M. 2002. Pathological diagnosis of variant Creutzfeldt–Jakob disease. *APMIS* 110:79–87.
123. Iwata N, Sato Y, Higuchi Y, Nohtomi K, Nagata N, Hasegawa H, Tobiume M, Nakamura Y, Hagiwara Ki, Furuoka H. 2006. Distribution of PrP^{Sc} in cattle with bovine spongiform encephalopathy slaughtered at abattoirs in Japan. *Jpn J Infect Dis* 59:100.
124. Orge L, Galo A, Machado C, Lima C, Ochoa C, Silva J, Ramos M, Simas JP. 2004. Identification of putative atypical scrapie in sheep in Portugal. *J Gen Virol* 85:3487–3491.
125. Hirogari Y, Kubo M, Kimura KM, Haritani M, Yokoyama T. 2003. Two different scrapie prions isolated in Japanese sheep flocks. *Microbiol Immunol* 47:871–876.
126. Hill AF, Zeidler M, Ironside J, Collinge J. 1997. Diagnosis of new variant Creutzfeldt–Jakob disease by tonsil biopsy. *Lancet* 349:99–100.
127. Hill A, Butterworth R, Joiner S, Jackson G, Rossor M, Thomas D, Frosh A, Tolley N, Bell J, Spencer M. 1999. Investigation of variant Creutzfeldt–Jakob disease and other human prion diseases with tonsil biopsy samples. *Lancet* 353:183–189.
128. Schreuder B, Somerville R. 2003. Bovine spongiform encephalopathy in sheep? *Rev Sci Tech* 22:103–120.
129. Schreuder B, Van Keulen L, Vromans M, Langeveld J, Smits M. 1996. Preclinical test for prion diseases. *Nature* 381:563.
130. Van Keulen L, Schreuder B, Vromans M, Langeveld J, Smits M. 2000. *Pathogenesis of Natural Scrapie in Sheep.* Springer, Vienna, Austria.

131. Wild MA, Spraker TR, Sigurdson CJ, O'Rourke KI, Miller MW. 2002. Preclinical diagnosis of chronic wasting disease in captive mule deer (*Odocoileus hemionus*) and white-tailed deer (*Odocoileus virginianus*) using tonsillar biopsy. *J Gen Virol* 83:2629–2634.
132. Hilton DA, Fathers E, Edwards P, Ironside JW, Zajicek J. 1998. Prion immunoreactivity in appendix before clinical onset of variant Creutzfeldt–Jakob disease. *Lancet* 352:703–704.
133. Hilton DA, Ghani AC, Conyers L, Edwards P, McCardle L, Penney M, Ritchie D, Ironside JW. 2002. Accumulation of prion protein in tonsil and appendix: Review of tissue samples. *BMJ* 325:633–634.
134. Hilton DA, Ghani AC, Conyers L, Edwards P, McCardle L, Ritchie D, Penney M, Hegazy D, Ironside JW. 2004. Prevalence of lymphoreticular prion protein accumulation in UK tissue samples. *J Pathol* 203:733–739.
135. Ironside JW, Hilton DA, Ghani A, Johnston NJ, Conyers L, McCardle LM, Best D. 2000. Retrospective study of prion-protein accumulation in tonsil and appendix tissues. *Lancet* 355:1693–1694.
136. Saá P, Castilla J, Soto C. 2006. Presymptomatic detection of prions in blood. *Science* 313:92–94.
137. Saborio GP, Permanne B, Soto C. 2001. Sensitive detection of pathological prion protein by cyclic amplification of protein misfolding. *Nature* 411:810–813.
138. Jones M, Peden A, Prowse C, Gröner A, Manson J, Turner M, Ironside J, MacGregor I, Head M. 2007. *In vitro* amplification and detection of variant Creutzfeldt–Jakob disease PrP^{Sc}. *J Pathol* 213:21–26.
139. Kurt TD, Perrott MR, Wilusz CJ, Wilusz J, Supattapone S, Telling GC, Zabel MD, Hoover EA. 2007. Efficient *in vitro* amplification of chronic wasting disease PrP^{res}. *J Virol* 81:9605–9608.
140. Murayama Y, Yoshioka M, Yokoyama T, Iwamaru Y, Imamura M, Masujin K, Yoshiba S, Mohri S. 2007. Efficient *in vitro* amplification of a mouse-adapted scrapie prion protein. *Neurosci Lett* 413:270–273.
141. Atarashi R, Moore RA, Sim VL, Hughson AG, Dorward DW, Onwubiko HA, Priola SA, Caughey B. 2007. Ultrasensitive detection of scrapie prion protein using seeded conversion of recombinant prion protein. *Nat Methods* 4:645–650.
142. Brown P, Cervenáková L, Diringer H. 2001. Blood infectivity and the prospects for a diagnostic screening test in Creutzfeldt–Jakob disease. *J Lab Clin Med* 137:5–13.
143. Yang WC, Schmerr MJ, Jackman R, Bodemer W, Yeung ES. 2005. Capillary electrophoresis-based noncompetitive immunoassay for the prion protein using fluorescein-labeled protein A as a fluorescent probe. *Anal Chem* 77:4489–4494.
144. Giese A, Bieschke J, Eigen M, Kretzschmar HA. 2000. Putting prions into focus: Application of single molecule detection to the diagnosis of prion diseases. *Arch Virol Suppl* 16:161–171.
145. Paramithiotis E, Pinard M, Lawton T, LaBoissiere S, Leathers VL, Zou W-Q, Estey LA, Lamontagne J, Lehto MT, Kondejewski LH. 2003. A prion protein epitope selective for the pathologically misfolded conformation. *Nat Med* 9:893–899.
146. Šerbec VČ, Bresjanac M, Popović M, Hartman KP, Galvani V, Rupreht R, Černilec M, Vranac T, Hafner I, Jerala R. 2004. Monoclonal antibody against a peptide of human prion protein discriminates between Creutzfeldt–Jacob's disease-affected and normal brain tissue. *J Biol Chem* 279:3694–3698.
147. Korth C, Stierli B, Streit P, Moser M, Schaller O, Fischer R, Schulz-Schaeffer W, Kretzschmar H, Raeber A, Braun U. 1997. Prion (PrP^{Sc})-specific epitope defined by a monoclonal antibody. *Nature* 390:74–77.
148. Rhie A, Kirby L, Sayer N, Wellesley R, Disterer P, Sylvester I, Gill A, Hope J, James W, Tahiri-Alaoui A. 2003. Characterization of 2′-fluoro-RNA aptamers that bind preferentially to disease-associated conformations of prion protein and inhibit conversion. *J Biol Chem* 278:39697–39705.
149. Sayer NM, Cubin M, Rhie A, Bullock M, Tahiri-Alaoui A, James W. 2004. Structural determinants of conformationally selective, prion-binding aptamers. *J Biol Chem* 279:13102–13109.
150. Saa P, Castilla J, Soto C. 2006. Ultra-efficient replication of infectious prions by automated protein misfolding cyclic amplification. *J Biol Chem* 281:35245–35252.
151. Legname G, Baskakov IV, Nguyen H-OB, Riesner D, Cohen FE, DeArmond SJ, Prusiner SB. 2004. Synthetic mammalian prions. *Science* 305:673–676.
152. Anderson RM, Donnelly CA, Ferguson NM, Woolhouse M, Watt C, Udy H, MaWhinney S, Dunstan S, Southwood T, Wilesmith J. 1996. Transmission dynamics and epidemiology of BSE in British cattle. *Nature* 382:779–788.
153. Andreoletti O, Litaise C, Simmons H, Corbiere F, Lugan S, Costes P, Schelcher F, Vilette D, Grassi J, Lacroux C. 2012. Highly efficient prion transmission by blood transfusion. *PLoS Pathog* 8:e1002782.
154. Houston F, McCutcheon S, Goldmann W, Chong A, Foster J, Siso S, Gonzalez L, Jeffrey M, Hunter N. 2008. Prion diseases are efficiently transmitted by blood transfusion in sheep. *Blood* 112:4739–4745.

155. Gill ON, Spencer Y, Richard-Loendt A, Kelly C, Dabaghian R, Boyes L, Linehan J, et al. 2013. Prevalent abnormal prion protein in human appendixes after bovine spongiform encephalopathy epizootic: Large scale survey. *BMJ* 347:f5675.
156. McCutcheon S, Alejo Blanco AR, Tan BC, Gonzalez L, Martin S, Mallinson G, Appleford NE, Turner ML, Manson JC, Houston EF. 2015. A prion reduction filter does not completely remove endogenous prion infectivity from sheep blood. *Transfusion* 55:2123–2133.
157. Sakudo A, Ano Y, Onodera T, Nitta K, Shintani H, Ikuta K, Tanaka Y. 2011. Fundamentals of prions and their inactivation (review). *Int J Mol Med* 27:483–489.
158. Rutala WA, Weber DJ, Society for Healthcare Epidemiology of A. 2010. Guideline for disinfection and sterilization of prion-contaminated medical instruments. *Infect Control Hosp Epidemiol* 31:107–117.
159. Rutala WA, Gergen MF, Weber DJ. 2008. Impact of an oil-based lubricant on the effectiveness of the sterilization processes. *Infect Control Hosp Epidemiol* 29:69–72.

Index

I

K

L

M

N

O

P

Q

R

S

T

U

V

W